COMBINATORIAL DESIGNS

PURE AND APPLIED MATHEMATICS

A Program of Monographs, Textbooks, and Lecture Notes

EXECUTIVE EDITORS

Earl J. Taft
Rutgers University
New Brunswick, New Jersey

Zuhair Nashed
University of Delaware
Newark, Delaware

CHAIRMEN OF THE EDITORIAL BOARD

S. Kobayashi
University of California, Berkeley
Berkeley, California

Edwin Hewitt
University of Washington
Seattle, Washington

EDITORIAL BOARD

M. S. Baouendi
Purdue University

Donald Passman
University of Wisconsin-Madison

Jack K. Hale
Brown University

Fred S. Roberts
Rutgers University

Marvin Marcus
University of California, Santa Barbara

Gian-Carlo Rota
Massachusetts Institute of Technology

W. S. Massey
Yale University

David Russell
University of Wisconsin-Madison

Leopoldo Nachbin
Centro Brasileiro de Pesquisas Físicas
and University of Rochester

Jane Cronin Scanlon
Rutgers University

Anil Nerode
Cornell University

Walter Schempp
Universität Siegen

Mark Teply
University of Wisconsin-Milwaukee

MONOGRAPHS AND TEXTBOOKS IN PURE AND APPLIED MATHEMATICS

1. K. Yano, Integral Formulas in Riemannian Geometry (1970) *(out of print)*
2. S. Kobayashi, Hyperbolic Manifolds and Holomorphic Mappings (1970) *(out of print)*
3. V. S. Vladimirov, Equations of Mathematical Physics (A. Jeffrey, editor; A. Littlewood, translator) (1970) *(out of print)*
4. B. N. Pshenichnyi, Necessary Conditions for an Extremum (L. Neustadt, translation editor; K. Makowski, translator) (1971)
5. L. Narici, E. Beckenstein, and G. Bachman, Functional Analysis and Valuation Theory (1971)
6. D. S. Passman, Infinite Group Rings (1971)
7. L. Dornhoff, Group Representation Theory (in two parts). Part A: Ordinary Representation Theory. Part B: Modular Representation Theory (1971, 1972)
8. W. Boothby and G. L. Weiss (eds.), Symmetric Spaces: Short Courses Presented at Washington University (1972)
9. Y. Matsushima, Differentiable Manifolds (E. T. Kobayashi, translator) (1972)
10. L. E. Ward, Jr., Topology: An Outline for a First Course (1972) *(out of print)*
11. A. Babakhanian, Cohomological Methods in Group Theory (1972)
12. R. Gilmer, Multiplicative Ideal Theory (1972)
13. J. Yeh, Stochastic Processes and the Wiener Integral (1973) *(out of print)*
14. J. Barros-Neto, Introduction to the Theory of Distributions (1973) *(out of print)*
15. R. Larsen, Functional Analysis: An Introduction (1973) *(out of print)*
16. K. Yano and S. Ishihara, Tangent and Cotangent Bundles: Differential Geometry (1973) *(out of print)*
17. C. Procesi, Rings with Polynomial Identities (1973)
18. R. Hermann, Geometry, Physics, and Systems (1973)
19. N. R. Wallach, Harmonic Analysis on Homogeneous Spaces (1973) *(out of print)*
20. J. Dieudonné, Introduction to the Theory of Formal Groups (1973)
21. I. Vaisman, Cohomology and Differential Forms (1973)
22. B.-Y. Chen, Geometry of Submanifolds (1973)
23. M. Marcus, Finite Dimensional Multilinear Algebra (in two parts) (1973, 1975)
24. R. Larsen, Banach Algebras: An Introduction (1973)
25. R. O. Kujala and A. L. Vitter (eds.), Value Distribution Theory: Part A; Part B: Deficit and Bezout Estimates by Wilhelm Stoll (1973)
26. K. B. Stolarsky, Algebraic Numbers and Diophantine Approximation (1974)
27. A. R. Magid, The Separable Galois Theory of Commutative Rings (1974)
28. B. R. McDonald, Finite Rings with Identity (1974)
29. J. Satake, Linear Algebra (S. Koh, T. A. Akiba, and S. Ihara, translators) (1975)

30. *J. S. Golan*, Localization of Noncommutative Rings (1975)
31. *G. Klambauer*, Mathematical Analysis (1975)
32. *M. K. Agoston*, Algebraic Topology: A First Course (1976)
33. *K. R. Goodearl*, Ring Theory: Nonsingular Rings and Modules (1976)
34. *L. E. Mansfield*, Linear Algebra with Geometric Applications: Selected Topics (1976)
35. *N. J. Pullman*, Matrix Theory and Its Applications (1976)
36. *B. R. McDonald*, Geometric Algebra Over Local Rings (1976)
37. *C. W. Groetsch*, Generalized Inverses of Linear Operators: Representation and Approximation (1977)
38. *J. E. Kuczkowski and J. L. Gersting*, Abstract Algebra: A First Look (1977)
39. *C. O. Christenson and W. L. Voxman*, Aspects of Topology (1977)
40. *M. Nagata*, Field Theory (1977)
41. *R. L. Long*, Algebraic Number Theory (1977)
42. *W. F. Pfeffer*, Integrals and Measures (1977)
43. *R. L. Wheeden and A. Zygmund*, Measure and Integral: An Introduction to Real Analysis (1977)
44. *J. H. Curtiss*, Introduction to Functions of a Complex Variable (1978)
45. *K. Hrbacek and T. Jech*, Introduction to Set Theory (1978)
46. *W. S. Massey*, Homology and Cohomology Theory (1978)
47. *M. Marcus*, Introduction to Modern Algebra (1978)
48. *E. C. Young*, Vector and Tensor Analysis (1978)
49. *S. B. Nadler, Jr.*, Hyperspaces of Sets (1978)
50. *S. K. Segal*, Topics in Group Rings (1978)
51. *A. C. M. van Rooij*, Non-Archimedean Functional Analysis (1978)
54. *L. Corwin and R. Szczarba*, Calculus in Vector Spaces (1979)
53. *C. Sadosky*, Interpolation of Operators and Singular Integrals: An Introduction to Harmonic Analysis (1979)
54. *J. Cronin*, Differential Equations: Introduction and Quantitative Theory (1980)
55. *C. W. Groetsch*, Elements of Applicable Functional Analysis (1980)
56. *I. Vaisman*, Foundations of Three-Dimensional Euclidean Geometry (1980)
57. *H. I. Freedman*, Deterministic Mathematical Models in Population Ecology (1980)
58. *S. B. Chae*, Lebesgue Integration (1980)
59. *C. S. Rees, S. M. Shah, and C. V. Stanojević*, Theory and Applications of Fourier Analysis (1981)
60. *L. Nachbin*, Introduction to Functional Analysis: Banach Spaces and Differential Calculus (R. M. Aron, translator) (1981)
61. *G. Orzech and M. Orzech*, Plane Algebraic Curves: An Introduction Via Valuations (1981)
62. *R. Johnsonbaugh and W. E. Pfaffenberger*, Foundations of Mathematical Analysis (1981)
63. *W. L. Voxman and R. H. Goetschel*, Advanced Calculus: An Introduction to Modern Analysis (1981)
64. *L. J. Corwin and R. H. Szcarba*, Multivariable Calculus (1982)
65. *V. I. Istrătescu*, Introduction to Linear Operator Theory (1981)
66. *R. D. Järvinen*, Finite and Infinite Dimensional Linear Spaces: A Comparative Study in Algebraic and Analytic Settings (1981)

67. *J. K. Beem and P. E. Ehrlich*, Global Lorentzian Geometry (1981)
68. *D. L. Armacost*, The Structure of Locally Compact Abelian Groups (1981)
69. *J. W. Brewer and M. K. Smith, eds.*, Emmy Noether: A Tribute to Her Life and Work (1981)
70. *K. H. Kim*, Boolean Matrix Theory and Applications (1982)
71. *T. W. Wieting*, The Mathematical Theory of Chromatic Plane Ornaments (1982)
72. *D. B. Gauld*, Differential Topology: An Introduction (1982)
73. *R. L. Faber*, Foundations of Euclidean and Non-Euclidean Geometry (1983)
74. *M. Carmeli*, Statistical Theory and Random Matrices (1983)
75. *J. H. Carruth, J. A. Hildebrant, and R. J. Koch*, The Theory of Topological Semigroups (1983)
76. *R. L. Faber*, Differential Geometry and Relativity Theory: An Introduction (1983)
77. *S. Barnett*, Polynomials and Linear Control Systems (1983)
78. *G. Karpilovsky*, Commutative Group Algebras (1983)
79. *F. Van Oystaeyen and A. Verschoren*, Relative Invariants of Rings: The Commutative Theory (1983)
80. *I. Vaisman*, A First Course in Differential Geometry (1984)
81. *G. W. Swan*, Applications of Optimal Control Theory in Biomedicine (1984)
82. *T. Petrie and J. D. Randall*, Transformation Groups on Manifolds (1984)
83. *K. Goebel and S. Reich*, Uniform Convexity, Hyperbolic Geometry, and Nonexpansive Mappings (1984)
84. *T. Albu and C. Năstăsescu*, Relative Finiteness in Module Theory (1984)
85. *K. Hrbacek and T. Jech*, Introduction to Set Theory, Second Edition, Revised and Expanded (1984)
86. *F. Van Oystaeyen and A. Verschoren*, Relative Invariants of Rings: The Noncommutative Theory (1984)
87. *B. R. McDonald*, Linear Algebra Over Commutative Rings (1984)
88. *M. Namba*, Geometry of Projective Algebraic Curves (1984)
89. *G. F. Webb*, Theory of Nonlinear Age-Dependent Population Dynamics (1985)
90. *M. R. Bremner, R. V. Moody, and J. Patera*, Tables of Dominant Weight Multiplicities for Representations of Simple Lie Algebras (1985)
91. *A. E. Fekete*, Real Linear Algebra (1985)
92. *S. B. Chae*, Holomorphy and Calculus in Normed Spaces (1985)
93. *A. J. Jerri*, Introduction to Integral Equations with Applications (1985)
94. *G. Karpilovsky*, Projective Representations of Finite Groups (1985)
95. *L. Narici and E. Beckenstein*, Topological Vector Spaces (1985)
96. *J. Weeks*, The Shape of Space: How to Visualize Surfaces and Three-Dimensional Manifolds (1985)
97. *P. R. Gribik and K. O. Kortanek*, Extremal Methods of Operations Research (1985)
98. *J.-A. Chao and W. A. Woyczynski, eds.*, Probability Theory and Harmonic Analysis (1986)
99. *G. D. Crown, M. H. Fenrick, and R. J. Valenza*, Abstract Algebra (1986)
100. *J. H. Carruth, J. A. Hildebrant, and R. J. Koch*, The Theory of Topological Semigroups, Volume 2 (1986)

101. *R. S. Doran and V. A. Belfi*, Characterizations of C*-Algebras: The Gelfand-Naimark Theorems (1986)
102. *M. W. Jeter*, Mathematical Programming: An Introduction to Optimization (1986)
103. *M. Altman*, A Unified Theory of Nonlinear Operator and Evolution Equations with Applications: A New Approach to Nonlinear Partial Differential Equations (1986)
104. *A. Verschoren*, Relative Invariants of Sheaves (1987)
105. *R. A. Usmani*, Applied Linear Algebra (1987)
106. *P. Blass and J. Lang*, Zariski Surfaces and Differential Equations in Characteristic $p > 0$ (1987)
107. *J. A. Reneke, R. E. Fennell, and R. B. Minton*. Structured Hereditary Systems (1987)
108. *H. Busemann and B. B. Phadke*, Spaces with Distinguished Geodesics (1987)
109. *R. Harte*, Invertibility and Singularity for Bounded Linear Operators (1988).
110. *G. S. Ladde, V. Lakshmikantham, and B. G. Zhang*, Oscillation Theory of Differential Equations with Deviating Arguments (1987)
111. *L. Dudkin, I. Rabinovich, and I. Vakhutinsky*, Iterative Aggregation Theory: Mathematical Methods of Coordinating Detailed and Aggregate Problems in Large Control Systems (1987)
112. *T. Okubo*, Differential Geometry (1987)
113. *D. L. Stancl and M. L. Stancl*, Real Analysis with Point-Set Topology (1987)
114. *T. C. Gard*, Introduction to Stochastic Differential Equations (1988)
115. *S. S. Abhyankar*, Enumerative Combinatorics of Young Tableaux (1988)
116. *H. Strade and R. Farnsteiner*, Modular Lie Algebras and Their Representations (1988)
117. *J. A. Huckaba*, Commutative Rings with Zero Divisors (1988)
118. *W. D. Wallis*, Combinatorial Designs (1988)
119. *W. Więsław*, Topological Fields (1988)

Other Volumes in Preparation

COMBINATORIAL DESIGNS

W. D. WALLIS
Southern Illinois University at Carbondale
Carbondale, Illinois

MARCEL DEKKER, INC.   New York and Basel

Wallis, W. D.
    Combinatorial designs / W. D. Wallis.
      p. cm. -- (Pure and applied mathematics ; v. 118)
    Includes index.
    ISBN 0-8247-7942-8
    1. Combinatorial designs and configurations. I. Title.
II. Series: Monographs and textbooks in pure and applied mathematics;
    v. 118.
    QA166.25.W34 1988
    511'.6--dc19   87-37967
                                                              CIP

COPYRIGHT©1988 by MARCEL DEKKER, INC. ALL RIGHTS RESERVED

Neither this book nor any part may be reproduced or transmitted in any form or by any means, electronic or mechanical, including photocopying, microfilming, and recording, or by any information storage and retrieval system, without permission in writing from the publisher.

MARCEL DEKKER, INC.
270 Madison Avenue, New York, New York 10016

Current printing (last digit):
10 9 8 7 6 5 4 3 2 1

PRINTED IN THE UNITED STATES OF AMERICA

Preface

It has been suggested that combinatorial theory is the fastest-growing area of modern mathematics. The largest part of the subject is graph theory, as evidenced by the ever-growing number of research publications in that part of the discipline. Accordingly, there have appeared a good many textbooks devoted to graph theory, suitable for use at junior, senior, and graduate levels. A second major part of combinatorial mathematics is enumeration. The third—the one we treat here—is the theory of combinatorial designs.

Design theory began with Euler's invention of the Latin square 200 years ago. The next area of particular interest was the balanced incomplete block design and in particular the finite projective and affine planes. This interest was fanned by the statistical interpretation of designs and by the geometrical nature of planes. Most writers of textbooks assume that these—what we might call "classical designs"—are the only combinatorial designs. However, over the last 10 to 15 years, combinatorial researchers have discussed a wider range of designs: one-factorizations, Room squares, and other designs based on unordered pairs, various tournament designs, nested designs, and so on.

Like several other mathematicians, I have taught courses at the senior/graduate level in combinatorial designs primarily by two methods:

(i) Use an elementary book as a text, supplemented by more advanced notes;
(ii) Use no textbook, but refer to various advanced books and research monographs.

Neither is satisfactory. The present book is intended for use as a text in such a course. In writing it my aims were:

To enable instructors to present combinatorial designs to undergraduate and beginning graduate students in more than a cursory fashion;

To give a good groundwork in the classical areas of design theory: block designs, finite geometries, and Latin squares;
To introduce some modern extensions of design theory;
To lead students toward the current boundaries of the subject—we hope they will be capable of understanding current research papers after studying the book;
To present a well-rounded textbook, providing motivation, instructive examples, theorems, and exercises for all its topics.

I assume the reader to have some mathematical background, including familiarity with matrices. Some abstract algebra occurs, but the reader will need little more than the definition of a group. The mathematical notations are standard, except for the consistent use of "k-set" to mean "set with k elements," and the notation (a $\cdots$ b), borrowed from Pascal, to denote $\{a, a + 1, \ldots, b\}$. The end of a theorem or of a proof is denoted ∎.

The earlier parts (at least half the book) are reasonably accessible; for example, it could be read by a senior in computer science with a moderately strong mathematical background, without a teacher. Some fairly difficult material is included in the later chapters, and the breadth and depth are such that a satisfactory graduate course that goes close to the boundaries of present research could be taught from the book without any need for extensive outside reference, although in such a course, most instructors will wish to replace some topics by material in their own favorite areas.

Many people have contributed to my knowledge of, and continued interest in, combinatorial designs, and therefore have contributed to the writing of this book. I acknowledge this with gratitude, but I think mention of individual names is unnecessary. However, I would like to thank Larry Cummings, who has read and taught from several parts of the manuscript; Liz Billington, Curt Lindner, Doug Stinson, Ralph Stanton, and Anne Street, for many helpful discussions; Linda Macak, who typed the manuscript; and a number of students who have suffered through various versions in courses.

W. D. Wallis

Contents

PREFACE		iii
1	BASIC CONCEPTS	1
	1.1 Combinatorial Designs	1
	1.2 Some Examples of Designs	3
	1.3 Block Designs	5
	1.4 Systems of Distinct Representatives	10
	Bibliographic Remarks	13
2	BALANCED DESIGNS	14
	2.1 Pairwise Balanced Designs	14
	2.2 Balanced Incomplete Block Designs	21
	2.3 t-Designs	32
	Bibliographic Remarks	34
3	SOME FINITE ALGEBRA	36
	3.1 Finite Fields	36
	3.2 Quadratic Elements	42
	3.3 Sums of Squares	46
	Bibliographic Remarks	53
4	DIFFERENCE SETS AND DIFFERENCE METHODS	55
	4.1 Difference Sets	55
	4.2 Properties of Difference Sets	60
	4.3 More General Methods	62
	Bibliographic Remarks	69

5	FINITE GEOMETRIES	70
	5.1 Finite Affine Planes	70
	5.2 Construction of Finite Affine Geometries	73
	5.3 Finite Projective Geometries	77
	5.4 Singer Difference Sets	83
	5.5 Ovals in Projective Planes	88
	5.6 The Desargues Configuration	90
	Bibliographic Remarks	94
6	MORE ABOUT BLOCK DESIGNS	95
	6.1 Residual and Derived Designs	95
	6.2 The Main Existence Theorem	98
	6.3 Another Proof of the Existence Theorem	105
	6.4 Resolvability	109
	Bibliographic Remarks	116
7	t-DESIGNS	118
	7.1 Contractions and Extensions	118
	7.2 Some Examples of 3-Designs	121
	7.3 Extensions of Symmetric Designs	128
	7.4 Derived t-Designs and Affine t-Designs	131
	Bibliographic Remarks	133
8	HADAMARD MATRICES	134
	8.1 Basic Ideas	134
	8.2 Hadamard Matrices and Block Designs	139
	8.3 Kronecker Product Constructions	143
	8.4 Williamson's Method	152
	8.5 Regular Hadamard Matrices	156
	Bibliographic Remarks	160
9	THE VARIABILITY OF HADAMARD MATRICES	163
	9.1 Equivalence of Hadamard Matrices	163
	9.2 Integral Equivalence	165
	9.3 Profiles	174
	Bibliographic Remarks	178
10	LATIN SQUARES	179
	10.1 Latin Squares and Subsquares	179
	10.2 Orthogonality	182

Contents

10.3 Idempotent Latin Squares	187
10.4 Transversal Designs	192
10.5 Spouse-Avoiding Mixed-Doubles Tournaments	198
10.6 Three Orthogonal Latin Squares	204
Bibliographic Remarks	207

11 ONE-FACTORS AND ONE-FACTORIZATIONS — 209

11.1 Basic Ideas	209
11.2 Starters	212
11.3 One-Factorizations of Complete Graphs	214
11.4 One-Factorizations of K_8	223
11.5 An Application to Projective Planes	233
Bibliographic Remarks	234

12 TRIPLE SYSTEMS — 236

12.1 Construction of Triple Systems	236
12.2 Subsystems	241
12.3 Simple Triple Systems	249
12.4 Difference Triples and Cyclic Triple Systems	252
12.5 Kirkman Triple Systems	254
Bibliographic Remarks	261

13 ROOM SQUARES — 263

13.1 Basic Ideas	263
13.2 Starter Constructions	265
13.3 Subsquare Constructions	269
13.4 The Existence Theorem	273
13.5 Subsquares of Room Squares	276
Bibliographic Remarks	282

14 ASYMPTOTIC RESULTS ON BALANCED INCOMPLETE BLOCK DESIGNS — 283

14.1 Introduction	283
14.2 Designs with Large λ	285
14.3 Designs in the Principal Fiber for $\lambda = 1$	293
14.4 Designs in Other Fibers for $\lambda = 1$	300
Bibliographic Remarks	304

ANSWERS, SOLUTIONS, AND HINTS	306
AUTHOR INDEX	319
SUBJECT INDEX	323

1
Basic Concepts

1.1 COMBINATORIAL DESIGNS

Informally, one may define a <u>combinatorial design</u> to be a way of selecting subsets from a finite set in such a way that some specified conditions are satisfied. Although we shall not define the type of condition precisely, we intend that they involve <u>incidence</u>: set membership, set intersection, and so on.

As an example, suppose that it is required to select 3-sets from the seven objects $\{0, 1, 2, 3, 4, 5, 6\}$ in such a way that each object occurs in three of the 3-sets and every intersection of two 3-sets has precisely one member. The solution of this problem—the way of selecting the 3-sets—is a combinatorial design. One possible solution is

$$\{012, 034, 056, 135, 146, 236, 245\}$$

where 012 represents $\{0, 1, 2\}$ and so on.

The example above does not use the arithmetical properties of the numbers 0, 1, 2, 3, 4, 5, 6. In fact, we do not even know what these symbols stand for. They could be the corresponding integers; they could be the integers modulo 7, or modulo some larger base; or they could be labels attached to any seven objects. The elements of the set are not viewed from an arithmetical viewpoint (although arithmetic will sometimes be very useful in showing that a design exists or in solving other problems).

Our example involved showing that a certain design exists. Other problems can arise. For example, suppose that in our problem there are n 3-sets. Then they have between them 3n entries. But there are 7 objects, and each occurs in three of the subsets. So when they are listed, there are 21 entries. Thus 3n = 21 and n = 7. In other words, the number of subsets is determined; we can prove that: If a solution exists, it must contain precisely 7 subsets. So one can prove properties of possible solutions.

We now prove the uniqueness of the given solution of our example, in the following sense. If we have any one solution, we can obviously obtain another by relabeling the objects in some way—for example, the relabeling of "1" as "2" and "2" as "1" in the solution above leads to the different-looking answer

$$\{012, 034, 056, 235, 246, 136, 145\}$$

But this is essentially the same as the original. We shall consider two solutions to be <u>isomorphic</u> if the list of 3-sets in one can be obtained from the list of 3-sets of the other by carrying out some permutation of the symbols consistently. What we now show is that the solution of the problem is unique, up to isomorphism.

Assume that we have a solution to the problem. There must be three 3-sets that contain the object 0. Suppose that they are 0ab, 0cd, and 0ef, where a, b, c, d, e, and f are chosen from the numbers 0 through 6. Since the intersection of any two sets is to have one entry, and since 0 is common to 0ab and 0cd, we must have

$$a \neq c, \quad a \neq d, \quad b \neq c, \quad b \neq d.$$

Moreover, since 0ab is to have three members, we know that

$$a \neq b, \quad a \neq 0, \quad b \neq 0.$$

By similar observations we see that a, b, c, d, e, and f must all be different from each other and must all be nonzero. So

$$\{a, b, c, d, e, f\} = \{1, 2, 3, 4, 5, 6\}.$$

Whichever ordering is used, we may permute the symbols at will, so we may take

$$a = 1, \quad b = 2, \quad c = 3, \quad d = 4, \quad e = 5, \quad f = 6$$

up to isomorphism, and the design contains the sets

012, 034, 056.

Now 1 belongs to three sets. One of them is 012. Suppose that 1gh is another. Since two sets have one common member, we have

$$g \neq 0, \quad g \neq 2, \quad h \neq 0, \quad h \neq 2$$

(consider 012 and 1gh). Also,

1.2 Some Examples of Designs

$|\{g,h\} \cap \{3,4\}| = 1.$

If h is the common member, we can swap the labels g and h; if 4 is the common member, we can swap the labels 3 and 4. So up to isomorphism, we can assume that

$g = 3, \quad h \neq 3, \quad h \neq 4.$

If h = 6, we can exchange labels 5 and 6. So we can assume that h = 5. Therefore, two of the sets containing 1 are 012 and 035. It is easy to see that the third set cannot contain 0, 2, 3, or 5, so it is 146.

We next consider the sets containing 2. In the same way we see that they must be

012, 236, 245.

So the design is uniquely determined, up to isomorphism.

We could also show that our problem has no solution unless the number of objects is precisely 7. More generally, suppose that we replaced "3" by "k": Each object is to belong to precisely k k-sets. Then we could prove that exactly $k^2 - k + 1$ objects are needed. But this necessary condition is not sufficient: Even with the correct number of elements, there is no solution for k = 7.

EXERCISES

1.1.1 Show that there exists a combinatorial design on the nine symbols $\{0,1,2,3,4,5,6,7,8\}$, which consists of 12 3-sets with the property that any two symbols occur together in exactly one 3-set. Show that the design is uniquely defined up to isomorphism.

1.1.2 A combinatorial design consists of the selection of 2-sets and 3-sets from the set $\{0,1,2,3,4,5\}$, such that:

(a) Every pair of symbols occur together in exactly one subset;
(b) There are at least three 3-sets.

Prove that no two 3-sets can be disjoint. Show that the number of 3-sets can be either 3 or 4, but no larger; and show that there are precisely two nonisomorphic solutions, one with three and one with four 3-sets.

1.2 SOME EXAMPLES OF DESIGNS

Many examples of combinatorial designs have been discussed in the literature. We shall define some of these, to provide examples of the concept.

Linked Design. A linked design is a way of selecting subsets from a set in such a way that any two subsets have intersection size μ, where μ is a constant for the design. Such a constant is called a parameter of the design; one might say that a linked design on v objects, in which every pair of blocks has intersection size μ, is a "linked design with parameters (v, μ)." The designs discussed in Section 1.1 were linked designs with parameters $(7,1)$, but with certain other conditions (constant subset size, and every element belonging to the same number of subsets).

One-Factorization or System of Pairs. A one-factorization on a set S of n symbols [such as S = (1 $\cdots$ n)] is a way of organizing the $n(n-1)/2$ unordered pairs of members of S into classes such that every pair belongs to precisely one such class and such that every symbol belongs to precisely one member of each class. A one-factorization is a combinatorial design; one may interpret the phrase "way of selecting subsets" in the definition to refer to the selection of [all $n(n-1)/2$] unordered pairs from the set S, and interpret the arrangement of these pairs into classes as a "specified conditions"; alternatively, one could say that the universal "finite set" is the set of unordered pairs on S. A one-factorization has only one parameter, the number n, which is the size of the symbol set S.

Latin Square. A Latin square L of side n is an n × n array whose entries all come from some n-set S, with the property that each row and each column is a permutation of S. One can either interpret the rows of L as subsets (of size n) which are ordered, or interpret the entries of L as subsets (of size 1) which are subjected to a two-dimensional positioning rule. Examples of Latin squares are

| 1 2 |
| 2 1 |

| 1 3 2 |
| 3 2 1 |
| 2 1 3 |

| 1 3 2 |
| 2 1 3 |
| 3 2 1 |

Graph. A graph $G = G(V, E)$ consists of a finite set V of objects called vertices (or points), and a set E of unordered pairs of members of V called edges. This definition is consistent with the conventions that the edges of a graph are not directed, that the two vertices constituting an edge (its "endpoints") are distinct, and that two vertices belong together to no edge or to one edge ("are adjacent") but to no more. However, the reader should be aware that the word "graph" is sometimes used for objects with directed edges, with loops on vertices, with multiple edges or with infinite vertex sets.

1.3 Block Designs

Graph theory is a large area of study in combinatorial mathematics, and the reader has probably already encountered it. It is not our intention to claim that graph theory is a part of design theory, and very often the methods of approach of a graph theorist and a design theorist are quite different. However, it should be noted that a graph is a form of design. Moreover, the terminology of graph theory is occasionally useful to us.

EXERCISES

1.2.1 Prove that the parameter of a one-factorization must be even.

1.2.2 Prove that there are precisely two Latin squares of side 3 with first row

namely, those shown in this section.

1.2.3 Exhibit a Latin square of side 4.

1.2.4 How many different graphs are there with vertex set $\{1, 2, 3\}$?

1.2.5 Invent a definition of "isomorphism" for graphs.

1.3 BLOCK DESIGNS

The members of the universal set S in a combinatorial design are usually called <u>treatments</u>, or <u>varieties</u>, and the subsets chosen are called <u>blocks</u>. We shall say that a design of a certain type is a <u>block design</u> if in the definition of the type of design, the blocks are simply unordered sets of treatments and there is no structural ordering or pattern in the blocks. This vague definition (containing the undefined phrase "structural ordering or pattern") is really a statement of intention: If the blocks are ordered internally (as in a Latin square) or arranged in a pattern (as in a one-factorization), we shall usually study them in a different way from a linked design, for example.

We now define an important class of block design. A <u>regular design</u> based on a v-set S is a collection of k-sets from S such that every member of S belongs to r of the k-sets. It is usual to write b for the number of blocks in a design. So a regular design has four parameters: v, b, r, and k. However, they are not independent.

THEOREM 1.1 In any regular design,

$$bk = vr. \tag{1.1}$$

Proof: One counts, in two different ways, all the ordered pairs (x, y) such that treatment x belongs to block y. Since every treatment belongs to r blocks, there are r pairs for each treatment, so the number is vr. Similarly, each block contributes k pairs, so the summation yields bk. Therefore,

bk = vr. ∎

The number of blocks that contain a given treatment is called the replication number or frequency of that treatment. So the defining characteristics of a regular design are that all elements have the same replication number and that all blocks have the same size.

If all v treatments occur in a block of a design, that block is called complete. If a regular design has that property, then k = v and obviously r = b. Such a design is a complete design, and it has very little interest unless some further structure is imposed (as, for example, in a Latin square. We say that a design is incomplete if at least one block is incomplete. If v = b, the design is called symmetric.

If x and y are any two different treatments in an incomplete design, we shall refer to the number of blocks that contain both x and y as the covalency of x and y, and write it as λ_{xy}. Many important properties of block designs are concerned with this covalency function. The one that has most frequently been studied is the property of balance: A balanced incomplete block design, or BIBD, is a regular incomplete design in which λ_{xy} is a constant, independent of the choice of x and y. It is usual to write λ for which constant value, which is called the index of the design. One often refers to a balanced incomplete block design by using the five parameters (v, b, r, k, λ); for example, we say "a $(9, 12, 4, 3, 1)$-design" or "a $(9, 12, 4, 3, 1)$-BIBD" to mean a balanced incomplete block design with 12 blocks of size 3, based on 9 treatments, with replication number 4 and covalency 1.

A balanced design with $\lambda = 0$, or a null design, is often called "trivial," and so is a complete design. We shall demand that a balanced incomplete block design be not trivial; since completeness is already outlawed, the added restriction is that $\lambda > 0$.

THEOREM 1.2 In a (v, b, r, k, λ)-BIBD,

$$r(k - 1) = \lambda(v - 1). \tag{1.2}$$

Proof: Consider the blocks of the design that contain a given treatment, x say. There are r such blocks. Because of the balance property, every treatment other than x must occur in λ of them. So if we list all entries of the blocks, we list

x, r times;
every other treatment, λ times.

1.3 Block Designs

The list contains rk entries. So

$$rk = r + \lambda(v - 1),$$
$$r(k - 1) = \lambda(v - 1). \qquad \blacksquare$$

The word <u>incidence</u> is used to describe the relationship between blocks and treatments in a design. One can say block B is incident with treatment t, or treatment t is incident with block B, to mean that t is a member of B. A block design may be specified by its <u>incidence matrix</u>: If the design has b blocks $B_1, B_2, \ldots, B_b$ and v treatments $t_1, t_2, \ldots, t_v$, define a $v \times b$ matrix A with (i, j) entry a_{ij} as follows:

$$a_{ij} = \begin{cases} 1 & \text{if } t_i \text{ belongs to } B_j, \\ 0 & \text{otherwise.} \end{cases}$$

This matrix A is called the incidence matrix of the design. The definition means that each block corresponds to a column of the incidence matrix, and each treatment corresponds to a row. For example, the regular design with treatments t_1, t_2, t_3, t_4 and blocks $B_1 = \{t_1, t_2\}$, $B_2 = \{t_3, t_4\}$, $B_3 = \{t_1, t_4\}$, $B_4 = \{t_2, t_3\}$ has incidence matrix

$$A = \begin{bmatrix} 1 & 0 & 1 & 0 \\ 1 & 0 & 0 & 1 \\ 0 & 1 & 0 & 1 \\ 0 & 1 & 1 & 0 \end{bmatrix}.$$

As another example, observe that the following blocks form a $(6, 10, 5, 3, 2)$-design:

123, 124, 135, 146, 156,
236, 245, 256, 345, 346.

This design has incidence matrix

$$\begin{bmatrix} 1 & 1 & 1 & 1 & 1 & 0 & 0 & 0 & 0 & 0 \\ 1 & 1 & 0 & 0 & 0 & 1 & 1 & 1 & 0 & 0 \\ 1 & 0 & 1 & 0 & 0 & 1 & 0 & 0 & 1 & 1 \\ 0 & 1 & 0 & 1 & 0 & 0 & 1 & 0 & 1 & 1 \\ 0 & 0 & 1 & 0 & 1 & 0 & 1 & 1 & 1 & 0 \\ 0 & 0 & 0 & 1 & 1 & 1 & 0 & 1 & 0 & 1 \end{bmatrix}.$$

It is often useful to discuss designs that lie within other designs. The generic term "subdesign" is used. However, in some cases it is not clear what the appropriate definition should be: for example, does one require that each block be a subset of a block in the original design, or that the set of blocks be a subset of the original set? Usually, we take the latter view—a subdesign of a block design is the block design formed from a subset of the block set and a (sufficiently large) subset of the treatment set. Observe that—for example—a subdesign of a balanced incomplete block design will not necessarily be a balanced incomplete block design.

As was pointed out in Section 1.1, we do not wish to consider two designs to be different if it is possible to relabel the treatments and blocks of one in such a way that the other design is obtained. To formalize the earlier discussion, block designs D and E are defined to be <u>isomorphic</u> if there is a one-to-one map from the set of treatments of D onto the set of treatments of E such that when the mapping is applied to a block of D, the result is a block of E, and such that every block of E can be derived from some block of D in this way. A map with this property is called an <u>isomorphism.</u>

Alternatively, one could define an isomorphism to consist of two one-to-one maps, one mapping the set of treatments of D onto the set of treatments of E, and the other mapping the set of blocks of D onto the set of blocks of E, with the property that if treatment t maps to treatment $t\varphi$ and block B maps to block $B\varphi$, then t belongs to B if and only if $t\varphi$ belongs to $B\varphi$:

$$t \in B \iff t\varphi \in B\varphi.$$

This property is described by saying that "isomorphisms preserve incidence."

In terms of incidence matrices, two block designs are isomorphic if the incidence matrix of the other by row and column permutations. As an example of this, consider the design that was discussed in Section 1.1. It is easy to verify that it is a balanced incomplete block design with parameters $(7, 7, 3, 3, 1)$. The first representation of the design had blocks

$$\{012, 034, 056, 135, 146, 236, 245\}.$$

The corresponding incidence matrix is

$$A = \begin{bmatrix} 1 & 1 & 1 & 0 & 0 & 0 & 0 \\ 1 & 0 & 0 & 1 & 1 & 0 & 0 \\ 1 & 0 & 0 & 0 & 0 & 1 & 1 \\ 0 & 1 & 0 & 1 & 0 & 1 & 0 \\ 0 & 1 & 0 & 0 & 1 & 0 & 1 \\ 0 & 0 & 1 & 1 & 0 & 0 & 1 \\ 0 & 0 & 1 & 0 & 1 & 1 & 0 \end{bmatrix}.$$

When we interchanged treatments 1 and 2 we obtained blocks

$$\{012, 034, 056, 235, 246, 136, 145\},$$

with incidence matrix

1.3 Block Designs

$$B = \begin{bmatrix} 1 & 1 & 1 & 0 & 0 & 0 & 0 \\ 1 & 0 & 0 & 0 & 0 & 1 & 1 \\ 1 & 0 & 0 & 1 & 1 & 0 & 0 \\ 0 & 1 & 0 & 1 & 0 & 1 & 0 \\ 0 & 1 & 0 & 0 & 1 & 0 & 1 \\ 0 & 0 & 1 & 1 & 0 & 0 & 1 \\ 0 & 0 & 1 & 0 & 1 & 1 & 0 \end{bmatrix}.$$

Obviously, B can be converted into the original matrix A by interchanging the second and third rows (which correspond to treatments 1 and 2, respectively). There are, in fact, other row and column permutations that have the same effect: If the last two rows are interchanged and then the permutation (4 6)(5 7) is carried out on the columns of B, we obtain A.

EXERCISES

1.3.1 Prove that there are exactly three nonisomorphic regular designs with $v = 4$, $b = 6$, $r = 3$, and $k = 2$.

1.3.2 Prove that there is a $(7, 7, 4, 4, 2)$-design and that it is unique up to isomorphism.

1.3.3 In each row of the following table, fill in the blanks so that the parameters are possible parameters for a balanced incomplete block design, or else show that this is impossible.

v	b	r	k	λ
	35		3	1
		6	4	2
14	7	4		
		13	6	1
21	28		3	
17		8	5	
21	30		7	
17			7	1

1.3.4 A design is constructed in the following way. The names of 16 treatments are written in a 4 × 4 array: for example,

```
 1  2  3  4
 5  6  7  8
 9 10 11 12
13 14 15 16.
```

There are 16 blocks: To form a block, one chooses a row and a column of the array, deletes the common treatment, and takes the other six elements. (For example, row 1 and column 2 yield the block $\{1,3,4,6,10,14\}$; treatment 2 is deleted.)
  (i) Prove that the design is a balanced incomplete block design. What are its parameters?
  (ii) Does the construction above work if the array is of a size different from 4×4?

1.4 SYSTEMS OF DISTINCT REPRESENTATIVES

Suppose that D is a block design based on $S = (1 \cdots v)$ with blocks B_1, B_2, ..., B_b. We define a <u>system of distinct representatives</u> (SDR) for D to be a way of selecting a member x_i from each block B_i such that $x_1, x_2, \ldots, x_v$ are all different.

As an example, consider the blocks

   123, 124, 134, 235, 246, 1256.

One system of distinct representatives for them is

   1, 2, 3, 5, 4, 6.

There are several others. On the other hand, the blocks

   124, 124, 134, 23, 24, 1256

have no SDR.

If the design D is to have an SDR, it is clearly necessary that D have at least as many treatments as blocks. The example above shows that this is not sufficient: If we consider the first five blocks of the design, they constitute a design with $b = 5$ and $v = 4$. If there were an SDR for the six-block design, then the first five elements would be an SDR for the five-block subdesign, which is impossible. The implied necessary condition—that D contain no subdesign with $v < b$—is in fact sufficient, as we now prove.

THEOREM 1.3 A block design D has a system of distinct representatives if and only if it never occurs that some n blocks contain between them less than n treatments.

1.4 Systems of Distinct Representatives

Proof: We proceed by induction on the number of blocks. If D has one block, the result is obvious. Suppose that D has b blocks $B_1, B_2, \ldots, B_b$ and the v treatments (1 $\cdots$ v), and that D satisfies the hypothesis that the union of any n blocks has size at least n, for $1 \leq n \leq b$. By induction, any set of n blocks has an SDR, provided that $n < b$. We distinguish two cases.

(i) Suppose that no set of n blocks contains less than $n + 1$ treatments in its union, for $n < b$. Select any element x_1 in B_1, and write $B_i^* = B_i \setminus \{x_1\}$, for $i \in (1 \cdots b)$. Then the union of any n of the B_i^* has at least n elements, for $n = 1, 2, \ldots, b - 1$. By the induction hypothesis, there is an SDR $x_2, x_3, \ldots, x_b$ for $B_2, B_3, \ldots, B_b$, so $x_1, x_2, x_3, \ldots, x_b$ is an SDR for the original design D.

(ii) Suppose that there is a set of n blocks whose union has precisely n elements, for some n less than b. Without loss of generality, take these blocks as $B_1, B_2, \ldots, B_n$. For $i > n$, write B_i^* to mean B_i with all members of $B_1 \cup B_2 \cup \cdots \cup B_n$ deleted. It is easy to see that the design with blocks $B_{n+1}^*, B_{n+2}^*, \ldots, B_b^*$ satisfies the conditions of the theorem (if $B_{n+i_1}^*, B_{n+i_2}^*, \ldots, B_{n+i_k}^*$ were k blocks whose union has less than k elements, then $B_1, B_2, \ldots, B_n, B_{n+i_1}, B_{n+i_2}, \ldots, B_{n+i_k}$ would be $n + k$ blocks of D whose union has less than $n + k$ elements, which is impossible). From the induction hypothesis both sets have SDRs, and clearly they are disjoint, so together they comprise an SDR for D. ∎

We give two important applications of this result. Another important application occurs later, as Theorem 10.1.

THEOREM 1.4 Suppose that a set of tasks are to be performed by a set of workers; no task may be assigned to a worker who is not qualified to perform it, and no worker may be required to perform more than one task. A necessary and sufficient condition that all tasks can be assigned is that for any set of n tasks, there are at least n workers capable of performing at least one of them each.

Proof: Number the tasks; for task i, write B_i for the set of workers qualified to perform it. Then what is required is an SDR for the B_i, and the necessary and sufficient condition is precisely that of Theorem 1.3.

THEOREM 1.5 Suppose that M is a square matrix of zeros and ones with the property that every row and every column sums to r. Then there are permutation matrices $P_1, P_2, \ldots, P_r$ such that

$$M = P_1 + P_2 + \cdots + P_r.$$

Proof: We proceed by induction on r. If r = 1, the theorem is clearly true. Suppose that it holds for r < k; let M be a v × v matrix of zeros and ones with constant row and column sums k. Define

$$B_i = \{j : m_{ij} = 1\}.$$

These B_i form a design with blocks of constant size k.

We show that the union of any n blocks contains at least n elements. Without loss of generality, suppose that they are blocks $B_1, B_2, \ldots, B_n$. Consider

$$\{(i,j) : j \in B_i, 1 \leq i \leq n\}.$$

This set has kn elements. But no integer j can occur as right-hand element in more than k of the pairs (since column j has only k nonzero entries), so at least n different values of j must be represented.

Now the blocks B_i have a system of distinct representatives. If x_i is the representative of B_i, write P_k for the permutation matrix with (i, j) entry if and only if $j = x_i$. Then $M - P_k$ is a zero-one matrix. Moreover, it has constant row-sum and column-sum k - 1. So by the induction hypothesis there are permutation matrices $P_1, P_2, \ldots, P_{k-1}$ such that

$$M - P_k = P_1 + P_2 + \cdots + P_{k-1},$$

giving the result. ■

EXERCISES

1.4.1 Show that the blocks

123, 124, 134, 235, 246, 1256

have precisely 17 SDRs.

1.4.2 Suppose that the design D has an SDR, and that each of the b blocks has at least t elements. Prove that if $t \leq b$, then D has at least t! SDRs.

1.4.3 S and T are sets of v elements each. P is a set of ordered pairs (s, t), where s is in S and t is in T. Define

$$A(s) = \{t : t \in T, (s,t) \in P\}$$
$$B(t) = \{s : s \in S, (s,t) \in P\}.$$

Prove that the following two properties are equivalent:
(a) For every k-subset X of S, $1 \leq k \leq v$, $\cup_{s \in X} A(s)$ has at least k elements;
(b) For every k-subset Y of T, $1 \leq k \leq v$, $\cup_{s \in Y} B(t)$ has at least k elements.

BIBLIOGRAPHIC REMARKS

References to the various types of designs will be given in later chapters when we discuss them in more detail. A selection of books that discuss designs in general, and which we have used frequently, is [3], [6], [7], [9]; but these are certainly not the only such books, and not necessarily the best. A recent volume that discusses block designs in great detail is [1].

The literature of graph theory is even more extensive than that of designs. Some good introductory texts are [2] and [5].

Various examples of balanced incomplete block designs have appeared in the literature. However, the first general definition is due to Yates [10]. Some confusion has arisen over the word "incomplete," and several authors have defined a complete design to be the set of all k-sets on a v-set, for some k and v; an incomplete design, for them, is one in which not all k-sets appear. This interpretation is certainly not what was intended by Yates, who specifically refers to "incomplete blocks"; moreover, it is not applicable to cases where the design is not regular. Accordingly, we assert that our definition of "complete" is the correct one. For a further discussion of this point (and one that is worded far more strongly than ours), the reader should consult the paper by Stanton [8].

There is an excellent discussion of systems of distinct representatives in [7]. Theorem 1.3 is due to P. Hall [4].

1. T. Beth, D. Jungnickel, and H. Lenz, Design Theory (Cambridge University Press, Cambridge, England, 1986).
2. J. A. Bondy and U. S. R. Murty, Graph Theory with Applications (Elsevier North-Holland, New York, 1976).
3. M. Hall, Combinatorial Theory, 2nd ed. (Wiley, New York, 1986).
4. P. Hall, "On representations of subsets," Journal of the London Mathematical Society 10(1935), 26-30.
5. F. Harary, Graph Theory (Addison-Wesley, Reading, Mass., 1969).
6. D. Raghavarao, Constructions and Combinatorial Problems in Design of Experiments (Wiley, New York, 1971).
7. H. J. Ryser, Combinatorial Mathematics (Wiley, New York, 1965).
8. R. G. Stanton, "The appropriateness of standard BIBD presentation," Ars Combinatoria 16A(1983), 289-296.
9. A. P. Street and W. D. Wallis, Combinatorics: A First Course (Charles Babbage Research Centre, Winnipeg, Manitoba, Canada, 1982).
10. F. Yates, "Incomplete randomized blocks," Annals of Eugenics 7(1936), 121-140.

2
Balanced Designs

2.1 PAIRWISE BALANCED DESIGNS

A <u>pairwise balanced design</u> of index λ is a way of selecting blocks from a set of treatments such that any two treatments have covalency λ. If there are v treatments and if every block size is a member of some set K of positive integers, the design is designated a PB(v;K;λ). So the parameters consist of two positive integers and one set of positive integers. The number of blocks is <u>not</u> normally treated as a parameter: One can have two pairwise balanced designs with the same parameters but with different numbers of blocks. Both

123, 145, 24, 25, 34, 35

and

123, 14, 15, 24, 25, 34, 35, 45

are PB(5;$\{3,2\}$;1)'s, but they have six and eight blocks, respectively.

It should be observed that we do not require every member of K to be a block size. For instance, the two examples are also PB(5;$\{4,3,2\}$;1)'s. More generally, any PB(v;K;λ) is also a PB(v;L;λ) whenever L contains K.

Various easy results may be proven about pairwise balanced designs. For example, two copies of a PB(v;K;λ), taken together, constitute a PB(v;K;2λ). Just as obviously, there is a PB(v;$\{v\}$;λ) for all positive integers v and λ (it is the complete design with λ sets, each containing all v elements). The following theorem is more interesting.

THEOREM 2.1 Suppose that there exists a PB(v;K;λ); and suppose that for every element k of K there exists a PB(k;L;μ). Then there exists a PB(v;L;$\lambda\mu$).

Proof: Assume that we have a PB(v;K;λ) based on a v-set V. Suppose its blocks are B_1, B_2, ..., B_n, where B_i has k_i elements. We replace each

2.1 Pairwise Balanced Designs

block by a new collection of blocks. Given B_i, form a $PB(k_i; L; \mu)$, but instead of taking the numbers 1, 2, ..., k_i as treatments, use the elements of B_i. This is done for every i. If x and y are any two treatments, then $\{x, y\}$ was contained in λ of the B_i, and in each case B_i has been replaced by a collection of blocks, μ of which contain $\{x, y\}$. So x and y occur together $\lambda \mu$ times in total. Therefore, the total collection of new blocks is a pairwise balanced design of index $\lambda \mu$, based on V. The size of any block of the new design is a member of L, so the design is a $PB(v; L; \lambda \mu)$. ∎

As an example, we construct a $PB(15; \{4, 3\}; 3)$ from the $PB(15; \{5, 3\}; 1)$ with blocks

```
01234     56789     ABCDE
05A   06C   07E   08B   09D
16B   17D   18A   19C   15E
27C   28E   29B   25D   26A
38D   39A   35C   36E   37B
49E   45B   46D   47A   48C
```

We need a $PB(5; \{4, 3\}; 3)$ and a $PB(3; \{4, 3\}; 3)$. The former has blocks

$$1234 \quad 1235 \quad 1245 \quad 1345 \quad 2345 \tag{2.1}$$

and the latter is the complete design

$$123 \quad 123 \quad 123.$$

The block 01234 is replaced by a copy of the set of blocks (2.1) in which the treatments have been relabeled using the mapping

$$(1, 2, 3, 4, 5) \longmapsto (0, 1, 2, 3, 4).$$

The new blocks are

$$0123 \quad 0124 \quad 0134 \quad 0234 \quad 1234. \tag{2.2}$$

Similarly, 56789 and ABCDE are replaced by

$$5678 \quad 5679 \quad 5689 \quad 5789 \quad 6789 \tag{2.3}$$

and

$$ABCD \quad ABCE \quad ABDE \quad ACDE \quad BCDE, \tag{2.4}$$

respectively. The block 05A is replaced by

05A 05A 05A

and similarly for every other 3-block. So the required design consists of the blocks listed in (2.2), (2.3), and (2.4) and three copies of every 3-set in the original design.

The PB(15; $\{5,3\}$;1) which we used above is an example of a useful class of designs. We now give a more general construction.

THEOREM 2.2 There exists a PB(3k; $\{3,k\}$;1) whenever k is odd.

Proof: We construct a design with treatment set $\{a_1, a_2, \ldots, a_k, b_1, b_2, \ldots, b_k, c_1, c_2, \ldots, c_k\}$. There are three blocks of size k (which we call "big blocks"), namely,

$$a_1 a_2 \cdots a_k \quad b_1 b_2 \cdots b_k \quad c_1 c_2 \cdots c_k.$$

The blocks of size 3 are the blocks

$$\{a_i b_{i+x} c_{i+2x} : 1 \leq i \leq k, \ 1 \leq x \leq k\}$$

where subscripts that exceed k are reduced modulo k.

The pairs $\{a_i, a_j\}$, $\{b_i, b_j\}$, and $\{c_i, c_j\}$ occur once each, in the big blocks. The pair $\{a_i, b_j\}$ occurs in $a_i b_{i+x} c_{i+2x}$ if and only if $j \equiv i + x$ (mod k); this will happen for only one value of x, either $x = j - i$ (if $j > i$) or $x = k + j - i$ (if $j \leq i$). So only one block contains $\{a_i, b_j\}$. A similar remark applies to $\{b_i, c_j\}$. If $\{a_i, c_j\}$ occurs in $a_i b_{i+x} c_{i+2x}$, then $i + 2x \equiv j$ (mod k), and this also uniquely defines x: since k is odd, the solution is $x \equiv (1/2) \times (k + 1)(j - 1)$ (mod k). ∎

The construction above does not generalize to even values of k, because $(1/2)(k + 1)(j - i)$ is not necessarily an integer in that case. However, we shall see in Chapter 10 that another construction is available for even k.

We now prove an interesting theorem concerning the number of blocks in a pairwise balanced design. We start with an easy remark.

LEMMA 2.3 In a pairwise balanced design with $\lambda = 1$, no two blocks have two common elements.

Proof: Suppose that $B_1 = \{x, y, \ldots\}$ and $B_2 = \{x, y, \ldots\}$. Then $\{x, y\}$ is a subset of B_1, and also of B_2. So $\lambda_{xy} \geq 2$. But this contradicts the property "$\lambda = 1$."

THEOREM 2.4 Suppose that there is a PB(v;K;1) with b blocks, where $b > 1$. Then $b \geq v$. If $b = v$, then either the PB(v;K;1) has one block of size $v - 1$ and the rest of size 2, or else $b = v = k^2 - k + 1$ for some integer k and all the blocks have size k.

2.1 Pairwise Balanced Designs

Proof: Suppose that a PB(v;K;1) has treatments $t_1, t_2, \ldots, t_v$ and b blocks $B_1, B_2, \ldots, B_b$. Say that k_i equals the number of elements in B_i, and that t_j belongs to r_j blocks. (We call r_j the frequency or replication number of t_j.) Then, counting all elements of all blocks,

$$\sum_{j=1}^{v} r_j = \sum_{i=1}^{b} k_i. \qquad (2.5)$$

If t_j does not belong to B_i, then t_j must belong to at least k_i blocks: for every element x of B_i there is a block that contains t_j and x, and these blocks are all disjoint because of Lemma 2.3. So

$$t_j \notin B_i \implies k_i \leq r_j. \qquad (2.6)$$

The blocks are incomplete, and we can assume that there are no blocks of size 1 (since we are seeking to minimize the number of blocks, and blocks of size 1 could be deleted). So

$$1 < k_i < v \quad \text{for} \quad i \in (1 \cdots b). \qquad (2.7)$$

There must be some treatment whose replication number is minimal. Suppose that it is t_v, and write $r_v = m$. Relabel the blocks so that those containing t_v are $B_1, B_2, \ldots, B_m$. We can select an element of each block, other than t_v, and clearly all these elements are different. Suppose (after relabeling) that $t_i \in B_i$, $t_i \neq t_v$. Then if $1 \leq i \leq m$, $1 \leq j \leq m$, and $i \neq j$, $t_j \notin B_i$; in particular,

$$t_1 \notin B_2, \ t_2 \notin B_3, \ \ldots, \ t_{m-1} \notin B_m, \ t_m \notin B_1,$$

so from (2.6)

$$k_2 \leq r_1, \ k_3 \leq r_2, \ \ldots, \ k_m \leq r_{m-1}, \ k_1 \leq r_m. \qquad (2.8)$$

Also, $t_v \notin B_i$ for $i > m$, so $k_i \leq r_v$ for $i > m$.

Adding these inequalities and comparing with (2.5), we obtain

$$\sum_{j=1}^{v} r_j = \sum_{i=1}^{b} k_i \leq \sum_{j=1}^{m} r_j + \sum_{j=m+1}^{b} r_v \leq \sum_{j=1}^{b} r_j$$

(since $r_v \leq r_j$ for all j), and this is impossible if b < v.

In particular, suppose that $b = v$. Then each of the inequalities in (2.8) must be an equality, and also $k_i = r_i$ for all $i > m$. If we relabel the treatments $t_1, t_2, \ldots, t_m$, we obtain

$$r_i = k_i, \quad \text{all} \quad i \in (1 \cdots v)$$

for some ordering of treatments and blocks. Moreover, t_v is unchanged. Let us further relabel the treatments (and simultaneously, the blocks) so that

$$r_1 \geq r_2 \geq \cdots \geq r_v.$$

(Since t_v had minimum frequency, it has still not been disturbed.)

We consider the various possibilities.

(i) Suppose that $r_1 > r_2$. Then $r_1 > r_j$ for all $j \geq 2$. So $k_1 = r_1 > r_j$ ($j \geq 2$). From (2.6), $t_j \in B_1$ for all $j > 1$. Of course, $t_1 \notin B_1$. So

$$B_1 = \{t_2, t_3, \ldots, t_v\},$$

and the other blocks must be

$$\{t_1, t_2\}, \{t_1, t_3\}, \ldots, \{t_2, t_v\}.$$

(ii) Suppose that $r_1 = r_2 = \cdots = r_{j-1} > r_j$, where $j > 2$. Then $t_j \in B_1$ and $t_j \in B_2$ [from (2.6)]; since $t_v \in B_1 \cap B_2$, the only possibility (according to Lemma 2.3) is $t_j = t_v$ and $j = v$. So we consider that case. Since $r_v < r_{v-1} = k_{v-1} < v$ [from (2.7)] there are at least two blocks not containing t_v. One might be B_v, but suppose the other is B_x, where $x \neq v$. Then from (2.6) we have $r_x = k_x \leq r_v$, which is a contradiction.

(iii) Finally, suppose that $r_1 = r_2 = \cdots = r_v$. We have constant block size and constant frequency—a balanced incomplete block design. From (1.1) and (1.2) we immediately deduce that $b = v = k^2 - k + 1$, where k is the common block size. ∎

Theorem 2.4 can be strengthened if further restrictions are put on the sizes of the blocks. Suppose that we write $g^K(v)$ for the smallest possible number of blocks in a PB(v;K;1); and, in particular, write $g^n(v)$ for $g^K(v)$ in the case where $K = (2 \cdots n)$.

THEOREM 2.5 For any v,

$$g^4(v) \geq \frac{v(6\left\lceil \frac{v-1}{3} \right\rceil - v + 1)}{12}.$$

2.1 Pairwise Balanced Designs

Proof: Suppose that there exists a PB(v;$\{2,3,4\}$;1) with b blocks B_1, B_2, ..., B_b and v treatments t_1, t_2, ..., t_v. As before we write k_i for the size of B_i and r_j for the frequency of t_j. Suppose that f_2 of the blocks have size 2. Since $\lambda = 1$,

$$v - 1 = \sum_{t_j \in B_i} (k_i - 1),$$

and since $k_i \leq 4$ for all i, we have

$$v - 1 \leq 3r_j,$$

so $r_j \geq \left\lceil \frac{v-1}{3} \right\rceil$ for all j. Summing over j, we have

$$\sum_{i=1}^{b} k_i = \sum_{j=1}^{v} r_j \geq v \left\lceil \frac{v-1}{3} \right\rceil.$$

The number of unordered pairs of treatments in B_i is $(1/2)k_i(k_i - 1)$. Since every unordered pair occurs exactly once,

$$\frac{1}{2} v(v - 1) = \frac{1}{2} \sum_{i=1}^{b} k_i(k_i - 1),$$

$$v(v - 1) = \sum_{i=1}^{b} k_i(k_i - 1). \tag{2.9}$$

Now observe that $(k_i - 3)(k_i - 4) = 0$ when $k_i = 3$ or $k_i = 4$, and equals 2 when $k_i = 2$. So

$$\sum_{i=1}^{b} (k_i - 3)(k_i - 4) = 2f_2 \geq 0.$$

On the other hand,

$$(k_i - 3)(k_i - 4) = k_i(k_i - 1) - 6k_i + 12,$$

so

$$0 \leq \sum_{i=1}^{b} (k_i - 3)(k_i - 4) = \sum_{i=1}^{b} k_i(k_i - 1) - 6 \sum_{i=1}^{b} k_i + 12b$$

$$\leq v(v - 1) - 6v \left\lceil \frac{v-1}{3} \right\rceil + 12b,$$

whence

$$b \geq \frac{6v \left\lceil \frac{v-1}{3} \right\rceil - v(v-1)}{12}$$

$$= \frac{v \left(6 \left\lceil \frac{v-1}{3} \right\rceil - v + 1 \right)}{12}. \qquad \blacksquare$$

Equation (2.9) can be useful in proving that certain designs cannot exist or in restricting the numbers of blocks of a given size. If f_j is the number of blocks of size j, (2.9) becomes

$$v(v-1) = \sum_{j \in K} j(j-1) f_j. \tag{2.10}$$

As an example of its use, we observe that no $PB(10; \{5,4\}; 3)$ exists: for if such a design existed, then

$$20 f_5 + 12 f_4 = 90;$$

the left-hand side is divisible by 4, but the right-hand side is not. Similarly, no $PB(6; \{4,3\}; 1)$ exists. For such a design would satisfy

$$12 f_4 + 6 f_3 = 30,$$

$$2 f_4 + f_3 = 5.$$

The only possible cases are $(f_4, f_3) = (0, 5), (1, 3),$ or $(2, 1)$.

(i) If $f_4 = 0$ and $f_3 = 5$, then the design is a balanced incomplete block design with $v = 6$, $k = 3$, and $\lambda = 1$. From (1.2), $5 = 2r$, which is impossible.
(ii) If $f_4 = 1$ and f_3, we can assume that the design contains the block 1234. The other three blocks are each of size 3. Since 1 has yet to appear in the same block as 5 and 6, 156 must be a block of the design; similarly, 256 must be one block, in order to satisfy the condition that 2 should

2.2 Balanced Incomplete Block Designs

appear precisely once with each other treatment. So 5 and 6 occur together more than once.

(iii) If $f_4 = 2$ and $f_3 = 1$, the design contains two sets of size 4 chosen from 6 objects. These sets must have two or more common elements; if $\{x,y\}$ is a subset of each, then x and y occur together more than once.

EXERCISES

2.1.1 Prove that no PB(8;$\{4,3\}$;1) can exist.

2.1.2 Assume that there exists a PB(v;$\{k,3\}$;1), and that v is congruent to 2 modulo 3. Prove that k is congruent to 2 modulo 3.

2.1.3 Suppose that there is a PB(7;$\{5,4,3\}$;1). Prove that the number of blocks of size 5 is divisible by 3, and that consequently no such blocks exist. Can there by any blocks of size 4?

2.1.4 Define $g^{(n)}(v)$ to be the smallest possible number of blocks in a PB(v;$\{2,3,\ldots,n\}$;1) which contains at least one block of size n. Prove that $g^{(4)}(7) = 10$, whereas $g^4(7) = 7$. Find $g^{(4)}(6)$ and $g^4(6)$.

2.1.5 Is it true that for any positive integer n, there is a positive integer V such that $g^{(n)}(v) = g^n(v)$ whenever $v \geq V$?

2.2 BALANCED INCOMPLETE BLOCK DESIGNS

One can consider a balanced incomplete block design as a kind of pairwise balanced design, so all the results of the preceding section apply to them. In particular, Theorem 2.1 can be applied. If the set L in that Theorem has only one element, the resulting design is regular. So we have the following corollary to Theorem 2.1.

THEOREM 2.6 Suppose that there exists a PB(v;K;λ), and suppose that for every member k of K there is a balanced incomplete block design on k treatments with block size ℓ and balance parameter μ. Then there exists a balanced incomplete block design with parameters

$$\left(v, \frac{\lambda\mu v(v-1)}{\ell(\ell-1)}, \frac{\lambda\mu(v-1)}{\ell(\ell-1)}, \ell, \lambda\mu\right).$$ ∎

In a balanced incomplete block design all the sets chosen are the same size, every treatment occurs equally often and the design is pairwise balanced. It is possible to relax the condition of pairwise balance but keep the other two conditions, and the result is a regular block design, which we discussed in Chapter 1. It is also possible to retain balance and retain the constant number of replications but have more than one block size; such designs are called (r,λ)-designs, where r and λ are the appropriate parameters.

However, we now prove that there are no examples where the restriction on the number of replications is the only one that is dropped.

THEOREM 2.7 A pairwise balanced design in which every set has the same size is a balanced incomplete block design.

Proof: If treatment x has frequency r_x, one obtains

$$(k - 1)r_x = \lambda(v - 1)$$

as in Theorem 1.2. But this means that all treatments have the same frequency. ∎

If one treatment is deleted from a balanced incomplete block design, the result is a pairwise balanced design with two block sizes, k and k - 1. These pairwise balanced designs will have constant replication number (unless k = 2, and the blocks of size 1 which arise are then deleted). This gives rise to many examples of (r, λ)-designs, but there are other examples which are not constructed in this way.

We shall now discuss the incidence matrices of balanced incomplete block designs. Incidence matrices belong to the class of (0, 1)-matrices, matrices each of whose entries is 0 or 1. It is interesting to observe that two equations characterize the incidence matrices of balanced incomplete block designs among the wider class of (0, 1)-matrices. To express these equations, we use two notations: an n × n identity matrix is denoted I_n, and an m × n matrix with every entry 1 is denoted $J_{m \times n}$. (If m = n, we simply write J_m for $J_{m \times n}$; and in any event, we usually omit the subscripts whenever possible.)

THEOREM 2.8 If A is the incidence matrix of a (v, b, r, k, λ) design, then

$$AA^T = (r - \lambda)I_v + \lambda J_v \qquad (2.11)$$

and

$$J_v A = kJ_{v \times b}. \qquad (2.12)$$

Conversely, if there is a v × b (0, 1)-matrix A which satisfies (2.11) and (2.12), then

$$v = \frac{r(k - 1)}{\lambda} + 1,$$

$$b = \frac{vr}{k},$$

2.2 Balanced Incomplete Block Designs

and provided that $k < v$, A is the incidence matrix of a (v, b, r, k, λ)-design.

Proof: First, suppose that A is the incidence matrix of a (v, b, r, k, λ)-design. As $a_{ij} = 1$ if and only if treatment i belongs to block j, the number of entries 1 in column j equals the number of members of block j, which is k. All other entries of A are zero, so the number of entries 1 in a column equals the sum of the entries in the column. But each entry of column j of JA equals the sum of the entries in column j of A. Therefore, (2.12) holds.

The (i, j) entry of AA^T is

$$\sum_{n=1}^{b} a_{in} a_{jn}. \tag{2.13}$$

Now $a_{in} a_{jn}$ equals 1 when both treatments i and j belong to block n, and 0 otherwise. So the sum (2.13) equals the number of blocks that contain both t_i and t_j; this is r when $i = j$ and λ when $i \neq j$. Therefore,

$$AA^T = rI + \lambda(J - I)$$
$$= (r - \lambda)I + \lambda J. \tag{2.11}$$

Conversely, suppose that A is a $(0, 1)$ matrix satisfying (2.11) and (2.12). Define a block design with treatments $t_1, t_2, \ldots, t_v$ and blocks $B_1, B_2, \ldots, B_b$ by

$$t_i \in B_j \text{ if and only if } a_{ij} = 1.$$

One sees easily that this is a balanced incomplete block design (provided that $k < v$). Consequently, v and b must satisfy the given equations. ∎

It is useful to know the determinant of the matrix AA^T. We prove a slightly more general result, in that we do not assume the parameters r and λ to be positive integers.

LEMMA 2.9 The determinant of the $v \times v$ matrix

$$M = (r - \lambda)I + \lambda J$$

is $(r - \lambda)^{v-1}[r + (v - 1)\lambda]$.

Proof: The matrix M is transformed as follows: first, subtract column 1 from every other column; second, add each of rows 2, 3, ..., n to row 1. The process is

$$M = \begin{bmatrix} r & \lambda & \lambda & \cdots & \lambda \\ \lambda & r & \lambda & \cdots & \lambda \\ \lambda & \lambda & r & \cdots & \lambda \\ \cdot & \cdot & \cdot & \cdots & \cdot \\ \lambda & \lambda & \lambda & \cdots & r \end{bmatrix} \longrightarrow \begin{bmatrix} r & \lambda-r & \lambda-r & \cdots & \lambda-r \\ \lambda & r-\lambda & 0 & \cdots & 0 \\ \lambda & 0 & r-\lambda & \cdots & 0 \\ \cdot & \cdot & \cdot & \cdots & \cdot \\ \lambda & 0 & 0 & \cdots & r-\lambda \end{bmatrix}$$

$$\longrightarrow \begin{bmatrix} r+(v-1)\lambda & 0 & 0 & \cdots & 0 \\ \lambda & r-\lambda & 0 & \cdots & 0 \\ \lambda & 0 & r-\lambda & \cdots & 0 \\ \cdot & \cdot & \cdot & \cdots & \cdot \\ \lambda & 0 & 0 & \cdots & r-\lambda \end{bmatrix}.$$

These transformations do not change the determinants; since the final matrix is zero above the diagonal, the determinant equals the product of the diagonal elements, which is $(r-\lambda)^{v-1}[r+(v-1)\lambda]$.

THEOREM 2.10 In a balanced incomplete block design, $b \geq v$.

Proof: Suppose that A is the incidence matrix of a (v,b,r,k,λ)-BIBD. Since $k < v$, the equation

$$\lambda(v-1) = r(k-1)$$

implies that $r > \lambda$. So the determinant of AA^T, which equals $(r-\lambda)^{v-1}[r+(v-1)\lambda]$, is nonzero. Therefore, AA^T has rank v, and

$$v = \text{rank}(AA^T) \leq \text{rank } A \leq \min(v,b).$$

So $v \leq b$. ∎

Theorem 2.10 is called **Fisher's inequality**. The original proof did not involve matrices, but rather used an ingenious method that takes its inspiration from statistics. If n readings $f_1, f_2, \ldots, f_n$ are given, their mean $\bar{f}$ is defined as $n^{-1} \Sigma f_i$, and their variance v is $n^{-1} \Sigma (f_i - \bar{f})^2$. It is obvious that $\Sigma (f_i - \bar{f})^2$ is nonnegative. In calculating variances, it is observed that

$$\sum_i (f_i - \bar{f})^2 = \left(\sum_i f_i^2\right) - n\bar{f}^2.$$

Now write $n = b - 1$. Define f_i to be the size of the intersection $B_i \cap B_b$ of blocks i and b in a (v,b,r,k,λ)-BIBD. We count the occurrences of pairs of members of B_b in the other blocks. Each pair must occur $\lambda - 1$ more times, so

2.2 Balanced Incomplete Block Designs

$$\sum \frac{1}{2} f_i(f_i - 1) = \frac{1}{2} k(k - 1)(\lambda - 1).$$

On the other hand,

$$\sum_i f_i = k(r - 1)$$

whence we have

$$\bar{f}_i = (b - 1)^{-1} k(r - 1)$$

and also

$$\sum_i f_i^2 = k(k - 1)(\lambda - 1) + k(r - 1).$$

Therefore,

$$\sum_i f_i^2 - (b - 1)\bar{f}^2 = k(k - 1)(\lambda - 1) + k(r - 1) - (b - 1)^{-1} k^2 (r - 1)^2.$$

Several identities are now used. The least obvious is the observation that

$$k^2(r - 1)^2 - r^2(k - 1)^2 = [k(r - 1) - r(k - 1)][k(r - 1) + r(k - 1)]$$
$$= (r - k)(2kr - k - r),$$

so that

$$k^2(r - 1)^2 = r^2(k - 1)^2 + (r - k)(2kr - k - r)$$
$$= \lambda(v - 1)r(k - 1) + (r - k)(2kr - k - r)$$

[using (1.2)]. Others needed are

$$k(b - 1) = vr - k = r(v - 1) + (r - k),$$
$$\lambda v = \lambda - rk + r,$$

which use (1.1) and (1.2), respectively. We can then proceed:

$$(b-1)\left[\sum f_1^2 - (b-1)\bar{f}^2\right]$$

$$= (b-1)k(k-1)(\lambda-1) + (b-1)k(r-1) - k^2(r-1)^2$$

$$= (b-1)k[(k-1)(\lambda-1) + r - 1] - \lambda(v-1)r(k-1) + (r-k)(k+r-2kr)$$

$$= r(v-1)(k\lambda - k - \lambda + r) + (r-k)(k\lambda - k - \lambda + r) - \lambda(v-1)r(k-1)$$
$$+ (r-k)(k+r-2kr)$$

$$= r(v-1)(k\lambda - k - \lambda + r - k\lambda + \lambda) + (r-k)(k\lambda - k - \lambda + r + k + r - 2kr)$$

$$= r(v-1)(r-k) + (r-k)(k\lambda - \lambda + 2r - 2kr)$$

$$= (r-k)(rv + r + k\lambda - \lambda - 2kr)$$

$$= (r-k)(rv + k\lambda - 2kr - \lambda v)$$

$$= (r-k)(v-k)(r-\lambda).$$

Now $v - k$ and $r - \lambda$ are both positive, so

$$r - k \geq 0,$$

and it follows from (1.1) that $b \geq v$.

If A is the incidence matrix of a regular design with parameters v, b, r, and k, then A^T is the incidence matrix of a regular design with parameters b, v, k, and r. The design is easy to construct. If the original design had blocks $B_1, B_2, \ldots, B_b$ and treatments $t_1, t_2, \ldots, t_v$, the new design has blocks $C_1, C_2, \ldots, C_v$ and treatments $u_1, u_2, \ldots, u_b$, and u_i belongs to C_j if and only if t_j belongs to B_i. The new design is called the dual of the original.

It is clear that the dual of an incomplete design is an incomplete design. We shall, however, prove that balance is preserved if and only if $v = b$. We would need to show that a $v \times v$ matrix of zeros and ones which satisfies

$$AA^T = (k-\lambda)I + \lambda J \tag{2.14}$$

$$JA = kJ \tag{2.15}$$

[which are the results of substituting $r = k$ in (2.11) and (2.12)] will also satisfy

$$A^T A = (k-\lambda)I + \lambda J \tag{2.16}$$

$$JA^T = kJ. \tag{2.17}$$

In fact, we shall prove the following, stronger theorem.

2.2 Balanced Incomplete Block Designs

THEOREM 2.11 If A is a nonsingular matrix of side v which satisfies one of (2.14), (2.16) and one of (2.15), (2.17), then it satisfies all four equations, and

$$k(k - 1) = \lambda(v - 1).$$

Proof: All matrices in the proof are $v \times v$. Since A is nonsingular, both AA^T and A^TA have nonzero determinant. So either (2.14) or (2.16) implies that $(k - \lambda)I + \lambda J$ has nonzero determinant; from Lemma 2.11, we have

$$k - \lambda \neq 0, \quad \lambda(v - 1) + k \neq 0.$$

First suppose that (2.14) and (2.15) hold and A is nonsingular. From (2.14) we deduce that

$$JAA^T = (k - \lambda)J + \lambda J^2$$

and substituting kJ for JA [from (2.15)] and vJ for J^2 yields

$$kJA^T = (k - \lambda)J + \lambda vJ.$$

So

$$kJA^T J = (k - \lambda + \lambda v)J^2$$
$$kJ(JA)^T = (k - \lambda + \lambda v)J^2$$
$$kJ(kJ)^T = (k - \lambda + \lambda v)J^2$$
$$k^2 J^2 = (k - \lambda + \lambda v)J^2,$$

whence $k^2 = k - \lambda + \lambda v$ and $k(k - 1) = \lambda(v - 1)$. Also, from $kJA^T = (k - \lambda + \lambda v)J$ we can now deduce

$$kJA^T = k^2 J;$$

since $k^2 = k - \lambda + \lambda v = \lambda(v - 1) + k$, which we know to be nonzero, k is nonzero, so we can divide by k to get

$$JA^T = kJ \qquad (2.17)$$

and

$$A^T A = A^{-1}(AA^T)A = (k - \lambda)I + \lambda A^{-1}JA; \tag{2.18}$$

but we have $AJ = kJ = JA$, so $A^{-1}JA = A$, and (2.18) becomes (2.16).

Next, suppose that (2.14) and (2.17) hold for the nonsingular A. Since $JA^T = kJ$ we have $AJ = kJ$, so $J = kA^{-1}J$, whence k must be nonzero and $A^{-1}J = k^{-1}J$. Now

$$\begin{aligned}
A^T &= A^{-1}(AA^T) \\
&= (k - \lambda)A^{-1} + \lambda A^{-1}J \\
&= (k - \lambda)A^{-1} + \lambda k^{-1}J;
\end{aligned} \tag{2.19}$$

$$\begin{aligned}
kJ = JA^T &= (k - \lambda)JA^{-1} + \lambda k^{-1}J^2 \\
&= (k - \lambda)JA^{-1} + \lambda k^{-1}vJ.
\end{aligned}$$

Therefore (division being possible because we know $k - \lambda \ne 0$),

$$JA^{-1} = \frac{k - \lambda k^{-1}v}{k - \lambda} J, \tag{2.20}$$

$$JA^{-1}J = \frac{k - \lambda k^{-1}v}{k - \lambda} (vJ).$$

But

$$\begin{aligned}
vk^{-1}J &= k^{-1}J^2 \\
&= k^{-1}J(kA^{-1}J) \\
&= JA^{-1}J.
\end{aligned}$$

So we have

$$vk^{-1} = \frac{(k - \lambda k^{-1}v)v}{k - \lambda},$$

$$k - \lambda = k^2 - \lambda v,$$

which is $k(k - 1) = \lambda(v - 1)$. We can substitute k^{-1} for the right-hand coefficient in (2.20):

2.2 Balanced Incomplete Block Designs

$$JA^{-1} = k^{-1}J,$$

whence $JA = kJ$, which is (2.15). Finally, returning to (2.19) and multiplying on the right by A, we get

$$A^T A = (k - \lambda)A^{-1}A + \lambda k^{-1}J$$
$$= (k - \lambda)I + \lambda k^{-1}JA$$
$$= (k - \lambda)I + \lambda k^{-1}(kJ)$$
$$= (k - \lambda)I + \lambda J. \qquad (2.16)$$

Finally, suppose that we start with (2.16) and (2.15). We see that these are just (2.14) and (2.17) with A replaced by A^T; so A^T satisfies (2.15) and (2.16) and $k(k-1) = \lambda(v-1)$, which means that A satisfies (2.14) and (2.17) and $k(k-1) = \lambda(v-1)$. A similar remark applies to the case of (2.16) and (2.17). ∎

COROLLARY 2.11.1 The dual of a balanced incomplete block design is a balanced incomplete block design if and only if the design is symmetric: $v = b$.

Proof: (i) If $b > v$ in a balanced incomplete block design, its dual design will have more treatments than blocks, and by Theorem 2.10 it cannot be a balanced incomplete block design.

(ii) If $b = v$, Theorem 2.11 applies. ∎

COROLLARY 2.11.2 The intersection of two distinct blocks of a symmetric balanced incomplete block design always contains λ elements.

Proof: Suppose that a symmetric balanced incomplete block design has incidence matrix A. The number of treatments common to block i and block j equals the number of places where column i and column j both have entry 1. As A is a (0,1)-matrix, this number must equal the scalar product of column i with column j, which in turn is the (i, j) entry of $A^T A$; this always equals λ, by Corollary 2.11.1. ∎

Suppose that there is a balanced incomplete block design with parameters (v, b, r, k, λ), having blocks $B_1, B_2, \ldots, B_b$. Write S for the v-set of all treatments in the design. Then the sets

$$S \backslash B_1, S \backslash B_2, \ldots, S \backslash B_b$$

form a design with treatment set S, which is called the <u>complementary design</u> or <u>complement</u> of the original.

THEOREM 2.12 The complementary design of a (v, b, r, k, λ)-design is a balanced incomplete block design with parameters

$$(v, b, b - r, v - k, b - 2r + \lambda),$$

provided that $b - 2r + \lambda$ is nonzero.

Proof: There are v treatments and b blocks in either design. Since each treatment belonged to r blocks in the original design, it will belong to the other $b - r$ blocks in the complement; since B_i has k elements, S has v elements, and B_i is a subset of S, $S \backslash B_i$ has $v - k$ elements. So we have a regular design and the first four parameters are verified. Finally, consider two distinct treatments t_1 and t_2 in the original design. They occur together in λ blocks. At t_1 belongs to r blocks, there will be $r - \lambda$ which contain t_1 but not t_2. Similarly, $r - \lambda$ blocks contain t_2 but not t_1. So the number of blocks that contain neither t_1 nor t_2 is

$$b - (r - \lambda) - (r - \lambda) - \lambda = b - 2r + \lambda,$$

and this will be the number of complementary blocks which contain both of them. The constancy of this number implies that the design is balanced. ∎

As an example, here are the blocks of a $(9, 12, 4, 3, 1)$-BIBD A and its complement, a $(9, 12, 8, 6, 5)$-BIBD C.

```
    1 2 3           4 5 6 7 8 9
    4 5 6           1 2 3 7 8 9
    7 8 9           1 2 3 4 5 6
    1 4 7           2 3 5 6 8 9
    2 5 8           1 3 4 6 7 9
    3 6 9           1 2 4 5 7 8
    1 6 8           2 3 4 5 7 9
    2 4 9           1 3 5 6 7 8
    3 5 7           1 2 4 6 8 9
    1 5 9           2 3 4 6 7 8
    2 6 7           1 3 4 5 8 9
    3 4 8           1 2 5 6 7 9

       A                  C
(original design)    (its complement)
```

EXERCISES

2.2.1 A is a $(0, 1)$-matrix. Prove that A is the incidence matrix of a regular block design if and only if A satisfies the equations

2.2 Balanced Incomplete Block Designs

$$J_v A = kJ_{v \times b}$$
$$AJ_b = rJ_{v \times b}$$

for some integers v, b, r, k.

2.2.2 S is a set and B is the set of all k-subsets of S. Prove that if B is considered as a set of blocks on the treatment set S, the result is a balanced incomplete block design. If S has v elements, what are the parameters of the design?

2.2.3 Are there any balanced incomplete block designs with $k = 2$, $\lambda = 1$, other than those which arise in Exercise 2.2.2?

2.2.4 Suppose that D is a balanced incomplete block design with parameters (v, b, r, k, λ) whose blocks are all different. Write S for the set of treatments on which D is based; and write T for the collection of all k-sets on S other than the blocks of D. Prove that if $b < \binom{v}{k}$, the set T of blocks forms a balanced incomplete block design based on S, and find its parameters.

2.2.5 D is the balanced incomplete block design with parameters $(6, 10, 5, 3, 2)$ and blocks

$$B_0 = \{0, 1, 2\}, \quad B_5 = \{1, 2, 5\},$$
$$B_1 = \{0, 1, 3\}, \quad B_6 = \{1, 3, 4\},$$
$$B_2 = \{0, 2, 4\}, \quad B_7 = \{1, 4, 5\},$$
$$B_3 = \{0, 3, 5\}, \quad B_8 = \{2, 3, 4\},$$
$$B_4 = \{0, 4, 5\}, \quad B_9 = \{2, 3, 5\}.$$

Construct the blocks of the dual of D, and verify that it is not balanced.

2.2.6 D is the $(15, 15, 7, 7, 3)$-design with blocks

$$B_0 = \{0, 1, 2, 3, 4, 5, 6\}, \quad B_4 = \{0, 3, 4, 9, 10, 13, 14\},$$
$$B_1 = \{0, 1, 2, 7, 8, 9, 10\}, \quad B_5 = \{0, 5, 6, 7, 8, 13, 14\},$$
$$B_2 = \{0, 1, 2, 11, 12, 13, 14\}, \quad B_6 = \{0, 5, 6, 9, 10, 11, 12\},$$
$$B_3 = \{0, 3, 4, 7, 8, 11, 12\}, \quad B_7 = \{1, 3, 5, 7, 9, 11, 13\},$$

$B_8 = \{1,3,6,7,10,12,14\}$, $\quad B_{12} = \{2,3,6,8,9,11,14\}$,

$B_9 = \{1,4,5,8,10,11,14\}$, $\quad B_{13} = \{2,4,5,7,9,12,14\}$,

$B_{10} = \{1,4,6,8,9,12,13\}$, $\quad B_{14} = \{2,4,6,7,10,11,13\}$.

$B_{11} = \{2,3,5,8,10,12,13\}$,

(i) Write down the blocks of the dual design D^* of D.

(ii) A *triplet* is a set of three blocks whose mutual intersection has size 3. (For example, $\{B_0, B_1, B_2\}$ is a triplet in D.) By considering triplets, prove that D and D^* are not isomorphic.

2.2.7 (Another proof of Fisher's inequality.) Suppose that $V_1, V_2, \ldots, V_v$ are the rows of the incidence matrix of a (v,b,r,k,λ)-BIBD. Define

$K_1 = V_1$;

$K_i = V_i - \dfrac{\lambda}{r+(i-2)\lambda}(V_1 + V_2 + \cdots + V_{i-1})$, $\quad 2 \leq i \leq v$.

Prove that the vectors $K_1, K_2, \ldots, K_v$ are v orthogonal, and therefore independent, vectors of length b, whence $v \leq b$.

2.2.8 Suppose that D is a balanced incomplete block design with parameters (v,b,r,k,λ). Form a new design by replacing each block of D by n identical copies of itself. Prove that this new structure is a balanced incomplete block design (called the n-multiple of D). What are its parameters?

2.3 t-DESIGNS

Suppose that t, v, k, and λ are positive integers with $1 < k < \lambda$. We define a t-design with parameters (v,k,λ), or $t-(v,k,\lambda)$ design to be a way of selecting blocks of size k from a v-set so that any set of t treatments appears as a subset of exactly λ blocks. A t-design is a generalization of a balanced incomplete block design; in fact, a balanced incomplete block design is a 2-design. We define b and r in the usual way.

LEMMA 2.13 In a $t-(v,k,\lambda)$ design,

$$b = \lambda \frac{\binom{v}{t}}{\binom{k}{t}}.$$

2.3 t-Designs

Proof: We count all the ordered pairs whose first element is a block and whose second element is a t-set contained in that block. Since there are b blocks and each contains $\binom{k}{t}$ t-sets, these are $b\binom{k}{t}$ pairs. On the other hand, there are $\binom{v}{t}$ t-sets in all, and each must appear in λ blocks, so the number is $\lambda\binom{v}{t}$. Equating these we get the result. ∎

Instead of computing r, we shall prove a more general theorem. Suppose that S is any set of s treatments of a t-design D. We define the <u>derivative</u> D_S of D with regard to S to be the block design whose treatments are the treatments of D other than S, and whose blocks are the B\S where B was a block that contained S. If S is singleton, D_S is called the <u>contraction</u> at S.

LEMMA 2.14 The derivative of a t-design with regard to a subset S of size s, s ≤ t, is a (t - s)-design with parameters (v - s, k - s, λ).

Proof: Let X be any set of t - s treatments of D_S. The blocks of D_S containing X are precisely the blocks B\S where B is a block of D that contains the t-set S ∪ X. These are λ such blocks B. So the design is a (t - s) - (v - s, k - s, λ) design. ∎

THEOREM 2.15 There is a constant r_s such that every s-set of treatments of a t-(v,k,λ) design belongs to exactly r_s blocks, when 0 ≤ s ≤ t; and

$$r_s = \lambda \frac{\binom{v-s}{t-s}}{\binom{k-s}{t-s}}.$$

Proof: The number of blocks of D containing S equals the number of blocks of D_S. From Lemma 2.14, if S is an s-set, this number equals the number of blocks in a (t - s) - (v - s, k - s, λ)-design. The result now follows from Lemma 2.13. ∎

COROLLARY 2.15.1 A t-design is an s-design for s ≤ t. ∎

It follows from Theorem 2.15 that if there is a t-(v,k,λ)-design, the expression for r_s must be an integer for 0 ≤ s ≤ t. In other words,

$$\binom{k-s}{t-s} \text{ divides } \lambda\binom{v-s}{t-s}, \quad 0 \leq s \leq t. \tag{2.21}$$

If the numbers (v,k,λ,t) satisfy (2.21), we refer to them as an <u>admissible quadruple</u>. Admissibility is clearly a necessary condition for a t-(v,k,λ)-design, but we shall see that it is not sufficient.

EXERCISES

2.3.1 Describe 0-designs.

2.3.2 Prove that if t, v, and k are any positive integers with $t \leq k \leq v$, there is a t-(v, k, λ) design with $\lambda = \binom{v-t}{k-t}$.

2.3.3 V consists of all n-tuples $(v_1, v_2, \ldots, v_n)$ where the v_i are integers modulo 2. Addition is defined by the law

$$(u_1, u_2, \ldots) + (v_1, v_2, \ldots) = (w_1, w_2, \ldots),$$

where $w_i \equiv u_i + v_i \pmod{2}$. Prove that the blocks

$$\{x, y, z, w\} : x, y, z, w \in V, \quad x + y + z + w = (0, 0, \ldots, 0)$$

form a 3-$(2^n, 4, 1)$-design.

2.3.4 Which of the following are admissible quadruples?
(i) $(11, 4, 1, 3)$ (ii) $(28, 7, 1, 5)$
(iii) $(14, 4, 1, 3)$ (iv) $(17, 6, 2, 4)$

2.3.5 Given a positive integer k, what is the value of v such that the quadruple (v, k, 1, k - 1) is admissible?

2.3.6 Prove that if $v \equiv 2 \pmod{6}$, then $(v, 4, 1, 3)$ is an admissible quadruple.

BIBLIOGRAPHIC REMARKS

Most of the results on pairwise balanced designs are well known in combinatorics. We have borrowed from [6]. Theorem 2.4 was proven by de Bruijn and Erdos [1]. The discussion of $g^4(v)$, and Theorem 2.5, are from [5].

Fisher's inequality is of course due to Sir Ronald Fisher, the famous statistician [2]; Stanton [4] pointed out that the original proof did not involve matrix theory, and also outlined the alternative proof in Exercise 2.2.7. Theorem 2.11 is due to Ryser [3].

1. N. G. de Bruijn and P. Erdos, "On a combinatorial problem," Indagationes Mathematicae 10(1948), 1277-1279.
2. R. A. Fisher, "An examination of the different possible solutions of a problem in incomplete blocks," Annals of Eugenics 10(1940), 52-75.
3. H. J. Ryser, "Matrices with integer elements in combinatorial investigations," American Journal of Mathematics 74(1952), 769-773.
4. R. G. Stanton, "The appropriateness of standard BIBD presentation," Ars Combinatoria 16A(1983), 289-296.

5. R. G. Stanton and D. R. Stinson, "Perfect pair-coverings with block sizes two, three and four," Journal of Combinatorics, Information and System Sciences 8(1983), 21-25.
6. A. P. Street and W. D. Wallis, Combinatorics: A First Course (Charles Babbage Research Centre, Winnipeg, Manitoba, Canada, 1982).

3
Some Finite Algebra

Some readers will have met the following material in algebra and number theory courses. However, it is desirable, at least, to skim through the material to ensure that it is familiar.

3.1 FINITE FIELDS

A <u>field</u> $\{F, +, \times\}$ is a set F together with two binary operations + and × that satisfy the following properties:

(F1) For every and a b in F, a + b and a × b are also in F
(F2) $\{F, +\}$ is an abelian group;
(F3) $\{F \setminus \{0\}, \times\}$ is an abelian group;
(F4) For any elements a, b, and c in F,

$$a \times (b + c) = (a \times b) + (a \times c).$$

We shall often omit the sign ×, and omit brackets as in ordinary arithmetic; as an example, the displayed equation in (F4) might be written

a(b + c) = ab + ac.

We write 0 and 1 for the additive and multiplicative identity elements, respectively; -a and a^{-1} denote the inverses of a under the two operations. The set $F \setminus \{0\}$ is called the multiplicative group of F and is often denoted F^*. Notice that 1 is a member of F^*, so $1 \neq 0$; therefore, every field has at least two elements.

LEMMA 3.1 If a and b are members of a field F, then

(i) a0 = 0;
(ii) (-1)a = -a;
(iii) ab = 0 $\Longrightarrow$ a = 0 or b = 0. ∎

3.1 Finite Fields

We are primarily interested in <u>finite</u> fields—those with a finite number of elements. In a finite field F, write 1_n for the sum of n copies of 1: $1_0 = 0$, and for $n \geq 0$, $1_{n+1} = 1_n + 1$. Clearly,

$$1_a \times 1_b = 1_{ab}, \quad 1_a + 1_b = 1_{a+b}, \quad 1_a - 1_b = 1_{a-b}$$

(in the last case, b cannot be greater than a).

By the finiteness of F, the sequence $1_1, 1_2, 1_3, \ldots$ must be contain repetitions. Suppose that $1_k = 1_\ell$, where $k > \ell$. Then $1_{k-\ell} = 0$. So 0 occurs in the sequence. The <u>characteristic</u> of F is defined to be the smallest positive integer n such that $1_n = 0$.

THEOREM 3.2 The characteristic of a finite field is prime.

Proof: Suppose that F has characteristic ab, where a and b are both integers greater than 1. Then

$$0 = 1_{ab} = 1_a \times 1_b$$

so that either $a = 0$ or $b = 0$, by Lemma 3.1(iii). But $a < ab$ and $b < ab$, which contradicts the minimality of ab.

THEOREM 3.3 If x is a nonzero element of a field of characteristic n, the sum of k copies of x equals zero if and only if n divides k.

Proof: The sum of k copies of x satisfies

$$x + x + \cdots + x = x(1 + 1 + \cdots + 1) = x \times 1_k,$$

and $x \times 1_k = 0$ if and only if $1_k = 0$ [by Lemma 3.1(iii)]. Say b is the remainder on dividing k by n: $k = an + b$ for some a, and $0 \leq b < n$. Then

$$1_k = 1_{an} + 1_b = 1_a \times 1_n + 1_b.$$

Since the characteristic is minimal, 1_b equals zero if and only if $b = 0$, which is equivalent to saying n divides k. ∎

The most frequently encountered fields—the rational, real, and complex numbers—are infinite. The most familiar finite field is the set Z_p of integers modulo p, where p is a prime. This field has characteristic p. However, there are other finite fields: for example, the set $\{0, 1, A, B\}$ forms a field under the operations

+	0	1	A	B		×	0	1	A	B
0	0	1	A	B		0	0	0	0	0
1	1	0	B	A		1	0	1	A	B
A	A	B	0	1		A	0	A	B	1
B	B	A	1	0		B	0	B	1	A

and that field has characteristic 2.

A <u>subfield</u> is defined to be a subset of a field which is itself a field with the same identity elements 0 and 1. If F is a field and G is a subfield of F, it is not hard to see that F is a vector space over G, using the usual definition. If both F and G are finite, F is isomorphic to the set of all k-tuples $(x_1, x_2, \ldots, x_k)$, where each x_i ranges through G. So F has g^k elements, where g is the order of G.

THEOREM 3.4 A finite field F with characteristic p has p^k elements for some positive integer k.

Proof: Write

$$P = \{1, 1_2, 1_3, \ldots, 1_p\}.$$

It is easy to verify that P is a field with identity elements 1_p (= 0) and 1. So P is a subfield of F. But P has p elements. So F and p^k elements, where k is the dimension of F when considered as a vector space over P. ∎

The field P is called the <u>prime field</u> of F.

It may be shown that any two fields with p^k elements are isomorphic. For this reason there is no confusion if we use the notation $GF(p^k)$ to mean a field with p^k elements. (GF stands for "Galois field.") The field $GF(p)$, where p is prime, is Z_p.

We now outline a proof that there is a finite field of every prime power order. A <u>polynomial</u> over a field F is a function of the form

$$x \mapsto f(x) = a_0 + a_1 x + a_2 x^2 + \cdots + a_k x^k,$$

where the a_i are members of F; k is called the degree of the polynomial. One usually refers to the polynomial as $f(x)$. If, further,

$$g(x) = b_0 + b_1 x + b_2 x^2 + \cdots + b_n x^n,$$

then we define the sum and product in the obvious way:

3.1 Finite Fields

$$(f+g)(x) = (a_0 + b_0) + (a_1 + b_1)x + (a_2 + b_2)x^2 + \cdots;$$

$$(fg)(x) = a_0 b_0 + (a_0 b_1 + a_1 b_0)x + \cdots + \left(\sum_{i+j=h} a_i b_j\right) x^h + \cdots;$$

$f+g$ has degree $\max(k,n)$ and fg has degree $k+n$.

A polynomial is called <u>reducible</u> if it can be written as a product of two polynomials each of degree at least 1, and <u>irreducible</u> otherwise. One can prove the existence of at least one irreducible polynomial of every degree over the field $GF(p)$, where p is prime.

Select a prime p, and select an irreducible polynomial $f(x)$ of degree k over $GF(p)$. If $a(x)$ is any polynomial over $GF(p)$, there is a unique polynomial $a(x)$ such that

$$a(x) = f(x)g(x) + \overline{a}(x)$$

for some polynomial $g(x)$, and $\overline{a}(x)$ has degree less than k. We shall call $\overline{a}(x)$ the residue of $a(x)$ [modulo $f(x)$]. It is not difficult to verify that the residues modulo $f(x)$ form a field with p^k elements, so we have:

THEOREM 3.5 There is a field with n elements if and only if $n = p^k$ for some prime p and some positive integer k. ∎

The details of the proof—for example, a verification that the unique factorization exists as described—may be found in algebra textbooks. For our purpose the most important thing is to observe how a field can be constructed when it is needed. As an example, we show how to construct a 9-element field.

Construction. We need a field of order 3 and an irreducible polynomial of degree 2 over it. The 3-element field is $F = \{0, 1, 2\}$, with addition and multiplication modulo 3. To calculate the irreducible quadratic polynomials over F, we observe that we need only consider the <u>monic</u> polynomials—those with 1 as the coefficient of x^2—since, for example, $2x^2 + 2ax + 2b$ is irreducible if and only if $x^2 + ax + b$ is. If $x^2 + ax + b$ is reducible, then

$$x^2 + ax + b = (x + \alpha)(x + \beta)$$

for some α and β in F. Setting α and β to the various possible values, we find there are six reducible monic quadratics, namely,

$$x^2, \ x^2 + x, \ x^2 + 2x, \ x^2 + 2x + 1, \ x^2 + 2, \ x^2 + x + 1.$$

Hence the remaining monic quadratics over F, namely

$$x^2 + 1, \; x^2 + x + 2, \; x^2 + 2x + 2,$$

are all irreducible.

We could use any of the three irreducible polynomials to construct $GF(3^2)$; let us use $x^2 + 1$. The elements will be

$$\{0, 1, 2, x, x+1, x+2, 2x, 2x+1, 2x+2\}$$

and all calculations are reduced modulo 3 and modulo $x^2 + 1$. For example,

$$\begin{aligned}(x+2)(2x+1) &= 2x^2 + 5x + 2 \\ &= 2x^2 + 2x + 2 \quad \text{(modulo 3)} \\ &= 2(x^2 + 1) + 2x \\ &= 2x \quad \text{(modulo } x^2 + 1\text{).}\end{aligned}$$

If we abbreviate $ax + b$ to ab, we obtain Tables 3.1 and 3.2.

In the discussion preceding Theorem 3.5, it was not necessary that p be a prime; we only used the fact that a field of order p existed. In particular, if $q = p^r$, we could use an irreducible polynomial of degree s over $GF(q)$ to form a field of order $q^s = p^{rs}$. The new field would have $GF(q)$ as a subfield.

TABLE 3.1

+	00	01	02	10	11	12	20	21	22
00	00	01	02	10	11	12	20	21	22
01	01	02	00	11	12	10	21	22	20
02	02	00	01	12	10	11	22	20	21
10	10	11	12	20	21	22	00	01	02
11	11	12	10	21	22	20	01	02	00
12	12	10	11	22	20	21	02	00	01
20	20	21	22	00	01	02	10	11	12
21	21	22	20	01	02	00	11	12	10
22	22	20	21	02	00	01	12	10	11

3.1 Finite Fields

TABLE 3.2

×	00	01	02	10	11	12	20	21	22
00	00	00	00	00	00	00	00	00	00
01	00	01	02	10	11	12	20	21	22
02	00	02	01	20	22	21	10	12	11
10	00	10	20	02	12	22	01	11	21
11	00	11	22	12	20	01	21	02	10
12	00	12	21	22	01	10	11	20	02
20	00	20	10	01	21	11	02	22	12
21	00	21	12	11	02	20	22	10	01
22	00	22	11	21	10	02	12	01	20

But there is only one field of order p^{rs} (up to isomorphism). So the field $GF(p^{rs})$, constructed using an irreducible polynomial of degree rs over a prime field of order p, always contains a field of p^r elements. We can deduce the following theorem.

THEOREM 3.6 The field $GF(p^k)$ contains a subfield $GF(p^r)$ whenever r divides k.

EXERCISES

3.1.1 Prove Lemma 3.1.

3.1.2 Prove that the field with four elements, exhibited in this section, is in fact a field.

3.1.3 Prove by direct construction that any field with four elements must be isomorphic to the one mentioned in Exercise 3.1.1.

3.1.4 Show that the integers $\{0, 2, 4, 6, 8\}$ form a field under addition and multiplication modulo 10.

3.1.5 F is a finite field and G is a subfield of F. Prove that F and G have the same characteristic.

3.1.6 Construct fields with 4 and 8 elements.

3.1.7 Prove that there are exactly eight irreducible monic polynomials of degree 3 over GF(3).

3.1.8 Using the irreducible polynomial $x^3 + 2x + 1$, construct a finite field with 27 elements.

3.1.9 If G and H are two fields, then G × H is defined to be the set of all ordered pairs whose first element is in G and whose second element is in H, with two binary operations defined by

$$(x,y) + (z,t) = (x + z, y + t),$$

$$(x,y) \times (z,t) = (xz, yt),$$

where the operations between x and z are the addition and multiplication of G, and the operations between y and t are those of H.
  (i) Prove that G × H contains a zero element (i.e., an identity element for addition, which acts like 0 for multiplication) and a unit element (i.e., an identity element for multiplication).
  (ii) Prove that G × H contains two nonzero elements whose product is zero (whence G × H is not field).
  (iii) Suppose that G and H both have prime order: say $|G| = p$ and $|H| = q$. Say $p \neq q$. If $0 \leq x < p$, write $\bar{x}$ for the element (x_p, x_q) where x_p means the element of G obtained when x is reduced modulo p, and similarly for x_q. Prove that G × H is essentially the arithmetic of integers modulo pq, provided that $\bar{x}$ is interpreted as "x (mod pq)."

3.2 QUADRATIC ELEMENTS

THEOREM 3.7 If F is a finite field, then the multiplicative group F^* of its nonzero elements is cyclic.

Proof: Suppose that F has m + 1 elements. Write $f_1, f_2, \ldots, f_m$ for its nonzero elements, and define r_i to be the <u>multiplicative order</u> of f_i, the smallest positive integer such that $f_i^{r_i} = 1$. Such an r exists, because of finiteness. Write

$$R = \{r_1, r_2, \ldots, r_m\}.$$

We prove:
  (i) If $f_i^k = 1$, then r_i divides k.
  (ii) If f_j is an integer power of f_i, then r_j divides r_i; in particular, f_i and f_i^{-1} have the same multiplicative order.
  (iii) If r_i and r_j are coprime, then $f_i f_j$ has order $r_i r_j$.
  (iv) R contains the least common multiple of its members.

3.2 Quadratic Elements

(i) Suppose that $f_i^k = 1$ and $k = qr_i + s$, where $0 \leq s < r_i$. Then

$$1 = f_i^k = f_i^{qr_i+s} = (f_i^{r_i})^q f_i^s = f_i^s,$$

so $s = 0$ because r_i is minimal. So r_i divides k.

(ii) Suppose that $f_i^h = f_j$, and say that $r_i = qr_j + s$, where $0 \leq r_j < s$. Then

$$1 = (f_i^{r_i})^h = (f_i^h)^{r_i} = f_j^{r_i} = (f_j^{r_j})^q f_j^s = f_j^s,$$

and again $s = 0$. So r_j divides r_i. In particular, suppose that $f_j = f_i^{-1}$. Then r_j divides r_i, but by the same argument r_i divides r_j, so $r_i = r_j$.

(iii) Suppose that r_i and r_j are coprime members of R, and that $f_k = f_i f_j$. Then

$$(f_i f_j)^{r_i r_j} = (f_i^{r_i})^{r_j} (f_j^{r_j})^{r_i} = 1.$$

So by (i), r_k divides $r_i r_j$.

On the other hand, $(f_i f_j)^{r_k} = 1$, so $f_i^{r_k} = (f_j^{r_k})^{-1}$. So $f_i^{r_k}$ and $f_j^{r_k}$ have the same multiplicative order, by (ii); call that order s. By (ii), s divides r_i and r_j, the orders of f_i and f_j. Since r_i and r_j are coprime, s must equal 1. Therefore, $f_i^{r_k}$ has multiplicative order 1, so $f_i^{r_k} = 1$. Therefore, r_i divides r_k, by (i). Similarly, r_j divides r_k. Again using the fact that r_i and r_j are coprime, $r_i r_j$ divides r_k. So $r_k = r_i r_j$.

(iv) It is easy to see that R contains all divisors of all of its elements: if $r_i = st$, one readily verifies that f_i^s has order t. Now suppose that r is the least common multiple of the elements of R and has prime factor decomposition

$$r = p_1^{a_1} p_2^{a_2} \cdots p_n^{a_n}.$$

For each i, there must be some element r_j of R such that $p_i^{a_i}$ divides r_j. So $p_i^{a_i}$ is in R. By (iii), r belongs to R.

Now suppose that f is an element of multiplicative order r, the least common multiple. The elements $1, f, f^2, \ldots, f^{r-1}$ must all be different members of F^*, so $r \leq m$. On the other hand, since each r_i divides r, each

f_i is a root of the equation $x^r - 1 = 0$. Since this equation has at least m distinct roots, its degree must be at least m: $r \geq m$. So $r = m$, and the powers of f make up all of F^*. So F^* is cyclic with generator f. ∎

As an example, consider $GF(3^2)$. The powers of the element $x + 1$ are

$(x + 1)^1 = x + 1;$ $(x + 1)^5 = 2x + 2;$

$(x + 1)^2 = 2x;$ $(x + 1)^6 = x;$

$(x + 1)^3 = 2x + 1;$ $(x + 1)^7 = x + 2;$

$(x + 1)^4 = 2;$ $(x + 1)^8 = 1,$

so $x + 1$ generates the multiplicative group of $GF(3^2)$.

A generator of the multiplicative group of a finite field is called a <u>primitive element</u> of the field; if the field has prime order, the term "primitive root" is often used.

Let us consider the equation $x^2 = q$. It has at most two roots over any field; if $x = f$ is a solution, then $x = -f$ is also a solution. It follows that $x^2 = q$ can have <u>precisely one</u> solution only if there is a root f that satisfies $f = -f$. One possibility is $f = 0$, which is a solution if and only if $q = 0$. Provided that the field has odd characteristic, there are no other cases where $f = -f$. We shall say that a nonzero element q of a field is <u>quadratic</u> if the equation $x^2 = q$ has a solution in the field: the element 2 is not quadratic in $GF(3)$, but is quadratic in $GF(3^2)$.

Suppose that $F = GF(p^n)$, where p is an odd prime. The set of all squares of the $p^n - 1$ nonzero elements must equal the set of all quadratic elements. If we list these squares, each quadratic element arises twice: once as f^2 and once as $(-f)^2$. So F has $(p^n - 1)/2$ quadratic elements.

We define a map χ from a finite field F to the integers as follows:

$\chi(x) = 1$ if x is quadratic;

$\chi(0) = 0;$

$\chi(x) = -1$ otherwise.

The map χ is called the <u>quadratic character</u> on F. The properties of χ can be deduced from the cyclic nature of F^*. Suppose that a is a primitive element of F. Then the quadratic elements are a^2, a^4, a^6, ... If the number of elements of F is even, a has odd order, and every power of a is quadratic. So we consider fields of odd order. If F has $2k + 1$ elements, a has order $2k$. There are k quadratic elements: a^2, a^4, ..., a^{2k} (= 1). The elements a, a^3, ..., a^{2k-1} are nonquadratic. From this it follows immediately that:

3.2 Quadratic Elements

LEMMA 3.8 In any finite field,

(i) if $x \neq 0$, then $\chi(x^{-1}) = \chi(x)$;
(ii) $\chi(xy) = \chi(x)\chi(y)$. ∎

THEOREM 3.9 If k is any nonzero element of $GF(p^n)$, where p is odd, then

$$\sum_{x \in F} \chi(x)\chi(x+k) = -1.$$

Proof: If $x \neq 0$, then $\chi(x^{-1}) = \pm 1$, so $\chi(x^{-1})\chi(x^{-1}) = 1$. Therefore,

$$\chi(x)\chi(x+k) = \chi(x) 1 \chi(x+k)$$
$$= [\chi(x)\chi(x^{-1})][\chi(x^{-1})\chi(x+k)]$$
$$= \chi(xx^{-1})\chi(x^{-1}(x+k))$$
$$= 1 \chi(1 + x^{-1}k)$$
$$= \chi(1 + x^{-1}k).$$

Since $k \neq 0$, $x^{-1}k \neq 0$. If $x \neq y$, then $x^{-1}k \neq y^{-1}k$. Thus as x runs through F^*, the product $x^{-1}k$ also runs through F^*. So

$$\sum_{x \in F} \chi(x)\chi(x+k) = \sum_{x \in F^*} \chi(1 + x^{-1}k) = \sum_{x^{-1}k \in F^*} \chi(1 + x^{-1}k)$$

$$= \sum_{\substack{y \in F \\ y \neq 1}} \chi(y), \quad \text{where } y = 1 + x^{-1}k,$$

$$= \left[\sum_{y \in F} \chi(y)\right] - \chi(1).$$

Now $\chi(y) = 1$ for $(p^n - 1)/2$ values of y, $\chi(y) = -1$ for $(p^n - 1)/2$ values and $\chi(0) = 0$. Hence

$$\sum_{y \in F} \chi(y) = \frac{1}{2}(p^n - 1) - \frac{1}{2}(p^n - 1) + 0 = 0;$$

since $(1) = 1$, we have

$$\sum_{x \in F^*} \chi(1 + x^{-1}k) = -1.$$

Therefore,

$$\sum_{x \in F} \chi(x)\chi(x + k) = 0 + \sum_{x \in F^*} \chi(x)\chi(x + k) = -1. \qquad \blacksquare$$

THEOREM 3.10 In $GF(p^n)$,

(i) if $p^n \equiv 1 \pmod 4$, then -1 is quadratic;
(ii) if $p^n \equiv 3 \pmod 4$, then -1 is nonquadratic.

Proof: We can assume that p^n is odd, say $p^n = 2k + 1$. Consider the equation $y^2 = 1$. If a is a primitive element, then the roots of this equation are precisely a^0 and a^k. Since -1 and 1 are distinct roots of $y^2 = 1$, and since $a^0 = 1$, we have $a^k = -1$. So -1 is quadratic if k is even, and nonquadratic if k is odd. But "k even" and "k odd" are precisely "$p^n \equiv 1 \pmod 4$" and "$p^n \equiv 3 \pmod 4$", respectively. $\qquad \blacksquare$

EXERCISES

3.2.1 Prove (in the notation of Theorem 3.7) that if $r_i = st$, then f_i^s has multiplicative order t.

3.2.2 Prove Lemma 3.8.

3.2.3 If p is an odd prime, show that the product of all quadratic elements in $GF(p^n)$ is

$$\begin{cases} 1 & \text{if } p^n \equiv 3 \pmod 4 \\ -1 & \text{if } p^n \equiv 1 \pmod 4. \end{cases}$$

3.3 SUMS OF SQUARES

We shall now study the decomposition of positive integers into sums of integer squares. Theorems 3.18 and 3.22 will be needed later; the proofs may be omitted at a first reading.

We first ask: Which integers n have the form

$$n = x^2 + y^2$$

where x and y are integers? If

3.3 Sums of Squares

$$q = x^2 + y^2$$
$$r = z^2 + t^2,$$

then

$$qr = (x^2 + y^2)(z^2 + t^2) = (xz + yt)^2 + (xt - yz)^2.$$

So:

LEMMA 3.11 If each prime factor of the positive integer n can be written as the sum of two integer squares, then n can be written in that way. ∎

We now consider the primes.

LEMMA 3.12 Let p be a prime congruent to 1 (mod 4). Then there exist integers x, y, and m, with $0 < m < p$, such that

$$x^2 + y^2 = mp.$$

Proof: By Theorem 3.10, -1 is a quadratic element modulo p. Choose y such that $y^2 \equiv -1$ (mod p) and $0 < y < p/2$ [if $y > p/2$, replace y by p - y]. Then

$$1 + y^2 \equiv 0 \pmod{p},$$

so $1 + y^2 = mp$ for some positive m.

$$mp = 1 + y^2 < 1 + \frac{p^2}{4} < p^2,$$

where $m < p$. Therefore, the numbers x, y, m, where x = 1, satisfy the conditions. ∎

LEMMA 3.13 Let p be a prime congruent to 1 (modulo 4). If there exist integers x, y, and m, with $1 < m < p$, such that

$$x^2 + y^2 = mp,$$

then there exist integers X, Y, and M, with $1 \leq M < m$, such that

$$X^2 + Y^2 = Mp.$$

Proof: (i) If m is even, then $x \equiv y \pmod{2}$. Since $x^2 + y^2 = mp$, we have

$$\left[\frac{x+y}{2}\right]^2 + \left[\frac{x-y}{2}\right]^2 = \frac{m}{2}p$$

and we choose

$$X = \frac{x+y}{2}, \quad Y = \frac{x-y}{2}, \quad M = \frac{m}{2}.$$

(ii) If m is odd, there are unique integers a and c such that

$$x = am + c, \quad |c| < m/2.$$

Similarly, there are unique b and d satisfying

$$y = bm + d, \quad |d| < m/2.$$

Since $x^2 + y^2 = mp$,

$$mp = (am+c)^2 + (bm+d)^2 = c^2 + d^2 + 2(ac+bd)m + (a^2+b^2)m^2 \quad (3.1)$$

which implies that $m \mid (c^2 + d^2)$. Now let $c^2 + d^2 = Mm$, for some $M \geq 0$. Hence, by (3.1),

$$mp = Mm + 2(ac+bd)m + (a^2+b^2)m^2,$$

so that

$$p = M + 2(ac+bd) + (a^2+b^2)m$$

and

$$\begin{aligned}
Mp &= M^2 + 2(ac+bd)M + (a^2+b^2)Mm \\
&= M^2 + 2(ac+bd)M + (a^2+b^2)(c^2+d^2) \\
&= M^2 + 2(ac+bd)M + (ac+bd)^2 + (ad-bc)^2 \\
&= (M+ac+bd)^2 + (ad-bc)^2.
\end{aligned}$$

So Mp has the required form. If $M = 0$, then $c = d = 0$, which means that $m \mid x$, $m \mid y$, and hence $m^2 \mid (x^2 + y^2)$, or in other words, $m^2 \mid mp$. But now

3.3 Sums of Squares

$m|p$; since p is prime and $1 < m < p$, we have a contradiction. So $M \geq 1$. Also,

$$Mm = c^2 + d^2 < \frac{m^2}{4} + \frac{m^2}{4} < m, \quad \text{so } M < m.$$

Hence $X = M + ac + bd$, $Y = ad - bc$, and M satisfy the conditions. ∎

THEOREM 3.14 Let p be a prime congruent to 1 (modulo 4). Then p can be represented as the sum of two integer squares.

Proof: By Lemma 3.12, there exist integers x, y, and m, with $1 \leq m < p$, such that $x^2 + y^2 = mp$. Let $m = k$ be the smallest possible value of m such that this is true. If $k > 1$, then by Lemma 3.13 there is a smaller possible value of m, which is a contradiction. So $k = 1$ and $p = x^2 + y^2$. ∎

On the other hand, primes congruent to 3 modulo 4 do not have this property:

LEMMA 3.15 No number congruent to 3 (mod 4) is a sum of two integer squares.

Proof: If x is even, $x^2 \equiv 0$ (mod 4). If x is odd, $x^2 \equiv 1$ (mod 4). If any event, the sum of two squares can only possibly be congruent to

$$0 + 0 = 0 \quad \text{or} \quad 0 + 1 = 1 \quad \text{or} \quad 1 + 1 = 2$$

and is never congruent to 3 modulo 4. ∎

We define a <u>proper representation</u> of an integer n as a sum of two squares to be a representation

$$n = x^2 + y^2,$$

where x and y are <u>relatively prime.</u>

LEMMA 3.16 Let p be a prime congruent to 3 (modulo 4) and let n be a positive multiple of p. Then n has no proper representation as the sum of two squares.

Proof: Suppose the contrary: $n = x^2 + y^2$, where x and y are integers such that $(x,y) = 1$. Now $p|n$; if $p|x$, then $p|y$ also, so that $p|(x,y)$, which is a contradiction. Hence p does not divide x. Therefore, x has an inverse (mod p); suppose u is such that $ux \equiv 1$ (mod p). Then $uxy \equiv y$ (mod p), so

$$n = x^2 + y^2 \equiv x^2 + u^2x^2y^2 \equiv x^2(1 + u^2y^2) \pmod{p}.$$

But p divides n, so $x^2(1 + u^2y^2) \equiv 0 \pmod{p}$. Since p does not divide x, $1 + u^2y^2 \equiv 0 \pmod p$. Therefore, $-1 \equiv (uy)^2$, and -1 is quadratic modulo p. But this is impossible, since $p \equiv 3 \pmod 4$. ∎

LEMMA 3.17 If $n = p^c m$, where p is a prime congruent to 3 (modulo 4), c is odd and p does not divide m, then n has no representation as the sum of two squares.

Proof: Suppose otherwise: $n = x^2 + y^2$. Let $(x,y) = d$, so that $x = dX$, $y = dY$ and $(X,Y) = 1$. Then $n = Nd^2$, where $N = X^2 + Y^2$. Now p^c divides n and c is odd, so p must divide N. But N cannot have the proper representation $X^2 + Y^2$ by Lemma 3.16, so we have a contradiction. ∎

THEOREM 3.18 A positive integer n can be represented as the sum of two squares if and only if each of its prime factors congruent to 3 (modulo 4) appears to an even power.

Proof: By Lemma 3.17 the condition is necessary. Now suppose that the prime power decomposition of n is

$$n = 2^a p_1^{b_1} p_2^{b_2} \cdots p_r^{b_r} q_1^{2c_1} q_2^{2c_2} \cdots q_s^{2c_s},$$

where $p_1, p_2, \ldots, p_r$ are the primes congruent to 1 (modulo 4) which divide n and $q_1, q_2, \ldots, q_s$ are those congruent to 3. By Theorem 3.14, each p_i is a sum of two squares. Clearly, each q_i^2 is a sum:

$$q_i^2 = 0^2 + q_i^2.$$

Also,

$$2 = 1^2 + 1^2.$$

So the result follows from Lemma 3.11, except for the trivial case $n = 1$; but

$$1 = 0^2 + 1^2.$$ ∎

The following corollary is now easy to prove, but it is surprisingly useful.

COROLLARY 3.18.1 If the positive integer n can be expressed as the sum of two rational squares, it equals the sum of two integer squares.

3.3 Sums of Squares

Proof: Suppose that

$$n = \left[\frac{x}{a}\right]^2 + \left[\frac{y}{b}\right]^2$$

where x, y, a, and b are integers and a and b are nonzero. Then na^2b^2 is a sum of two integer squares:

$$na^2b^2 = (bx)^2 + (ay)^2.$$

So no prime congruent to 3 (mod 4) occurs to an odd power in the prime power decomposition of na^2b^2. Since a^2b^2 is a perfect square, n must also have this property. The result now follows from the theorem. ∎

It is easy to show, by an argument similar to Lemma 3.15, that no integer congruent to 7 (modulo 8) is a sum of three squares. But we shall show that every integer is a sum of four squares. If $q = a^2 + b^2 + c^2 + d^2$ and $r = e^2 + f^2 + g^2 + h^2$, then $qr = A^2 + B^2 + C^2 + D^2$, where

$$\begin{aligned} A &= ae + bf + cg + dh, & B &= af - be + ch - dg \\ C &= ag - bh - ce + df, & D &= ah + bg - cf - de \end{aligned} \tag{3.2}$$

So:

LEMMA 3.19 If every prime can be represented as the sum of four squares, then every positive integer can be expressed as the sum of four squares. ∎

LEMMA 3.20 If p is an odd prime, then there exist integers x, y, z, m with $0 < m < p$, such that

$$x^2 + y^2 + z^2 = mp.$$

Proof: Consider the congruence

$$x^2 + y^2 + z^2 \equiv 0 \pmod{p}. \tag{3.3}$$

It has the trivial solution $x = y = z = 0$. If $p \equiv 1 \pmod 4$, then by Theorem 3.14, we can find integers X and Y such that $p = X^2 + Y^2$, so (3.3) has also the nontrivial solution $x = X$, $y = Y$, $z = 0$.

Suppose that (3.3) has only the trivial solution. This means that $p \equiv 3 \pmod 4$ and hence that -1 is not quadratic modulo p. Let a be quadratic modulo p. If $-(a + 1)$ is also a quadratic element, we can choose x, y, and z such that $x^2 \equiv 1$, $y^2 \equiv a$, and $z^2 \equiv -(a+1) \pmod p$, giving a nontrivial solution to (3.3). So if a is quadratic, $-(a + 1)$ is not quadratic. So $a + 1 = (-1)[-(a + 1)]$ is quadratic. So the hypothesis that (3.3) has only the trivial solution implies that whenever a is quadratic, $a + 1$ is quadratic. But 1 is quadratic, implying that 2 is quadratic, and in turn that 3 is quadratic, and so on. Hence all the integers are quadratic modulo p, which is impossible. So (3.3) must have a nontrivial solution.

Let such a solution be given by x, y, z. We can assume that $x < p/2$, for if $x > p/2$, we can replace it by $p - x$. Similarly, $y < p/2$ and $z < p/2$, so since $0 < x^2 + y^2 + z^2 < 3p^2/4$, and $x^2 + y^2 + z^2 = mp$, we have $0 < m < p$, as required. ∎

LEMMA 3.21 If p is an odd prime and if there exist integers w, x, y, z, and m, with $1 < m < p$, such that

$$w^2 + x^2 + y^2 + z^2 = mp,$$

then there exist integers W, X, Y, Z, and M with $1 \leq M < m$, such that

$$W^2 + X^2 + Y^2 + Z^2 = Mp.$$

Proof: (i) If m is even, then w, x, y, and z are all odd, or all even, or two are odd and two are even. In any case we can assume that $w \equiv x \pmod{2}$ and $y \equiv z \pmod{2}$, without loss of generality. Then

$$\left[\frac{w+x}{2}\right]^2 + \left[\frac{w-x}{2}\right]^2 + \left[\frac{y+z}{2}\right]^2 + \left[\frac{y-z}{2}\right]^2 = \frac{m}{2}p \quad (3.4)$$

so

$$W = \frac{w+x}{2}, \quad X = \frac{w-x}{2}, \quad Y = \frac{y+z}{2}, \quad Z = \frac{y-z}{2}, \quad M = \frac{m}{2}$$

satisfy the conditions.

(ii) If m is odd, we can find a, b, c, d, e, f, g, h such that

$$w = am + e, \quad x = bm + f, \quad y = cm + g, \quad z = dm + h,$$

where $|e| < m/2$, $|f| < m/2$, $|g| < m/2$, $|h| < m/2$. Substituting into (3.4), we find that [in the notation of (3.2)]

$$e^2 + f^2 + g^2 + h^2 + 2Am + (a^2 + b^2 + c^2 + d^2)m^2 = mp, \quad (3.5)$$

which implies that m divides $e^2 + f^2 + g^2 + h^2$. Define M by letting $Mm = e^2 + f^2 + g^2 + h^2$, where $M \geq 0$.

If $M = 0$, then $e = f = g = h = 0$, so that $m^2 \mid (w^2 + x^2 + y^2 + z^2)$, which implies that $m^2 \mid mp$ and hence that $m \mid p$. But $1 < m < p$ and p is prime, which is a contradiction. Hence $M \geq 1$. Also, $e^2 + f^2 + g^2 + h^2 < 4 \cdot [m^2/4] = m^2$, so $M < m$, and altogether $1 \leq M < m$.

Substituting into (3.5) we have

$$Mm + 2Am + (a^2 + b^2 + c^2 + d^2)m^2 = mp;$$

multiplying by M/m gives

$$M^2 + 2AM + A^2 + B^2 + C^2 + D^2 = Mp,$$

in the notation of (3.2). But now

$$(M + A)^2 + B^2 + C^2 + D^2 = Mp,$$

so the integers $W = M + A$, $X = B$, $Y = C$, $Z = D$, and M satisfy the conditions. ∎

THEOREM 3.22 Every positive integer can be written as the sum of four squares.

Proof: One can apply the same proof as that of Theorem 3.14 to the results of Lemmas 3.20 and 3.21, to show that every odd prime is a sum of four squares. So the theorem follows from Lemma 3.19, provided that we note:

$$2 = 1^2 + 1^2 + 0^2 + 0^2$$
$$1 = 1^2 + 0^2 + 0^2 + 0^2. \qquad \blacksquare$$

BIBLIOGRAPHIC REMARKS

The theory of finite fields has been studied widely. Discussions of those parts that are needed in combinatorics will be found in many of the elementary algebra books; two good examples are [1] and [2].

Similarly, basic number theory books contain considerably more than we have mentioned here about quadratic residues and quadratic elements. Two good references are [3] and [4].

The theorem that every positive integer is the sum of four squares is due to Lagrange. There is a nice discussion of this result in [3].

1. G. Birkhoff and T. C. Bartee, <u>Modern Applied Algebra</u> (McGraw-Hill, New York, 1970).

2. R. A. Dean, Elements of Abstract Algebra (Wiley, New York, 1966).
3. G. H. Hardy and E. M. Wright, An Introduction to the Theory of Numbers (Clarendon, Oxford, 1938).
4. I. M. Vinogradov, Elements of Number Theory (Dover, New York, 1954).

4
Difference Sets and Difference Methods

4.1 DIFFERENCE SETS

Suppose that G is an abelian group of order v, written in additive notation, and suppose that B is a set of k elements of G. Then the <u>design generated from B</u> (in G) consists of all the blocks

$$B + g : g \in G.$$

It is a block design with $b = v$ and $r = k$.

LEMMA 4.1 If B contains precisely λ_i ordered pairs of elements whose difference is g_i, and if x and y are members of G such that $x - y = g_i$, then $\{x,y\}$ is a subset of precisely λ_i blocks.

Proof: Suppose that the pairs of elements of B with difference g_i are

$$\{b_1, c_1\}, \{b_2, c_2\}, \ldots, \{b_{\lambda_i}, c_{\lambda_i}\};$$

for $j = 1, 2, \ldots, \lambda_i$, $c_j - b_j = g_i$. If $x - y = g_i$, then $\{x,y\}$ will occur as a subset of $B + (x - c_1)$, $B + (x - c_2)$, ..., and $B + (x - c_{\lambda_i})$. All these blocks are different—if $x - c_j = x - c_h$, then $c_j = c_h$, $b_j = b_h$ and the two pairs are the same. So $\{x,y\}$ is a subset of at least λ_i blocks. But if $\{x,y\}$ is a subset of $B + d$, then $(x - d) - (y - d) = g_i$ and $x - d$ and $y - d$ are in B. So $x - d = c_j$ and $y - d = b_j$ for some j, and we have already listed the pair $\{x - d, y - d\}$ above. So no further cases can arise and $\{x,y\}$ is a subset of <u>precisely</u> λ_i blocks. ∎

To construct a balanced design, we must make all the λ_i equal. This is effected by the following definition.

56   4. Difference Sets and Difference Methods

DEFINITION A (v, k, λ)-difference set is a k-set B of elements of an abelian group G of order v with the property that given any nonzero member d of G, there are precisely λ ordered pairs of elements of B whose difference is d.

THEOREM 4.2 The design developed from a (v, k, λ)-difference set is a (v, k, λ)-symmetric BIBD. ∎

COROLLARY 4.2.1 If there is a (v, k, λ)-difference set, then $\lambda(v - 1) = k(k - 1)$. ∎

Table 4.1 gives some examples of (v, k, λ)-difference sets over the integers modulo v.

THEOREM 4.3 Suppose that v is a prime power congruent to 3 modulo 4; for convenience, write $v = 4t - 1$. Then there is a $(4t - 1, 2t - 1, t - 1)$-difference set.

Proof: We shall show that the set D of quadratic elements of GF(v), the elements which are perfect squares, form the required difference set. There are $(v - 1)/2$ quadratic elements. The product of a quadratic element with x is a quadratic element if and only if x is a quadratic element; since $v \equiv 3$ (mod 4), -1 is not quadratic. The inverse of a quadratic element is always quadratic.

Suppose that there are λ ordered pairs (x,y) of elements of D with difference 1. Then, if q is any quadratic element,

$(y - x) = 1 \implies (qy - qx) = q,$

and qx and qy are quadratic; different pairs (x,y) give different pairs (qx,qy). So we have λ ordered pairs of elements of D with difference q. But if $(z - t) = q$, then $(q^{-1}z - q^{-1}t) = 1$, so every pair giving difference q comes from a pair giving difference 1. We have covered the differences q where q is quadratic. But this is sufficient. Every nonzero nonquadratic element is the negative of a quadratic element, so all the nonzero differences occur in λ ways: the pairs with difference -q are all the pairs (qy,qx). So D is a (v, k, λ)-difference set, with $v = 4t - 1$ and $k = 2t - 1$. From Corollary 4.2.1,

$\lambda = t - 1.$ ∎

THEOREM 4.4 If p^n and $p^n + 2$ are both odd prime powers, then there is a difference set with parameters $[v, (v - 1)/2, (v - 3)/4]$, where $v = p^n(p^n + 2)$.

Proof: We work in the group $G = GF(p^n) \times GF(p^n + 2)$. This additive abelian group consists of all the pairs (x,y) where x is in $GF(p^n)$ and y is in $GF(p^n + 2)$; addition is defined as usual by

4.1 Difference Sets

Table 4.1

v	k	λ	Difference set
7	3	1	1, 2, 4
15	7	3	1, 2, 3, 5, 6, 9, 11
21	5	1	1, 2, 7, 9, 19
73	9	1	1, 2, 4, 8, 16, 32, 37, 55, 64
37	9	2	1, 7, 9, 10, 12, 16, 26, 33, 34

$$(x,y) + (z,t) = (x+z, y+t),$$

where the addition $x + z$ is carried out in $GF(p^n)$ and $y + t$ is carried out in $GF(p^n + 2)$. There is also a multiplication:

$$(x,y)(z,t) = (xz, yt),$$

where the products xz and yt are computed in the appropriate fields.

From now on, a and b denote primitive elements in $GF(p^n)$ and $GF(p^n + 2)$, respectively. It will be convenient to have a single symbol for p^n; we write $p^n = s$. It is also useful to have special names for certain elements of G; we write $c = (a, b)$, $d = (a, 0)$, so $c^i = (a^i, b^i)$, and so on. As expected, $c^0 = (1,1)$, the usual multiplicative identity. We write d^0 for $(1,0)$. The zero is $0 = (0,0)$. Note that the powers of c will eventually contain all the elements (a^i, b^j) where i and j are both even or both odd, so $\{c^0, c, c^2, \ldots\}$ is a multiplicative group of order $(s^2 - 1)/2$. We shall verify that

$$D = \{c^0, c, c^2, \ldots, c^{[(s^2-1)/2]-1}, 0, d^0, d, \ldots, d^{s-2}\} \qquad (4.1)$$

is a difference set with parameters $(s(s+2), (s(s+2)-1)/2, (s(s+2)-3)/4)$.

The differences between elements of D fall into five classes:

(i) $d^{i+j} - d^i = d^i(d^j - 1)$, for $0 \leq i \leq s - 2$, $0 < j \leq s - 2$;
(ii) $\pm(d^i - 0) = d^i$, for $0 \leq i \leq s - 2$;
(iii) $\pm(c^i - d^j)$, for $0 \leq i \leq (s^2 - 1)/2 - 1$, $0 \leq j \leq s - 2$;
(iv) $\pm(c^i - 0) = \pm c^i$, for $0 \leq i \leq (s^2 - 1)/2 - 1$;
(v) $(c^{i+j} - c^i) = c^i(c^j - 1)$, for $0 \leq i \leq (s^2 - 1)/2 - 1$, $0 < j \leq (s^2 - 1)/2 - 1$.

We shall show that each possibility occurs $(s(s + 2) - 3)/4$ times among these.

4. Difference Sets and Difference Methods

Differences of type (i) have the form $(a^i(a^j - 1), 0)$, where i ranges from 0 to s - 2 and j ranges from 1 to s - 2. For fixed $j \neq 0$, the elements $a^i(a^j - 1)$ range through all the values $a^0, a^1, \ldots, a^{s-2}$ once each: since the powers of a form the set of nonzero elements of the field $GF(p^n)$, and $a^j - 1$ is nonzero, we have $a^j - 1 = a^k$ for some k; so

$$\{a^i(a^j - 1) : 0 \leq i \leq s - 2\} = \{a^{i+k} : 0 \leq i \leq s - 2\},$$

and this set is just a rewritten version of $\{a^0, a, a^2, \ldots, a^{s-2}\}$. So, as j varies through its s - 2 values, the $a^i(a^j - 1)$ contain the powers of a, s - 2 times each; and the differences of type (i) contain the elements d^i, $0 \leq i \leq s - 2$, s - 2 times each.

The differences of type (ii) contain the d^i twice each.

Since $c^i - d^j = (a^i - a^j, b^i)$, the differences of type (iii) are the

$$(a^i(1 - a^{j-i}), b^i)$$

and their negatives. Consider i fixed. As j varies, $1 - a^{j-i}$ goes through s - 1 different values, including 0 but not including -1, so $a^i(a^{j-i} - 1)$ goes through the values $\{0, a^0, a, \ldots, a^{s-2}\}$ except that a^i is missing. So the differences with right-hand element b^i consist of all the (x, b^i) with $x \in GF(p^n)$ once each, except that (a^i, b^i) is missing. When the negatives are taken, the situation is the same; $(-a^i, -b^i)$ is missing from the elements with right-hand component $-b^i$. These omissions are precisely remedied by the differences of type (iv). To calculate how many times each difference appears overall requires a little more arithmetic, in order to take into account the fact that $b^i = b^{(s+1)+i} = b^{2(s+1)+i} = \cdots$; this is most easily done by observing that there are $s(s^2 - 1)$ differences of types (iii) and (iv) in total, and that each of the $s(s + 1)$ differences (x, b^i) occurs equally often, so the frequency must be $s(s^2 - 1)/s(s + 1) = s - 1$.

Now we consider the differences of type (v). They have the form

$$(a^i(a^j - 1), b^i(b^j - 1)) \tag{4.2}$$

for $0 \leq i \leq (s^2 - 1)/2 - 1$, $0 < j \leq (s^2 - 1)/2 - 1$. We first distinguish two special cases. If $j = h(s - 1)$ for some h, so that $a^j = 1$, then the difference is $(0, b^i(b^{h(s-1)} - 1))$. For a given value of h, all the elements $(0, y)$ occur $[(s^2 - 1)/2]/(s + 1) = (s - 1)/2$ times; as h ranges all through its $(s - 1)/2$ possible values, each difference $(0, y)$ will therefore occur $(s - 1)^2/4$ times. Such differences occurred s - 1 times among the differences of types (iii) and (iv), so in total they occur

$$\frac{1}{4}(s - 1)^2 + s - 1 = \frac{1}{4}(s^2 + 2s - 3) = \frac{s(s + 2) - 3}{4}$$

4.1 Difference Sets

times. Similarly, if $j = k(s + 1)$, the differences have the form $(x, 0)$, and it easy to show that each such difference appears $(s(s + 2) - 3)/4$ times among types (i), (ii), and (v).

Now assume that $j \neq h(s - 1)$ and $j \neq k(s + 1)$, so that (4.2) has neither element zero. For fixed j, the differences (4.2) are all different, and as i varies they range through $(s^2 - 1)/2$ different values. Also, the negatives of these differences will range through $(s^2 - 1)/2$ different values. We observe that these two sets are disjoint for given j: if they were not, then

$$-(a^i(a^j - 1), b^i(b^j - 1)) = (a^k(a^j - 1), b^k(b^j - 1))$$

for some i and k; since $b^j - 1$ is nonzero, we have

$$a^i + a^k = 0,$$
$$b^i + b^k = 0.$$

This will mean that $(-1) = a^{i-k}$ in $GF(p^n)$ and $(-1) = b^{i-k}$ in $GF(p^n + 2)$. Either $i - k$ is even, so that -1 is a quadratic element in both fields, or $i - k$ is odd, and -1 is quadratic in neither. Since one of p^n and $p^n + 2$ is congruent to 1 (modulo 4), and the other is congruent to 3, -1 cannot have the same quadratic character in both—a contradiction. So the elements $\pm(c^{i+j} - c^i)$, where j is fixed, contain no repetitions; since there are $s^2 - 1$ of them, they must include every (x, y) with $x \neq 0$ and $y \neq 0$ precisely once. There are $(s - 1)/2$ values of j of the form $h(s - 1)$, and $(s - 3)/2$ of the form $k(s + 1)$, so the number of values of j that give differences (x, y) with $x \neq 0$ and $y \neq 0$ is

$$\frac{1}{2}(s^2 - 1) - 1 - \frac{1}{2}(s - 1) - \frac{1}{2}(s - 3) = \frac{1}{2}(s - 1)^2.$$

So when we range through all possible i and j, the differences $\pm(e^{i+j} - c^i)$ will contain every (x, y) with $x \neq 0$ and $y \neq 0$ a total of $(s^2 - 1)/2$ times each. But these differences include all differences of type (v) twice (once with sign + and once with sign −), so each difference occurs $(s - 1)^2/4$ times, together with $s - 1$ occurrences of types (iii) and (iv), a total of $[s(s+2) - 3]/4$. times.

So λ equals the required constant, and D as shown in (4.1) is the required difference set. ■

As an example, suppose that $p = 5$ and $n = 1$, so that $p^n = 5$ and $p^n + 2 = 7$. The possible candidates for a are 2 and 3, the primitive roots modulo 5; while b can be 3 or 5. If we take $a = 2$, $b = 3$, we obtain $c = (2, 3)$, $d = (2, 0)$. Since p and p + 2 are primes, the arithmetic here is simply modular arithmetic modulo 35 (see Exercise 3.1.9). Now $(2, 3) = (17, 17)$, since $17 \equiv 2 \pmod 5$ and $17 \equiv 3 \pmod 7$, and $(2,0) = (7,7)$, so we can take c and d as 17 and 7 (mod 35); $d^0 = (1,0) = 21$. So

$$D = \{17^0, 17^1, 17^2, \ldots, 17^{11}, 0, 21, 7, 7^2, 7^3\}$$
$$= \{1, 17, 9, 13, 11, 12, 29, 3, 16, 27, 4, 33, 0, 21, 7, 14, 28\}$$
$$= \{0, 1, 3, 4, 7, 9, 11, 12, 13, 14, 16, 17, 21, 27, 28, 29, 33\},$$

which is the required $(35, 17, 8)$-difference set.

Observe that both families of difference sets which we have so far constructed have the form $v = 4t - 1$, $k = 2t - 1$, $\lambda = t - 1$. They are called "Hadamard difference sets." (The symmetric balanced incomplete block designs derived from them are called "Hadamard designs"; they will be very important in Chapter 8.) Some other difference sets—the "Singer difference sets"—will be seen in Chapter 5.

EXERCISES

4.1.1 Verify that the sets in Table 4.1 are, in fact, difference sets with the stated parameters.

4.1.2 Find a $(13, 4, 1)$-difference set.

4.1.3 Find a $(7, 4, 2)$-difference set.

4.1.4 Find difference sets with the parameters $(15, 7, 3)$ and $(143, 71, 35)$.

4.1.5 For what values of t up to $t = 50$ can one construct a $(4t - 1, 2t - 1, t - 1)$-difference set using Theorems 4.3 and 4.4?

4.1.6 G is an abelian group, written in additive notation. Let $G = A \cup B$, where $A \cap B = \emptyset$. We call $\{A, B\}$ a pair of <u>antidifference sets</u> if and only if every nonzero element of G occurs equally often in the set of differences $A - B = \{a - b : a \in A, b \in B\}$. Show that $\{A, B\}$ are antidifference sets if and only if A is a difference set.

4.1.7 Suppose that D is the $(4t - 1, 2t - 1, t - 1)$-difference set obtained from Theorem 4.3 for some prime power $4t - 1$. Prove that $D \cup \{0\}$ is a $(4t - 1, 2t, t)$-difference set.

4.2 PROPERTIES OF DIFFERENCE SETS

Consider the two sets

$$\{0, 1, 3\}, \quad \{2, 4, 5, 6\}$$

modulo 7. Both are difference sets; their parameters are $(7, 3, 1)$ and $(7, 4, 2)$, respectively. We observe that the two sets are complements, and ask: When

4.2 Properties of Difference Sets

is the complement of a difference set again a difference set? The answer is that this is always true.

THEOREM 4.5 The complement of a (v, k, λ)-difference set is a $(v, v - k, v - 2k + \lambda)$-difference set.

Proof: Suppose that D is a (v, k, λ)-difference set in an abelian group G. Consider the set of all ordered pairs of elements of G:

$$S = \{(g, h) : g \in G, h \in G\}.$$

S can be broken into four disjoint parts, $S = S_1 \cup S_2 \cup S_3 \cup S_4$, where

$$S_1 = \{(g, h) : g \in D, h \in D\},$$
$$S_2 = \{(g, h) : g \in D, h \notin D\},$$
$$S_3 = \{(g, h) : g \notin D, h \in D\},$$
$$S_4 = \{(g, h) : g \notin D, h \notin D\}.$$

Let us write $\delta_i(x)$ for the number of elements (g, h) of S_i such that $g - h = x$. Then $\delta_1(x) = \lambda$ for all $x \neq 0$, and we wish to show that $\delta_4(x) = v - 2k + \lambda$ for all $x \neq 0$.

For fixed g, the elements $g - h$ range through G as h ranges through G, so the collection of all differences arising from elements of S will contain every element of G v times:

$$\delta_1(x) + \delta_2(x) + \delta_3(x) + \delta_4(x) = v.$$

In the same way, if we check all the differences $g - h$ where $g \in D$ and $h \in G$, we get every group element k times (since D has k elements). So

$$\delta_1(x) + \delta_2(x) = k$$

for all x; and similarly,

$$\delta_1(x) + \delta_3(x) = k,$$

so $\delta_2(x) = \delta_3(x) = k - \lambda$ provided that x is not zero. Thus

$$\delta_4(x) = v - \delta_1(x) - \delta_2(x) - \delta_3(x)$$
$$= v - 2k + \lambda. \qquad \blacksquare$$

It may be verified that v − k is never equal to k when a (v,k,λ)-difference set exists. So Theorem 4.5 tells us that every difference set we have constructed gives rise to another difference set.

Suppose that D is a (v,k,λ)-difference set and consider the set D + s:

$$D + s = \{d + s : d \in D\}.$$

If the ordered pairs in D that have difference g are $\{x_1,y_1\}$, $\{x_2,y_2\}$, ..., $\{x_\lambda,y_\lambda\}$, then the pairs $\{x_j + s, y_j + s\}$, for $1 \leq j \leq \lambda$, also have difference g. Considering all g in G, we see that D + s is a (v,k,λ)-difference set. Such a set is called <u>shift</u> of D.

As an example, suppose that one wished to construct all possible $(7,3,1)$-difference sets. Such a set must contain two elements x and x + 1, to produce difference 1. Since $\{x, x+1, y\}$ is a shift of $\{0, 1, y-x\}$, one can start by assuming 0 and 1 to be in the difference set. Testing all the possibilities, one finds that the $(7,3,1)$-difference sets are $\{0,1,3\}$, $\{0,1,5\}$, and their shifts.

Although $\{0,1,5\}$ is not a shift of $\{0,1,3\}$, the two sets are related. If the elements $\{0,1,5\}$ are multiplied by three, the result is $\{0,3,15\}$, which reduces to $\{0,3,1\}$ modulo 7. This concept—the <u>multiple</u> of a difference set—is of course available only when the underlying additive group has a multiplicative structure as well. In theory, a difference set can arise in any abelian group, but in practice many examples come from the modular arithmetic of the integers or from finite fields, where a suitable multiplication is available. Multiples of difference sets are discussed in Exercise 4.2.3.

EXERCISES

4.2.1 Prove that there is no (v,k,λ)-difference set with $v = 2k$.

4.2.2 If D is a (v,k,λ)-difference set, denote by D' the set of additive inverses of elements of D. Prove that D' is a (v,k,λ)-difference set.

4.2.3 Suppose that D is a (v,k,λ)-difference set over the integers modulo v.
   (i) If m is an integer prime to v, define mD to be the set formed by multiplying the members of D by m and reducing modulo v. Prove that mD is a (v,k,λ)-difference set.
   (ii) If m is prime to v, and mD equals D or some shift of D, then m is called a <u>multiplier</u> of D. Prove that the set of multipliers of D form a group under multiplication modulo v.
   (iii) Prove that the condition "m is prime to v" in part (i) is necessary.

4.3 MORE GENERAL METHODS

Difference sets arise from the wish that design developed from a block should be balanced. One possible generalization is to develop two or more initial

4.3 More General Methods

TABLE 4.2

v	b	r	k	λ	Initial blocks
9	18	8	4	3	$\{1,2,3,5\}; \{1,2,5,7\}$
41	82	10	5	1	$\{1,10,16,18,37\}; \{5,8,9,21,39\}$
13	26	12	6	5	$\{1,2,4,7,8,12\}; \{1,2,3,4,8,12\}$
16	80	15	3	2	$\{1,2,4\};\{1,2,8\};\{1,3,13\};\{1,4,9\};\{1,5,10\}$
22	44	14	7	4	$\{1,7,12,16,19,21,22\}; \{1,6,8,9,10,14,20\}$

blocks and hope that the collection of all blocks formed will be a balanced design. If n initial blocks of size k are developed in a group G of order v, the resulting design will have $(v,b,r,k) = (v,nv,nk,k)$.

Suppose that blocks $B_1, B_2, \ldots, B_n$ are developed in G, and that the number of ordered pairs of elements of B_j with difference g_i is λ_{ij}. Then it follows that if $x - y = g_i$, then $\{x,y\}$ will be a subset of λ_{ij} of the blocks generated from B_j. So we shall get a balanced design if all the sums $\sum_{j=1}^{n} \lambda_{ij}$ are equal. In this case the initial blocks are called supplementary difference sets.

As an example, consider the development of the initial blocks $\{0,1,2,4\}$ and $\{0,1,4,6\}$ modulo 9. The differences arising in the first initial block are $\pm 1, \pm 2, \pm 4, \pm 1, \pm 3, \pm 2$, while the second block yields $\pm 1, \pm 4, \pm 6, \pm 3, \pm 5$, and ± 2. It is seen (using the facts that $\pm 6 = \pm 3$ and $\pm 5 = \pm 4$) that each difference arises three times, so the resulting design is balanced with $\lambda = 3$, and we have a $(9,18,8,4,3)$-design.

Table 4.2 shows some initial blocks that generate balanced incomplete block designs in cyclic groups. As a noncyclic example, the sets $\{00,01,10,44\}$ and $\{00,02,20,33\}$ may be developed over $Z_5 \times Z_5$ to form a $(25,50,8,4,1)$-design.

THEOREM 4.6 Let v be a prime power congruent to 1 modulo 4; for convenience, write $v = 4t + 1$. Then there exist two supplementary difference sets which can be developed to give a $(4t+1, 2(4t+1), 4t, 2t, 2t-1)$-design.

Proof: Let a be a primitive element of the field GF(v). We show that the sets

$$Q = \{a^{2b} \mid b = 1, 2, \ldots, 2t\}$$

and

$$R = \{a^{2b+1} \mid b = 1, 2, \ldots, 2t\}$$

of nonzero quadratic elements and nonquadratic elements, respectively, are the required supplementary difference sets. Certainly, $|Q| = |R| = 2t$, $\lambda = 2t - 1$.

Consider the ordered pairs of elements of Q whose difference is 1. Suppose that there are λ_1 such pairs, (q_{11}, q_{21}), (q_{12}, q_{22}), ..., $(q_{1\lambda_1}, q_{2\lambda_1})$, such that

$$q_{1i} - q_{2i} = 1$$

for all i. Then the λ_1 pairs of the form $(q_{1i}q, q_{2i}q)$ have the property that

$$q_{1i}q - q_{2i}q = q,$$

for any $q \in Q$, and all these pairs belong to $Q \times Q$. Conversely, if

$$q_{1*} - q_{2*} = q,$$

then $(q_{1*}q^{-1}, q_{2*}q^{-1})$ must be one of the pairs (q_{1i}, q_{1i}). So there are precisely λ_1 ordered pairs of members of Q whose difference is q, for any $q \in Q$. Similarly, the number of ordered pairs of elements of R whose difference is r equals λ_1, for any r in R: the pair (q_{1i}, q_{2i}) gives rise to the pair (rq_{1i}, rq_{2i}) which satisfies

$$rq_{1i} - rq_{2i} = r,$$

and conversely.

Similarly, if there are λ_2 pairs (q_{1i}, q_{2i}) of elements of Q that satisfy

$$q_{1i} - q_{2i} = x,$$

then there will be λ_2 ordered pairs of members of Q whose difference is r, and λ_2 ordered pairs of elements of R whose difference is q, for any $r \in R$ and for any $q \in Q$.

It follows that when we go through all the differences of pairs in $Q \times Q$ and in $R \times R$, every nonzero element of GF(v) will occur $\lambda_1 + \lambda_2$ times. So the two sets are supplementary difference sets. It follows immediately [from (1.1) and (1.2)] that $\lambda = 2t - 1$. ∎

It is not necessary that all blocks of a design be developed using the same arithmetical process. For example, consider the result of developing the following blocks modulo 15:

$\{0, 1, 4\}$, $\{0, 2, 8\}$, $\{0, 5, 10\}$.

4.3 More General Methods

One obtains 45 blocks; however, the last 15 blocks consist of the five blocks

$$\{0,5,10\},\ \{1,6,11\},\ \{2,7,12\},\ \{3,8,13\},\ \{4,9,14\},$$

each listed three times. If the repetitions are omitted, the resulting 35 blocks comprise a $(15, 35, 7, 3, 1)$ design.

We refer to "partial development" of a block. In particular, most examples of this technique have arisen in cyclic cases, where the group is the set of integers under addition modulo v, and we refer to "partial circulation": We write

$$\{0,5,10\}\ (\text{mod } 15)\ \text{PC}(5)$$

to mean the set of five blocks generated by $\{0,5,10\}$ by partially circulating through five steps in Z_{15}.

Another approach to this situation follows from the fact that Z_{15} is isomorphic to the direct product $Z_3 \times Z_5$. One isomorphism is

$$x \mapsto xx;$$

in other words, the integer x (mod 15) is mapped to the ordered pair (x (mod 3)), x (mod 5)). Under this mapping the blocks in the example map to

$$\{00, 11, 14\},\ \{00, 22, 23\},\ \{00, 20, 10\}.$$

We write (mod 3, 5) to mean that an initial block is developed through all 15 possible values; (mod -, 5) means that $\{ab, \ldots\}$ goes through the five possible values obtained by developing the right-hand components modulo 5, but leaving the left-hand components fixed. So we say that the design is obtained by developing

$$\{00, 11, 14\},\ \{00, 22, 23\}\ (\text{mod } 3, 5),\ \{00, 20, 10\}\ (\text{mod } -, 5).$$

Another generalization of difference sets and supplementary difference sets is the addition of an "infinity" element. Suppose that G is an abelian group with $v - 1$ elements. Append to G a new element ∞; call the new set G'. The addition in G' is constructed from that of G by the additional law

$$\infty + g = \infty \quad \text{for all } g \in G'.$$

If B is a k-set of elements of G', the design generated from B consists of the blocks

$$B + x : x \in G$$

(there is no block "B + ∞" because it would contain only k copies of the element ∞). For example, if G is Z_3, the block $\{\infty, 0, 1\}$ gives rise to $\{\infty, 0, 1\}$, $\{\infty, 1, 2\}$, and $\{\infty, 2, 0\}$.

One cannot obtain a balanced incomplete block design by developing one initial block in this way: If ∞ is a member of B, it will appear in all v - 1 blocks which are generated, and if it is not in B, then it never occurs; the other elements of G' appear in k - 1 or k blocks, according as ∞ ∈ B or ∞ ∉ B. Regularity is impossible. However, one can sometimes achieve a balanced incomplete block design by developing more than one initial block. For example, the blocks

$$\{\infty, 0, 1, 3, 7\}, \{0, 1, 2, 4, 5\}$$

form a (10, 18, 9, 5, 4)-design when developed over Z_9'.

Suppose that it is possible to develop n initial blocks over G', where G is a group of order v - 1, so as to produce a (v, b, r, k, λ)-design, and suppose that ∞ occurs in d initial blocks. Then ∞ lies in d(v - 1) of the blocks after development, so

$$r = d(v - 1). \tag{4.3}$$

Since d of the initial blocks contain k - 1 elements other than ∞, each noninfinite element will appear in d(k - 1) of the blocks developed from them, and they will each appear in (n - d)k of the other blocks, so

$$r = d(k - 1) + (n - d)k. \tag{4.4}$$

Comparing (4.3) and (4.4) we deduce that

$$dv = nk. \tag{4.5}$$

There are n sets of (v - 1) blocks. Since ∞ occurs in d(v - 1) blocks with k - 1 other elements in each, we must have

$$(v - 1)\lambda = d(v - 1)(k - 1)$$

or

$$\lambda = d(k - 1)$$

if λ is to be constant, so the design has parameters

$$(v, n(v - 1), d(v - 1), k, d(k - 1)). \tag{4.6}$$

4.3 More General Methods

TABLE 4.3

Parameters	Initial blocks
(10, 30, 9, 3, 2)	$\{\infty, 0, 5\}$, $\{0, 1, 4\}$, $\{0, 2, 3\}$, $\{0, 2, 7\}$ (mod 9)
(10, 18, 9, 5, 4)	$\{\infty, 0, 1, 3, 7\}$, $\{0, 1, 2, 4, 5\}$ (mod 9)
(12, 44, 11, 3, 2)	$\{\infty, 0, 3\}$, $\{0, 1, 3\}$, $\{0, 1, 5\}$, $\{0, 4, 6\}$ (mod 11)
(15, 42, 14, 5, 4)	$\{\infty, 0, 1, 2, 7\}$, $\{0, 1, 4, 9, 11\}$, $\{0, 1, 4, 10, 12\}$ (mod 14)

Given these restrictions, the method has proven very useful in constructing individual designs. Some examples are shown in Table 4.3.

THEOREM 4.7 If there exists a (4t - 1, 2t - 1, t - 1)-difference set, then there exists a

(4t, 8t - 2, 4t - 1, 2t, 2t - 1)-design.

Proof: Suppose that a (4t - 1, 2t - 1, t - 1)-difference set D exists in the abelian group G. Then $G \backslash D$ is a (4t - 1, 2t, t)-difference set, as in Theorem 4.5. Consider the initial blocks $G \backslash D$ and $D \cup \{\infty\}$ in G'. The parameter conditions (4.6) and (4.7) are satisfied. If x is a nonzero element of G, then x arises as the difference of an unordered pair t - 1 times in D and t times in $G \backslash D$, so it arises a total of 2t - 1 times in the combined differences, and $\lambda = 2t - 1$, as required. ∎

COROLLARY 4.7.1 There is a design with parameters

(4t, 8t - 2, 4t - 1, 2t, 2t - 1)

whenever 4t - 1 is a prime power. ∎

The corollary follows using Theorem 4.3.

A number of generalizations and adaptations of the methods of this section can be made, to construct specific designs. The reader should consult, for example, Exercises 4.3.6 and 4.3.7.

EXERCISES

4.3.1 Verify that the sets shown in Tables 4.2 and 4.3 generate designs as indicated.

4.3.2 Find two initial blocks whose development modulo 13 yields a (13, 26, 6, 3, 1)-design.

4.3.3 Find a (9, 18, 8, 4, 3)-design developed from two initial blocks in the group $Z_3 \times Z_3$, one of the initial blocks being $\{00, 01, 10, 11\}$.

4.3.4 A design is developed from n initial blocks of size k over G', where G is an abelian group of order v − 1. The element ∞ belongs to d blocks.
  (i) Prove from (4.5) and (4.6) that the necessary conditions (1.1) and (1.2) are satisfied.
  (ii) The ordered pairs of elements contribute $[k(k-1)]/2$ distinct differences in each of the blocks without ∞, and $[(k-1)(k-2)]/2$ noninfinite differences in each of the blocks containing ∞. Verify that the total number of differences is $\lambda \binom{v-1}{2}$.

4.3.5 Construct a (6, 10, 5, 3, 2)-design from two initial blocks over Z_5'.

4.3.6 Prove that when the following blocks are developed as indicated, the result is a (37, 111, 12, 4, 1) design based on $(Z_3 \times Z_{11}') \cup \{\infty\infty\}$: $\{00, 01, 12, 15\}$, $\{01, 03, 08, 10\}$, $\{0\infty, 07, 15, 21\}$ (mod(3, 11)), $\{\infty\infty, 00, 10, 20\}$ mod(−, 11), $\{0\infty, 1\infty, 2\infty, \infty\infty\}$.

4.3.7 Consider the following method of constructing a balanced incomplete block design with parameters

$$(pq + 1,\ b,\ r,\ p + 1,\ 1),$$

where p and q are positive integers. The design is based on Z_{pq}'; the blocks consist of npq blocks developed from n initial blocks in Z_{pq}, together with the q blocks

$$\{\infty,\ x,\ q + x,\ \ldots,\ (p-1)q + x\} : 0 \leq x < q.$$

  (i) Prove that p + 1 must divide q − 1; show that

$$r = q,\quad n = \frac{q-1}{p+1},\quad b = \frac{q(pq+1)}{p+1}.$$

  (ii) Suppose that p = 2. The initial blocks have the form $\{\infty, 0, q\}$, $\{0, a_1, b_1\}, \ldots, \{0, a_n, b_n\}$. The blocks developed from $\{0, a_i, b_i\}$ include three that contain 0: say they are $\{0, c, d\}$, $\{0, e, f\}$, and $\{0, g, h\}$. Prove that among $\{c, d, e, f, g, h\}$, either two belong to $\{1, 2, \ldots, q-1\}$ and four to $\{q, q+1, \ldots, 2q-1\}$ or else all six belong to the first set and none to the second. Deduce that there must exist nonnegative integers A and B such that

$$A + B = n, \quad 2A + 6B = q - 1, \quad 4A = q - 1.$$

Hence prove that $q \equiv 0 \pmod{12}$.

(iii) Prove that the following sets generate a $(27, 117, 13, 3, 1)$-design [according to part (ii), this is the smallest possible from this construction when $p = 2$]:

$$\{0, 1, 22\}, \quad \{0, 2, 8\}, \quad \{0, 3, 14\}, \quad \{0, 7, 17\}, \quad \{\infty, 0, 13\}.$$

(iv) Can a similar analysis be carried out when $p = 3$?

4.3.8 Formulate a definition for a "shift" of a set of supplementary difference sets and for a "multiplier" in the case where the underlying additive group is a modular arithmetic. Do the "multipliers" in your definition form a group?

BIBLIOGRAPHIC REMARKS

The rise of difference methods in the construction of balanced incomplete block designs is largely due to Bose; in particular, see his paper [2]. However, various examples of difference sets and difference methods have occurred before Bose's paper. The construction of Theorem 4.3 is due to Paley [4], and these difference sets are sometimes called "Paley difference sets." Theorem 4.4 is from [5].

The theory and properties of difference sets have been studied by many authors. Three important books on the subject are those of Baumert [1], Mann [3], and Storer [6].

1. L. D. Baumert, Cyclic Difference Sets (Springer-Verlag, Heidelberg, West Germany, 1971).
2. R. C. Bose, "On the construction of balanced incomplete block designs," Annals of Eugenics 9(1939), 353-399.
3. H. B. Mann, Addition Theorems (Wiley, New York, 1965).
4. R. E. A. C. Paley, "On orthogonal matrices," Journal of Mathematics and Physics 12(1933), 311-320.
5. R. G. Stanton and D. A. Sprott, "A family of difference sets," Canadian Journal of Mathematics 10(1958), 73-77.
6. T. Storer, Cyclotomy and Difference Sets (Markham, Chicago, 1967).

5
Finite Geometries

5.1 FINITE AFFINE PLANES

In this chapter we develop the theory of the finite analogs of Euclidean and projective geometry, primarily because these objects give rise to a number of balanced incomplete block designs. An <u>incidence structure</u> consists of two sets—a set P of points and a set L of lines—together with a binary relation of <u>incidence</u> between elements of P and elements of L. If a point p is incident with a line ℓ, one says "p lies on ℓ" or "ℓ contains p." With any line one can associate a subset of P, namely the set of all points that lie on the given line. We shall never wish to discuss geometrical situations where a line can contain no points or where two different lines can have the same set of points, so we can in fact <u>define</u> lines to be nonempty sets of points.

We define a <u>geometry</u> to consist of a set P of objects called points and a set L of nonempty subsets of P called lines which satisfies the two axioms (A1) and (A2):

(A1) Given any two points, there is one and only one line that contains them both;
(A2) There is a set of four points, no three of which belong to one common line.

In particular, an <u>affine plane</u> is a geometry that obeys the further axiom

(A3) Given any point p and given any line q that does not contain p, there is exactly one line that contains p and contains no point of q.

Axiom (A3) is the well-known parallel axiom, or "Euclid's fifth postulate"; if we were to add the distance properties of ordinary physical space (in a suitable form) to the axioms (A1), (A2), and (A3), we would have a set of axioms for ordinary Euclidean plane geometry. However, we shall instead impose a condition of finiteness which is inconsistent with Euclidean metrical geometry, and make the following definition.

5.1 Finite Affine Planes

DEFINITION A finite affine plane is a finite set P of objects called points, together with a set L of nonempty subsets of P called lines, which satisfy the axioms (A1), (A2), and (A3).

It is clear that (A1) is a "balance" axiom; a geometry whose point set is finite is a pairwise balanced design with $\lambda = 1$. From (A1) it is easy to deduce the following property: Given any two lines, there is at most one point contained in both. Disjoint lines in a finite affine plane are called parallel; we can interpret axiom (A1) a meaning that two points determine exactly one common line; the remark just made means that two nonparallel lines have exactly one common point; and axiom (A3) means that if p does not lie on q, there is exactly one line through p parallel to q.

We shall use geometric terminology. In particular, lines with a common point are called <u>concurrent</u> and points that all lie on the same line are called <u>collinear</u>. We say "the line ab" to mean the unique line containing points a and b.

Any finite affine plane must contain at least four points, by (A3). There is a four-point plane, denoted AG(2,2), which contains exactly six lines: Their point sets are the six possible unordered pairs from the four points. This geometry has the interesting property that one can find two lines which between them contain all four points. We prove that this characterizes AG(2,2).

LEMMA 5.1 If a finite affine plane has two lines ℓ and m which between them contain all the points of the plane, it is AG(2,2).

Proof: By (A2), the plane must contain four distinct points, a, b, c, and d, say, of which no three are collinear. Clearly, two must belong to each of ℓ and m: say

$$\ell = \{a, b, \ldots\}, \quad m = \{c, d, \ldots\}.$$

Suppose ℓ also contains another point, e. Then consider the lines ac, bd, and de. Do ac and bd meet? Since all points lie on $\ell \cup m$, any intersection point must be either on ℓ or on m. If it were on ℓ, it must be the point a, which is where ℓ and ac meet. But by the same argument it must be b. Since two lines cannot have two common points, a = b, which is a contradiction. So ac and bd are parallel. But similarly, we see that ac and de are parallel. But this means that there are two lines through d parallel to ac, which contradicts (A3).

It follows that neither ℓ nor m can contain a third point. So the plane has exactly four points. The six lines ab, ac, ad, bc, bd, and cd must all belong to the plane. So we have a copy of AG(2,2). But there can be no further lines: any such line would have exactly one point, and if (for example) {a} were a line, then {a} and ab would be two lines through a, parallel to cd—another contradiction. So the plane is precisely AG(2,2). ∎

LEMMA 5.2 In a finite affine plane there is a parameter n such that every line contains n points and every point lines on n + 1 lines.

Proof: If the points of the plane are all contained in two lines, then it is AG(2,2), and the lemma is satisfied with n = 2. So assume that no two lines contain all the points.

Now select any two lines ℓ and m, and select a point p that lies on neither of them. Suppose that ℓ contains points a_1, a_2, ..., a_n. Then p lies on exactly n lines that meet ℓ, namely pa_1, pa_2, ..., pa_n. (It is easy to see that these lines are distinct.) It also lies on one line, k say, parallel to ℓ [by (A3)].

Consider these n + 1 lines. One will be parallel to m, and the other n will meet m in n distinct points. These n points are all the points of m (since every point of m lies on some line through p), so m contains n points. Since ℓ and m were lines chosen at random, it follows that all lines contain n points for a fixed n.

Now p lies on n + 1 lines. Since p could have been chosen to be any point not on the arbitrary line ℓ, it follows that the number of lines through any point is greater by 1 than the common number of points per line. ∎

The constant n is called the <u>parameter</u> of the affine plane. (Some writers call it the "order," but it seems more appropriate to use that term for the number of points in the whole plane rather than in one line.) A finite affine plane with parameter n is called an AG(2, n).

THEOREM 5.3 Provided that "points" are identified with "treatments" and "lines" are identified with "blocks," a finite affine plane with parameter n is precisely a balanced incomplete block design with parameters

$$(n^2,\ n^2 + n,\ n + 1,\ n,\ 1). \tag{5.1}$$

Proof: Suppose that a finite affine plane is interpreted as a block design, as indicated. Lemma 5.2 says that r = n + 1 and k = n, and axiom (A1) says that the design is balanced and $\lambda = 1$. So we have a balanced incomplete block design. The equations (1.1) and (1.2) must be satisfied; these yield $v = n^2$, $b = n^2 + n$, and we have the parameters (5.1).

The converse is left as an exercise (see Exercise 5.1.2). ∎

EXERCISES

5.1.1 A trivial affine geometry is defined to be a structure that obeys (A1) and (A3) but not (A2). Describe all finite trivial affine geometries.

5.2 Construction of Finite Affine Geometries

5.1.2 Prove that a balanced incomplete block design with parameters

$$(n^2, n^2 + n, n + 1, n, 1)$$

is a finite affine plane.

5.1.3 A <u>finite incidence geometry</u> consists of a finite set P of points and a set of nonempty subsets of P called lines, which satisfy the axioms:

(F1) Two points determine exactly one line joining them;
(F2) There are at least two lines and every line contains at least two points;
(F3) There is a constant n such that every line contains n points.

  (i) Prove that in any finite incidence geometry, two lines can have at most one common point.
  (ii) Prove that every finite incidence geometry is a balanced incomplete block design.
  (iii) Prove that in a finite incidence geometry, there is a number t such that for any point p and any line ℓ not through p there are exactly t lines containing p which have no common point with ℓ.

5.1.4 Prove that an AG(2,n) is never a 3-design (except in the trivial case n = 2).

5.2 CONSTRUCTION OF FINITE AFFINE GEOMETRIES

It is well known that Euclidean geometry can be constructed algebraically, using the properties of equations over the real field. In this section we examine the analogous construction using a finite field.

Suppose that F is the field GF(n) with n elements. Let V be the set of all ordered pairs of members of F. We write L for the set of all linear equations

$$ax + by + c = 0,$$

where a, b, and c are elements of F and x and y are indeterminates, with the case a = b = 0 omitted. Two equations are equivalent if one can be obtained from the other by multiplying by a nonzero field member throughout. If ℓ is any member of L, we write "the line ℓ" to mean the set of all pairs (x,y) that satisfy the equation ℓ; equivalent equations give rise to the same line. We shall verify that if we interpret the members of V as points and the "lines" derived from T as lines, we have constructed an affine plane of parameter n.

First, suppose that (a,b) and (c,d) are any two points. Then there is a unique line containing both: it is the line

$$(x - a)(d - b) = (c - a)(y - b).$$

(Since $a = c$ and $b = d$ cannot both be true for two distinct points, this is always a line. It is easy to show that no other line contains both points.) So (A1) is satisfied.

To prove that (A2) is true, it is sufficient to find four points of which no three are collinear. But the points $(0,0)$, $(0,1)$, $(1,0)$, and $(1,1)$ always have this property (see Exercise 5.2.1).

Next consider (A3). We show that there is a unique line through (e,f) and parallel to $ax + by + c = 0$. By definition, two lines have a common point if and only if their equations have a common solution. So the lines parallel to $ax + by + c = 0$ would be the lines $ax + by + d = 0$, where $d \neq c$. If (e,f) is to lie on this line,

$$ae + bf + d = 0.$$

This will be true for exactly one value of d, namely

$$d = -ae - bf.$$

So (A3) is satisfied.

Finally, the number of points on a line is precisely n: on the line $ax + by + c = 0$, where $b \neq 0$, the n points are

$$\{(x, -b^{-1}(c + ax)) : x \in F\}$$

and on the line $ax + c = 0$ they are

$$\{(a^{-1}c, y) : y \in F\}.$$

(In the latter case, $a = 0$ is impossible.)

Since there is a field with n elements whenever n is a prime power, we have proven the following result.

THEOREM 5.4 There is a finite affine plane $AG(2,n)$, or equivalently a balanced incomplete block design with parameters

$$(n^2, n^2 + n, n + 1, n, 1),$$

whenever n is a prime power.

5.2 Construction of Finite Affine Geometries

As an example we again construct an AG(2,2). We write F = GF(2) = $\{0,1\}$. The set V consists of the points

$$\{(0,0), (0,1), (1,0), (1,1)\},$$

which we abbreviate to

$$\{00, 01, 10, 11\}.$$

The possible choices of a, b, and c are 010, 011, 100, 101, 110, 111. These six choices lead to the following equations:

(i) y = 0, with points 00, 10;
(ii) y + 1 = 0, with points 01, 11;
(iii) x = 0, with points 00, 01;
(iv) x + 1 = 0, with points 10, 11;
(v) x + y = 0, with points 00, 11;
(vi) x + y + 1 = 0, with points 01, 10.

So the AG(2,2) is the (4,6,3,2,1)-design with blocks

$$\{00,10\}, \{01,11\}, \{00,01\},$$
$$\{10,11\}, \{00,11\}, \{01,10\}.$$

The geometry developed from a finite field is called a finite Euclidean geometry and denoted EG(2,n). We have not proven that all finite affine planes are Euclidean, and in fact this is not the case. In Section 5.6 we give an example of a finite affine plane that is not derived from a field.

Since every field has prime power order, an EG(2,n) exists if and only if n is a prime power. The status of the existence of finite affine planes is not decided. In all known examples the parameter n is a prime power, but it has not been proven that other cases are impossible. The situation is discussed in Chapter 6.

Just as higher-dimensional Euclidean geometry can be constructed from real equations in more than two variables, so we can construct higher-dimensional finite affine geometries. We shall use this algebraic approach for the actual definition of the geometries rather than proceeding from axioms.

DEFINITION The finite affine geometry AG(d,n) of dimension d over GF(n) consists of the n^d vectors of length d over GF(n), which are called points. If V is any k-dimensional subspace of AG(d,n) and p is any member of AG(d,n), then p + V, defined by

$$p + V = \{p + v : v \in V\},$$

is called a k-flat. In particular, (k - 1)-flats are called primes and 1-flats are called lines.

THEOREM 5.5 Given two points of AG(d,n), there is a unique line that contains them both.

Proof: Suppose that q and r are two points of AG(d,n). Write V for the set of all points which are multiples of r - q. Then q + V is a line that contains both q and r:

$q = q + 0(r - q),$

$r = q + 1(r - q).$

Moreover, suppose that p + W is another line which contains both q and r; say that W is the set of all multiples of a nonzero vector w. Then there are field elements a and b such that

$q = p + aw,$

$r = p + bw.$

So q - r is a multiple of w. It follows that V and W are the same space. It remains to show that p + W and q + W are the same set. But if x is any member of p + W, say x = p + cw, then x = q + (c - a)w and x ∈ q + W, and conversely. So the line is unique. ∎

COROLLARY 5.5.1 The points of an AG(d,n), interpreted as treatments, and the lines, interpreted as blocks, form a balanced incomplete block design with parameters

$$\left(n^d,\ \frac{n^{d-1}(n^d - 1)}{n - 1},\ \frac{n^d - 1}{d - 1},\ n,\ 1\right).$$

Proof: Balance follows from Theorem 5.5; and the points of the line p + V are the n vectors

p + av : a ∈ GF(n)

where v is a generator of V, so the block size is a constant n. So by Theorem 2.9 we have a balanced incomplete block design, and the other parameters follow from (1.1) and (1.2). ∎

One can derive a great many other balanced incomplete block designs from the finite affine geometries. If we interpret k-flats, rather than lines, as blocks, we obtain a design with parameters

5.3 Finite Projective Geometries

$$\left(n^d, \frac{\lambda n^{d-k}(n^d - 1)}{n^k - 1}, \frac{\lambda(n^d - 1)}{n^k - 1}, n^k, \lambda \right), \tag{5.2}$$

where

$$\lambda = \frac{n^{d-1} - 1}{n - 1} \cdot \frac{n^{d-2} - 1}{n^2 - 1} \cdot \ldots \cdot \frac{n^{d-k+1} - 1}{n^{k-1} - 1}. \tag{5.3}$$

(The proof is left as an exercise.)

EXERCISES

5.2.1 Verify that in any finite affine plane constructed over a field, no three of the points $(0,0)$, $(1,0)$, $(0,1)$, and $(1,1)$ are collinear.

5.2.2 Prove that if the k-flats of AG(d,n) are interpreted as blocks and the points as treatments, the result is a balanced incomplete block design with parameters defined by (5.2) and (5.3).

5.2.3 Suppose that A is any 2-flat of AG(d,n). By a line of A we mean a line of the AG(d,n) all of whose points lie in A. Prove that the points and lines of A form an AG(2,n).

5.2.4 Prove that AG(d,n) is never a 3-design when $n > 2$. (Hint: There are two sorts of 3-sets of points—those which are collinear and those which are not.)

5.2.5 Prove that the balanced incomplete block design of Exercise 5.2.2 is a 3-design when $n = 2$ and $k \geq 2$.

5.3 FINITE PROJECTIVE GEOMETRIES

LEMMA 5.6 Suppose that ℓ is a line in a finite affine plane AG(2,n). Then there are exactly $n - 1$ lines parallel to ℓ.

Proof: Say ℓ contains the points $p_1, p_2, \ldots, p_n$. There are n lines other than ℓ which pass through p_i, for $i \in (1 \cdots n)$; call these lines ℓ_{i1}, $\ell_{i2}, \ldots, \ell_{in}$. If $i \neq j$, there is exactly one line that contains both p_i and p_j, by (A1), and that line is ℓ. So $\ell_{ix} \neq \ell_{jy}$ unless $i = j$ and $x = y$, and there are n^2 distinct lines ℓ_{ix}. The lines parallel to ℓ—those which do not meet ℓ—are precisely those lines not equal to ℓ or to any of the ℓ_{ix}; their number must be

$$(n^2 + n) - (n^2 + 1),$$

which is $n - 1$. ∎

COROLLARY 5.6.1 Lines that are parallel to the same line ℓ are parallel to one another.

Proof: The number of points on each of the $n-1$ lines parallel to ℓ is n, so their union contains at most $n^2 - n$ points, and will contain less if any of the two lines have a common point. But (A3) tells us that each of the $n^2 - n$ points which do not lie on ℓ must be contained on a line parallel to ℓ, so all the points must be in the union. Therefore, the lines parallel to ℓ have no common points: they are parallel to each other. ∎

COROLLARY 5.6.2 The lines of AG(2,n) may be partitioned into $n+1$ subsets of size n, called parallel classes, such that two lines meet if and only if they are in different parallel classes. ∎

THEOREM 5.7 There exists an AG(2,n), or balanced incomplete block design with parameters

$$(n^2, n^2 + n, n + 1, n, 1),$$

if and only if there exists a symmetric balanced incomplete block design with parameters

$$(n^2 + n + 1, n + 1, 1). \tag{5.4}$$

Proof: Suppose that there is an AG(2,n) with point set P and line set L. The line set can be partitioned into $n+1$ subsets $L_1, L_2, \ldots, L_{n+1}$, where each L_i is a parallel class of lines. Since lines in a parallel class do not intersect, no point belongs to two lines in the same subset L_i; simple arithmetic shows that each point belongs to exactly one line in each L_i.

We form a new design whose treatments are the elements of P together with $n+1$ new "points" $p_1, p_2, \ldots, p_{n+1}$. The blocks are formed from the lines in L: for each line ℓ in L, there is a unique subset L_i such that $\ell \in L_i$; we define a line $\ell^* = \ell \cup \{p_i\}$. The blocks of the new design are the $n^2 + n$ blocks ℓ^*, together with ℓ_∞, where

$$\ell_\infty = \{p_1, p_2, \ldots, p_{n+1}\}.$$

The new design has $n^2 + n + 1$ treatments and $n^2 + n + 1$ blocks which are $(n+1)$-sets of treatments. It remains to prove that it is balanced, with $\lambda = 1$. Now every pair of members of P belongs to precisely one member of L; they will belong to the corresponding new block, and no other. If p belongs to P, then $\{p, p_j\}$ is a subset of ℓ^*, where ℓ is the unique line passing through p which lies in L_j, while $\{p_i, p_j\}$ is a subset of ℓ_∞ alone. So the new design is balanced, with $\lambda = 1$; from (1.1) we find that $r = n + 1$.

5.3 Finite Projective Geometries

Conversely, if a symmetric design with parameters $(n^2 + n + 1, n + 1, 1)$ exists, delete from it one block and all its members. An $AG(2,n)$ is obtained: the details are left to the reader (see Exercise 5.3.1). ■

The construction just given is the finite analog of the way in which real projective plane geometry may be derived from real Euclidean plane geometry by appending "points at infinity," where parallel lines meet, and a "line at infinity," which consists of all the points at infinity. For this reason, a symmetric design with the parameters (5.4) is called a <u>finite projective plane</u>. Alternatively, one may make a formal definition by axioms, such as the following.

DEFINITION A finite projective plane consists of a finite set P of points and a set of subsets of P called lines, which satisfies the axioms (P1), (P2), and (P3):

(P1) Given two points, there is exactly one line that contains both;
(P2) Given two lines, there is exactly one point that lies in both;
(P3) There are four points, of which no three are collinear.

THEOREM 5.8 In a finite projective plane, as defined by the axioms (P1), (P2), (P3), every line contains $n + 1$ points for some parameter n. Such a plane is a symmetric balanced incomplete block design with parameters

$(n^2 + n + 1, n + 1, 1)$. ■

The proof is left as an exercise. A finite projective plane with parameter n is denoted $PG(2,n)$.

As a first example we exhibit the geometry $PG(2,2)$. In the preceding section we derived $AG(2,2)$ from the field $GF(2)$; its points were

$\{00, 01, 10, 11\}.$

The six lines fall into three parallel classes, which we shall denote as follows:

$L_1: \begin{cases} y = 0, & \text{points } 00, 10; \\ y + 1 = 0, & \text{points } 01, 11; \end{cases}$

$L_2: \begin{cases} x = 0, & \text{points } 00, 01; \\ x + 1 = 0, & \text{points } 10, 11; \end{cases}$

$L_3: \begin{cases} x + y = 0, & \text{points } 00, 11; \\ x + y + 1 = 0, & \text{points } 01, 10. \end{cases}$

We append new points p_1, p_2, p_3 and obtain the following lines:

$\{00, 10, p_1\}$, $\{01, 11, p_1\}$,

$\{00, 01, p_2\}$, $\{10, 11, p_2\}$,

$\{00, 11, p_3\}$, $\{01, 10, p_3\}$,

$\{p_1, p_2, p_3\}$.

If we perform the substitution

$$(00, 01, 10, 11, p_1, p_2, p_3) \mapsto (0, 3, 1, 6, 2, 4, 5),$$

we obtain the blocks

```
012   362
034   164
056   315,
245
```

which are the sets we first obtained for a $(7, 3, 1)$-design in Section 1.1.

As before, we can generalize to projective geometries of higher dimension; and as before, the easiest technique is the algebraic one. We define a <u>finite projective geometry of dimension</u> d over GF(n), or PG(d,n), to be a set of points which are $(d+1)$-vectors over GF(n); however, the zero vector is not allowed, and two points are considered equal if the vector of one is a multiple of the vector of another. Alternatively, the points correspond to the vectors of the form

$$(0, 0, \ldots, 0, 1, *, \ldots, *),$$

where asterisks denote any field elements, and where the leading 1 must occur (the zero vector is not allowed). In this interpretation, a prime consists of all the points x,

$$x = (x_1, x_2, \ldots, x_{d+1}),$$

such that

$$a_1 x_1 + a_2 x_2 + \cdots + a_{d+1} x_{d+1} = 0$$

for some constants $a_1, a_2, \ldots, a_{d+1}$, not all zero—in other words, all the points whose vectors are in the subspace

5.3 Finite Projective Geometries

ax = 0.

When these same rules (delete zero; all scalar multiples of a vector are equivalent) are applied to a (k+1)-dimensional space, we obtain a k-flat; a 1-flat is a line.

It is easy to see that taking the k-flats of a PG(d,n) as blocks and the points as treatments, we obtain a balanced incomplete block design with parameters

$$\left(\frac{n^{d+1}-1}{n-1}, \; \frac{(n^{d+1}-n)(n^{d+1}-n)}{(n^{k+1}-1)(n^{k+1}-n)} \lambda, \; \frac{n^{d+1}-n}{n^{k+1}-n} \lambda, \; \frac{n^{k+1}-1}{n-1}, \; \lambda \right) \quad (5.5)$$

where

$$\lambda = \frac{(n^{d+1}-n^2)(n^{d+1}-n^3) \cdots (n^{d+1}-n^k)}{(n^{k+1}-n^2)(n^{k+1}-n^3) \cdots (n^{k+1}-n^k)}. \quad (5.6)$$

Not every finite affine plane can be constructed algebraically from a field, as we shall prove; so the field construction is not the only one for PG(2,n). However, it may be shown that when $n \le 8$, the only examples of AG(2,n) are the EG(2,n); so there is no AG(2,6) and the other AG(2,n) with $n \le 8$ are uniquely determined. It follows that PG(2,6) is impossible and that PG(2,n) is unique for other values of n up to 8. Those PG(2,n) that are constructed from fields are called field planes.

Field planes have various special properties which are not necessarily enjoyed by projective planes in general. We explore some of these in Section 5.6 to prove that planes other than field planes exist. For the moment we prove one simple but surprising fact about field planes whose parameter is a power of 2.

By a quadrangle we mean a set of four points, no three of which are collinear. In a quadrangle $\{a,b,c,d\}$ the lines ab and cd are called opposite sides, so that a quadrangle determines three pairs of opposite sides; and the intersection of such a pair of lines is called a diagonal point.

THEOREM 5.9 In a field plane over $GF(2^t)$, the three diagonal points of any quadrangle are collinear.

Proof: Consider the quadrangle $\{a,b,c,d\}$. Since a, b, and c are not collinear, they are independent [as vectors over $GF(2^t)$]; so one can find a matrix to map them simultaneously into any desired independent set. In particular, select a matrix M such that

$$aM = (1,0,0), \quad bM = (0,1,0), \quad cM = (0,0,1).$$

Write $dM = (p, q, r)$. Since abd are not collinear, it follows that aM, bM, and dM are not collinear, so $r \neq 0$. Similarly, p and q are nonzero.

It is easy to calculate the diagonal points of $\{aM, bM, cM, dM\}$:

aMbM $\cap$ cMdM = $(p, q, 0)$ = u, say;

aMcM $\cap$ bMdM = $(p, 0, r)$ = v, say;

aMdM $\cap$ bMcM = $(0, q, r)$ = w, say.

These three points are collinear: they lie on

$$\frac{x_1}{p} + \frac{x_2}{q} + \frac{x_3}{r} = 0.$$

(The equation makes sense because none of p, q, r can be zero; the points satisfy it because the characteristic of the field is two.) Now u, v, and w are dependent vectors, so the original diagonal points uM^{-1}, vM^{-1}, and wM^{-1} are dependent as vectors, and the points are collinear. ∎

The points of the quadrangle together with the diagonal points form a substructure isomorphic to PG(2,2). It is not surprising that the field planes PG(2, 2^t) contain subplanes isomorphic to PG(2,2); the strength of the theorem is the fact that every quadrangle forms a subplane.

EXERCISES

5.3.1 Prove that if any one line (and all its points) is deleted from a PG(2,n), the result is an AG(2,n).

5.3.2 Prove Theorem 5.8.

5.3.3 Consider the geometry PG(2,5).
(i) What are the points of the line

$$x_1 + x_2 + 3x_3 = 0?$$

(ii) What is the equation of the line joining $(1,1,1)$ and $(2,3,2)$?
(iii) What is the point of intersection of the lines

$$x_1 + x_2 + 3x_3 = 0, \quad 2x_1 + 3x_2 + x_3 = 0?$$

5.3.4 Prove that PG(d,n) has $(n^{d+1} - 1)/(n - 1)$ points.

5.3.5 Prove that by taking the k-flats of PG(d,n) as blocks and the points as treatments, one obtains a balanced incomplete block design with parameters defined by (5.5) and (5.6).

5.4 Singer Difference Sets

5.3.6 Prove that the field plane over GF(3) contains a quadrangle whose diagonal points are not collinear.

5.3.7 Prove that there is exactly one PG(2,2) and exactly one PG(2,3) (up to isomorphism).

5.4 SINGER DIFFERENCE SETS

Consider the equations (5.4) and (5.5) in the case where n is a prime power and the blocks are primes in $PG(d,n)$: $k = d - 1$, so

$$\lambda = \frac{(n^{d+1} - n^2)(n^{d+1} - n^3) \cdots (n^{d+1} - n^{d-1})}{(n^d - n^2)(n^d - n^3) \cdots (n^d - n^{d-1})}$$

$$= \frac{n^2 \cdot n^3 \cdot \cdots \cdot n^{d-1}(n^{d-1} - 1)(n^{d-2} - 1) \cdots (n^2 - 1)}{n^2 \cdot n^3 \cdot \cdots \cdot n^{d-1}(n^{d-2} - 1)(n^{d-3} - 1) \cdots (n - 1)}$$

$$= \frac{n^{d-1} - 1}{n - 1}$$

and

$$b = \frac{(n^{d+1} - 1)(n^{d+1} - n)}{(n^d - 1)(n^d - n)} \lambda$$

$$= \frac{(n^{d+1} - 1)n}{n(n^{d-1} - 1)} \frac{n^{d-1} - 1}{n - 1}$$

$$= \frac{n^{d+1} - 1}{n - 1}$$

$$= v,$$

so the design is symmetric.

We shall prove that design can in fact be constructed from a difference set in the cyclic group of order v. To prove this we need a fact about finite fields. Suppose that x is a primitive element of $GF(n^{d+1})$. Since $GF(n^{d+1})$ can be constructed using $(d + 1)$th-degree polynomials over $GF(n)$, there is an element of $GF(n)$ that satisfies an irreducible polynomial equation of degree $d + 1$ over $GF(n)$. The fact we need is that x satisfies such an equation. (This is not surprising, but it needs to be proven—see, for example, algebra textbooks.)

Suppose that f is an irreducible polynomial of degree $d + 1$ over $GF(n)$ which has x as a root: say

$$f(y) = c_0 + c_1 y + \cdots + c_d y^d + y^{d+1},$$

and $f(x) = 0$. Since f is irreducible, $c_0 \neq 0$. Then

$$x^{d+1} = -c_0 - c_1 x - \cdots - c_d x^d. \tag{5.7}$$

It follows that for any exponent i, we can find elements $a_{i0}, a_{id}, \ldots, a_{id}$ of $GF(n)$ such that

$$x^i = a_{i0} + a_{i1} x + \cdots + a_{id} x^d. \tag{5.8}$$

[Take $a_{ii} = 1$ and $a_{ij} = 0$ when $j \neq i$, if $i \leq d$; for larger i, use repeated substitutions of (5.7).] This establishes a correspondence between powers of x and vectors over $GF(n)$: the $n^{d+1} - 1$ different powers correspond to the $n^{d+1} - 1$ different nonzero vectors. Since different powers cannot give the same vector, we have a one-to-one correspondence; and the vectors can of course be interpreted as points of $PG(d,n)$.

If we write $v = (n^{d+1} - 1)/(n - 1)$, then $0, 1, x^v, x^{2v}, \ldots, x^{(n-1)v}$ are the n solutions of the equation $y^n = y$, so they are the elements of the subfield $GF(n)$ in $GF(n^{d+1})$. For any s, (5.8) yields

$$x^{i+sv} = (x^{sv} a_{i0}) + (x^{sv} a_{i1}) x + \cdots + (x^{sv} a_{id}) x^d.$$

So we see that if $a_i = (a_{i0}, a_{i1}, \ldots, a_{id})$ is the vector corresponding to x^i, then the vector corresponding to x^{i+sv} is $x^{sv} a_i$, which is a nonzero $GF(n)$-multiple of a_i, and the two points of $GF(d,n)$ are the same. That is, x^i and x^j correspond to the same point if and only if i and j are congruent modulo v.

The map "multiplication by x," which has the effect

$$0 \mapsto 0, \quad x^i \mapsto x^{i+1},$$

can be interpreted as a mapping of the points of $PG(d,n)$, taking a_i to a_{i+1}. Now

$$x^{i+1} = x(a_{i0} + a_{i1} x + \cdots + a_{id} x^d)$$

$$= a_{i0} x + a_{i1} x^2 + \cdots + a_{id} x^{d+1}$$

5.4 Singer Difference Sets

$$= a_{i0}x + a_{i1}x^2 + \cdots + a_{id}(-c_0 - c_1 x - \cdots - c_d x^d)$$

$$= -c_0 a_{id} + (a_{i0} - c_1 a_{id})x + \cdots + (a_{i,d-1} - c_d a_{id})x^d,$$

using (5.7), so

$$a_{i+1} = (-c_0 a_{id}, \, a_{i0} - c_1 a_{id}, \, \ldots, \, a_{i,d-1} - c_d a_{id}).$$

Therefore, the map takes the point p to pφ, where

$$p = (p_0, p_1, \ldots, p_d)$$

$$p\varphi = (-c_0 p_d, \, p_0 - c_0 p_d, \, \ldots, \, p_{d-1} - c_0 p_d).$$

Now consider a prime h of the PG(d, n). Since primes are vector spaces of dimension d − 1, they consist of all the points whose coordinates satisfy one linear equation: There exist elements $h_0, h_1, \ldots, h_d$ of GF(n) such that

$$h = \{p : \sum_j h_j p_j = 0\}.$$

Define numbers $k_0, k_1, \ldots, d_d$ by

$$k_1 = h_0,$$

$$k_2 = h_1,$$

$$\ldots$$

$$k_d = h_{d-1},$$

$$k_0 = -c_0^{-1}(h_d + c_1 h_0 + c_2 h_1 + \cdots + c_d h_{d-1}).$$

If p lies in h, then write q_i for ith coordinate of pφ.

$$k_j q_j = -k_0 c_0 p_d + \sum_{j=0}^{d-1} k_{j+1}(p_j - c_j p_d)$$

$$= (h_d + c_1 h_0 + \cdots + c_d h_{d-1}) p_d + \sum_{j=0}^{d-1} h_j(p_j - c_j p_d)$$

$$= \sum_{j=0}^{d} h_j p_j$$

$$= 0.$$

So $h\varphi$ is the prime

$$\left\{ d : \sum k_j q_j = 0 \right\}.$$

Therefore, the map φ takes primes into primes.

Suppose that we replace the point a_i by the integer i modulo v. Then we have shown that the map "plus 1" maps primes into primes. It is not hard to show that all v primes can be obtained from a given prime in this way. So all v primes can be obtained by developing an initial prime modulo v. This must mean that the prime is a difference set modulo v when the identification between a_i and i is made. We have:

THEOREM 5.10 If n is a prime power, there is a cyclic difference set with parameters

$$\left(\frac{n^{d+1} - 1}{n - 1},\ \frac{n^d - 1}{n - 1},\ \frac{n^{d-1} - 1}{n - 1} \right). \qquad \blacksquare$$

These difference sets are called Singer difference sets.

In particular, when $d = 2$, a "prime" is just a line in the projective plane, so we have shown that any finite projective plane constructed over a field is cyclic when considered as a balanced incomplete block design. So there is a cyclic $(n^2 + n + 1, n + 1, 1)$-design whenever n is a prime power.

To illustrate the theorem we use the case $d = 2$, $n = 3$; in other words, we construct a $(13, 4, 1)$-difference set. Then $d + 1 = 3$, and the polynomial f(y) must be an irreducible polynomial of degree 3 over GF(3). One example is

$$g(y) = 1 - y + y^3;$$

the reader should verify that it is irreducible. We assume that x is a root of $g(y) = 0$; in other words, x satisfies

$$x^3 = 2 + x.$$

The powers of x can now be expressed in terms of 1, x, and x^2 and made to correspond to points in PG(2, 3), as follows:

5.4 Singer Difference Sets

$$
\begin{aligned}
x^0 &= 1 & &\leftrightarrow (1,0,0) = 2(2,0,0) \leftrightarrow 2 & &= x^{13}\\
x^1 &= x & &\leftrightarrow (0,1,0) = 2(0,2,0) \leftrightarrow 2x & &= x^{14}\\
x^2 &= x^2 & &\leftrightarrow (0,0,1) = 2(0,0,2) \leftrightarrow 2x^2 & &= x^{15}\\
x^3 &= 2+x & &\leftrightarrow (2,1,0) = 2(1,2,0) \leftrightarrow 1+2x & &= x^{16}\\
x^4 &= 2x+x^2 & &\leftrightarrow (0,2,1) = 2(0,1,2) \leftrightarrow x+2x^2 & &= x^{17}\\
x^5 &= 2+x+2x^2 & &\leftrightarrow (2,1,2) = 2(1,2,1) \leftrightarrow 1+2x+x^2 & &= x^{18}\\
x^6 &= 1+x+x^2 & &\leftrightarrow (1,1,1) = 2(2,2,2) \leftrightarrow 2+2x+2x^2 & &= x^{19}\\
x^7 &= 2+2x+x^2 & &\leftrightarrow (2,2,1) = 2(1,1,2) \leftrightarrow 1+x+2x^2 & &= x^{20}\\
x^8 &= 2+2x^2 & &\leftrightarrow (2,0,2) = 2(1,0,1) \leftrightarrow 1+x^2 & &= x^{21}\\
x^9 &= 1+x & &\leftrightarrow (1,1,0) = 2(2,2,0) \leftrightarrow 2+2x & &= x^{22}\\
x^{10} &= x+x^2 & &\leftrightarrow (0,1,1) = 2(0,2,2) \leftrightarrow 2x+2x^2 & &= x^{23}\\
x^{11} &= 2+x+x^2 & &\leftrightarrow (2,1,1) = 2(1,2,2) \leftrightarrow 1+2x+2x^2 & &= x^{24}\\
x^{12} &= 2+x^2 & &\leftrightarrow (2,0,1) = 2(1,0,2) \leftrightarrow 1+2x^2 & &= x^{25}
\end{aligned}
$$

For example, the first line means that $x^0 = 1$ which corresponds to $(1,0,0)$ and $x^{13} = 2$ which corresponds to $(2,0,0)$; also, $(1,0,0) = 2(2,0,0)$ over $GF(3)$ [and since one is a multiple of the other, they represent the same point of $PG(2,3)$].

Now select any line: for example, we choose the line $x_3 = 0$, which contains the points

$(0,1,0), (1,0,0), (1,1,0), (1,2,0).$

These points correspond to

$x^1, x^0, x^9, x^3.$

So we have the cyclic difference set $\{1,0,9,3\}$, or equivalently $\{0,1,3,9\}$.

Any other line could have been chosen. For example, $x_1 + x_2 + x_3 = 0$ has points $(0,1,2), (1,0,2), (1,1,1), (1,2,0)$ which correspond to x^4, x^{12}, x^6, x^3, and give difference set $\{3,4,6,12\}$. $x_1 = x_2$ has points $(0,0,1)$, $(1,1,0), (1,1,1), (1,1,2)$, powers x^2, x^9, x^6, x^7, and difference set $\{2,6,7,9\}$.

EXERCISES

5.4.1 Prove that if h is a prime of $PG(d,n)$ and φ is the mapping defined in this section, then

$$h\varphi^i \neq h\varphi^j$$

unless $i \equiv j \pmod{v}$. Deduce that all primes can be constructed from h by mapping with the powers of φ.

5.4.2 Prove that if F is any finite field and f is a reducible cubic polynomial over F, then $f(a) = 0$ for some $a \in F$. Hence prove that $1 - y + y^3$ is irreducible over GF(3).

5.4.3 Find Singer difference sets to generate PG(2,2) and PG(2,5).

5.4.4 $GF(2^2)$ consists of the four elements 0, 1, w, and w + 1, where $1 + w + w^2 = 0$.
  (i) Verify that
  $$f(y) = w + wy + wy^2 + y^3$$
  is an irreducible polynomial over $GF(2^2)$.
  (ii) Assume that x is a primitive element of $GF(2^6)$ which satisfies $f(x) = 0$. Construct a Singer difference set that generates PG(2,4) (use the line $x_3 = 0$).

5.5 OVALS IN PROJECTIVE PLANES

Suppose that c is any set of points in a PG(2,n). A line ℓ is called an i-secant of c if ℓ and c have i common points. In particular a 2-secant is simply called a <u>secant</u>, a 1-secant is a <u>tangent</u>, and a 0-secant is an <u>external line</u> to c. A set c that has no i-secant for $i > k$ is called a k-<u>arc</u>.

Suppose that p is a point of a 2-arc c in a PG(2,n). If c has m elements, there are m - 1 lines through p to the other points of c; these lines are all different (for otherwise we would have a 3-secant). So p lies on exactly m - 1 secants, and consequently on n + 2 - m tangents. It follows that $m \le n + 2$; a 2-arc with n + 2 points is called a <u>maximal arc</u>. Also, the total number of tangents is m(n + 2 - m).

We are interested in 2-arcs with either n + 1 points or n + 2 points. These are called <u>ovals</u>; the two classes are called type I ovals and type II ovals, respectively. A type I oval has one tangent through each of its points, a total of exactly n + 1 tangents, and a type II oval has no tangents whatsoever.

Suppose that a PG(2,n) has a type II oval c. Select a point q outside the oval. Every line through q meets c in 0 or 2 points. But every point of c must lie on some line through q. It follows that c has an even number of points. So n is even.

LEMMA 5.11 If c is a type I oval in a projective plane PG(2,n), where n is even, then the tangents to c are concurrent.

Proof: If x is any point of c, there is a tangent at x. If x is not on c, count all the intersections of c with lines through x. The answer is n + 1, which is odd, but it also equals the number of tangents through x plus twice

5.5 Ovals in Projective Planes

the number of secants through x; so there is at least one tangent through x. Thus every point lies on at least one tangent to c.

Suppose that x and y lie on some secant to c. Then the tangents to c through x and y must be different (since otherwise the line xy would be both a tangent and a secant). So the tangents through points of a secant are all different. Since c has exactly n + 1 tangents, it follows that each of a secant lies on exactly one tangent.

Now consider the point m of intersection of two tangents. This point cannot lie on any secant. So all the lines joining m to the points of c are tangents, and m lies on all n + 1 tangents. ∎

The point m is called the <u>nucleus</u> of the oval c. It is clear that $c \cup \{m\}$ is a type II oval. On the other hand, if any point m is deleted from a type II oval, the result is a type I oval with m as nucleus.

COROLLARY 5.11.1 If n is odd, then the tangents of a type I oval in a PG(2,n) are not concurrent.

COROLLARY 5.11.2 A PG(2,n), where n is even, contains a type II oval if and only if it contains a type I oval. ∎

In particular, consider the field plane over $GF(2^t)$, and ket c consist of all the points that satisfy

$$x_1 x_2 + x_2 x_3 + x_3 x_1 = 0.$$

Clearly, c consists of the points (0,0,1), (0,1,0), and all the $(1, y, y(1+y)^{-1}$ for $y \neq 1$. So c contains n + 1 points, and it is easy to see that no three are collinear (see Exercise 5.5.1). Therefore, c is a type I oval. So we have:

COROLLARY 5.11.3 There is a $PG(2,2^t)$ containing a type II oval for every positive integer t. ∎

One reason for our interest in ovals is given in the following lemma.

LEMMA 5.12 Suppose that c is a type II oval in a PG(2,2k). Then there is a balanced incomplete block design with parameters

$$(2k^2 - k, 4k^2 - 1, 2k + 1, k, 1).$$

Proof: The treatments of the design will correspond to the external lines of c, and the blocks to the points of the plane that are not in c. A block contains a treatment if and only if the corresponding point lies on the corresponding point lies on the corresponding line. Since c has 2k + 2 points, it has

$$\binom{2k+2}{2} = (k+1)(2k+1)$$

secants, and therefore $4k^2 + 2k + 1 - (k+1)(2k+1)$ external lines. So the number of treatments is $4k^2 + 2k + 1 - (k+1)(2k+1) = 2k^2 - k$. Each point not on c lies on $2k+1$ lines, of which $k+1$ are secants, so each block of the design will contain $(2k+1) - (k+1) = k$ treatments. Any two external lines intersect in a unique point that is not on c, so the design is balanced with $\lambda = 1$. The other parameters follow from (1.1) and (1.2) (or they could be deduced directly). ∎

THEOREM 5.13 There is a design with parameters

$$(2k^2 - k,\ 4k^2 - 1,\ 2k+1,\ k,\ 1)$$

whenever k is a power of 2. ∎

This theorem cannot be reversed—for example, we know a $(15, 35, 7, 3, 1)$-design, but 3 is not a power of 2 and in fact we shall show in Chapter 6 that no $PG(2, 6)$ exists. But it is a useful theorem that provides an infinite class of block designs.

Ovals have other combinatorial applications. In Chapter 11 we use them, together with other tools, to prove the uniqueness of $PG(2, 4)$. They have also been used in other aspects of design construction and they are interesting in their own right, in finite geometry.

EXERCISE

5.5.5 Verify that no three of the points

$$(0, 0, 1),\ (0, 1, 0),\ \{(1,\ y,\ y(1+y)) : y \neq 1\}$$

are collinear over any field of characteristic 2.

5.6 THE DESARGUES CONFIGURATION

The configuration of Desargues consists of two disjoint <u>triangles</u>, or sets of three noncollinear points, abc and def say, in which the three lines ad, be, and cf meet in a seventh point p, with the further property that the three points ab ∩ de, ac ∩ df, and bc ∩ ef all lie on some line ℓ. The configuration is illustrated in Figure 5.1. One usually says that the two triangles abc and def are in perspective through the point p and also in perspective through the line ℓ.

We say a finite projective plane is Desarguesian if it satisfies the following law: Whenever the six points a, b, c, d, e, f are such that be and cf are

5.6 The Desargues Configuration

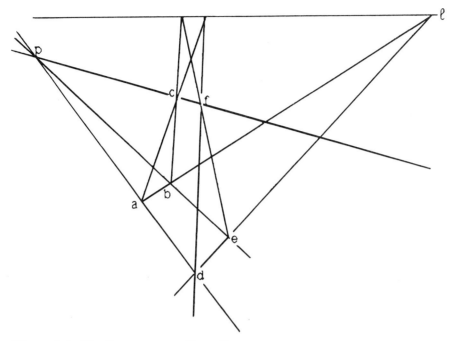

Figure 5.1 The Desargues configuration.

concurrent lines, and abc and def are not collinear sets, then ab ∩ de, ac ∩ df, and bc ∩ ef are collinear. That is, whenever two triangles are in perspective from a point, they form a Desargues configuration.

Suppose that p were to lie on the line ab. Then ab and de are the same line, and the intersection consists of all the points of the line. One could argue that no single point ab ∩ de is defined, so no Desargues configuration is possible; or one could say that the line joining ac ∩ df to bc ∩ ef certainly contains a point of ab ∩ de, and say that the triangles always yield a Desargues configuration. We shall take the latter view; if p lies on one of the sides of one of the triangles, then it lies on a common side of the two triangles and we have a trivial Desargues configuration.

THEOREM 5.14 Every field plane is Desarguesian.

Proof: Suppose that abc and def are two noncollinear sets, and that ad, be, and cf meet in the point p. There is no loss of generality in assuming that $a = (1, 0, 0)$, $b = (0, 1, 0)$, and $c = (0, 0, 1)$, because we can transform these into any three noncollinear points using a nonsingular matrix (compare with the proof of Theorem 5.9).

Write $p = (p_1, p_2, p_3)$. If p lies on ab, there is nothing to prove, so we can assume that $p_3 \neq 0$; similarly, $p_1 \neq 0$ and $p_2 \neq 0$. The points d, e, and f lie on ap, bp, and cp, respectively, and they are distinct from a, b, c, p, so we can write

$$d = (1 + gp_1,\ gp_2,\ gp_3)$$

$$e = (hp_1,\ 1 + hp_2,\ hp_3)$$

$$f = (ip_1,\ ip_2,\ 1 + ip_3)$$

for some nonzero g, h, and i.

The general point on de has the form

$$q_1 d + q_2 e, \qquad (5.9)$$

where q_1 and q_2 are members of the field. The points of ab are the points with third coordinate zero. So the point ab $\cap$ de will have the form (5.9) with

$$q_1 gp_3 + q_2 hp_3 = 0;$$

since points are invariant under nonzero scalar multiplication, we can take $q_1 = h$, $q_2 = -g$, and we obtain

$$\text{ab} \cap \text{de} = (h,\ -g,\ 0)$$

and similarly

$$\text{ac} \cap \text{df} = (-i,\ 0,\ g)$$

$$\text{bc} \cap \text{ef} = (0,\ i,\ -h).$$

All these points lie on the line

$$gx_1 + hx_2 + ix_3 = 0,$$

so the triangles are in perspective from that line. ∎

THEOREM 5.15 There is a non-Desarguesian PG(2,9).

Proof: We show how to construct a plane by a difference method. It has 91 points denoted A_i, B_i, C_i, D_i, E_i, F_i, G_i for $0 \leq i \leq 12$, and 91 lines denoted ℓ_i, m_i, n_i, p_i, q_i, r_i, s_i for $0 \leq i \leq 12$, where

5.6 The Desargues Configuration

$$\ell_0 = \{A_0, A_1, A_3, A_9, B_0, C_0, D_0, E_0, F_0, G_0\}$$
$$m_0 = \{A_0, B_1, B_8, D_3, D_{11}, E_2, E_5, E_6, G_7, G_9\}$$
$$n_0 = \{A_0, C_1, C_8, E_7, E_9, F_3, F_{11}, G_2, G_5, G_6\}$$
$$p_0 = \{A_0, B_7, B_9, D_1, D_8, F_2, F_5, F_6, G_3, G_{11}\}$$
$$q_0 = \{A_0, B_2, B_5, B_6, C_3, C_{11}, E_1, E_8, F_7, F_9\}$$
$$r_0 = \{A_0, C_7, C_9, D_2, D_5, D_6, E_3, E_{11}, F_7, F_8\}$$
$$s_0 = \{A_0, B_3, B_{11}, C_2, C_5, C_6, D_7, D_9, G_1, G_8\};$$

the other lines are derived by developing the subscripts modulo 13.

We exhibit two triangles in perspective from a point that does not form a Desargues configuration. The triangles are $A_1 B_1 C_8$ and $A_3 B_8 E_7$, which are in perspective through A_0:

$$\ell_0 = A_0 A_1 A_3, \quad m_0 = A_0 B_1 B_8, \quad n_0 = A_0 C_8 E_7.$$

One easily checks that

$$\begin{aligned}
A_1 B_1 &= \ell_1 = \{A_1, A_2, A_4, A_{10}, B_1, C_1, D_1, E_1, F_1, G_1\}, \\
A_3 B_8 &= q_3 = \{A_3, B_5, B_8, B_9, C_6, C_1, E_4, E_{11}, F_{10}, F_{12}\},
\end{aligned} \Biggr\} \ell_1 \cap q_3 = C_1,$$

$$\begin{aligned}
A_1 C_8 &= r_1 = \{A_1, C_8, C_{10}, D_3, D_6, D_7, E_4, E_{12}, F_8, F_9\} \\
A_3 E_7 &= \ell_7 = \{A_7, A_8, A_{10}, A_3, B_7, C_7, D_7, E_7, F_7, G_7\}
\end{aligned} \Biggr\} r_1 \cap \ell_7 = D_7,$$

$$\begin{aligned}
B_1 C_8 &= s_3 = \{A_3, B_6, B_1, C_5, C_8, C_9, D_{10}, D_{12}, G_4, G_{11}\}, \\
B_8 E_7 &= q_6 = \{A_6, B_8, B_{11}, B_{12}, C_9, C_4, E_7, E_1, F_0, F_2\},
\end{aligned} \Biggr\} s_3 \cap q_6 = C_9,$$

and C_1, D_7, and C_9 are not collinear:

$$C_1 D_7 = r_5, \quad C_1 C_9 = q_{11}, \quad D_7 C_9 = r_2. \qquad \blacksquare$$

COROLLARY 5.15.1 There are nonisomorphic planes PG(2,9).

EXERCISES

5.6.1 The dual of a balanced incomplete block design was defined in Section 2.2 to be the design obtained by exchanging the roles of blocks and treatments (or, equivalently, by transposing the incidence matrix). Prove that the dual of a Desarguesian PG(2,n) is a Desarguesian PG(2,n).

5.6.2 Verify that the lines given in the proof of Theorem 5.15 form a PG(2,9).

5.6.3 Find further pairs of triangles in the plane of Theorem 5.15 which are in perspective through a point but not through any line.

BIBLIOGRAPHIC REMARKS

The study of finite geometries started in the late nineteenth century. The most significant early study was that of Veblen and Bussey, starting in [9]. The relations between designs and finite geometries are discussed in [2], [6], and [8]. Two excellent introductory texts on finite geometries are [1] and [3], and more advanced books include [2], [4], and [5]. Theorem 5.10 is from [7].

1. A. A. Albert and R. F. Sandler, An Introduction to Finite Projective Planes (Holt, Rinehart and Winston, New York, 1968).
2. P. Dembowski, Finite Geometries (Springer-Verlag, New York, 1968).
3. H. L. Dorwart, The Geometry of Incidence (Prentice-Hall, Englewood Cliffs, N.J., 1966).
4. J. Hirschfeld, Projective Geometries over Finite Fields (Oxford University Press, Oxford, 1983).
5. D. R. Hughes and F. C. Piper, Projective Planes (Springer-Verlag, Heidelberg, West Germany, 1973).
6. D. R. Hughes and F. C. Piper, Design Theory (Cambridge University Press, Cambridge, 1985).
7. J. Singer, "A theorem in finite projective geometry and some applications to number theory," Transactions of the American Mathematical Society 43(1938), 377-385.
8. S. Vajda, Patterns and Configurations in Finite Spaces (Griffin, London, 1978).
9. O. Veblen and W. H. Bussey, "Finite projective geometries," Transactions of the American Mathematical Society 7(1906), 242-259.

6
More About Block Designs

6.1 RESIDUAL AND DERIVED DESIGNS

The relationship between affine and projective planes can be generalized to other block designs. If B_0 is any block of pairwise balanced design with index λ, then any two treatments that <u>do not</u> belong to B_0 must occur together in λ of the remaining blocks, while any two members of B_0 must be together in $\lambda - 1$ of the remaining blocks. It follows that the blocks $B \backslash B_0$ form a pairwise balanced design of index λ when B ranges through the remaining blocks, and the $B \cap B_0$ form a pairwise balanced design of index $\lambda - 1$. We shall refer to these as the <u>residual</u> and <u>derived</u> designs of the original with respect to the blocks B_0. (The derived design is a special case of the derivative that was defined in Section 2.3.)

In general the replication number of a residual or derived design is not constant, and the block sizes follow no particular pattern. To solve the first of these problems it suffices to make the original design equireplicate: If a design has constant frequency r, its residual and derived designs also have constant frequencies, r and r - 1, respectively. The block pattern can be predicted if the size of $B \cap B_0$ is constant; the derived blocks will be of that constant size, and the blocks of the residual design will be $|B \cap B_0|$ smaller than the blocks of the original. In particular, if we start with a symmetric balanced incomplete block design, we obtain constant frequencies and block sizes:

THEOREM 6.1 The residual design of a (v, k, λ)-design as a balanced incomplete block design with parameters

$$(v - k, \ v - 1, \ k, \ k - \lambda, \ \lambda), \tag{6.1}$$

provided that $\lambda \neq k - 1$. The derived design of a (v, k, λ)-design is a balanced incomplete block design with parameters

$$(k,\ v - 1,\ k - 1,\ \lambda,\ \lambda - 1), \tag{6.2}$$

provided that $\lambda \neq 1$. ∎

The two exceptions are made so as to exclude "trivial" designs.

Affine planes are the residual designs of projective planes; the derived designs of projective planes are trivial. As an example where both designs are nontrivial, consider the $(11, 6, 3)$-design B in Table 6.1, whose residual design R and derived design D with respect to the first block are also shown. They have parameters $(5, 10, 6, 3, 3)$ and $(6, 10, 5, 3, 2)$, respectively.

Suppose that a design D has the parameters $(v - k,\ v - 1,\ k,\ k - \lambda,\ \lambda)$ for some v, k, and λ. One can ask whether D is embeddable—that is, does there exist a (v, k, λ)-design of which D is a residual? In the case $\lambda = 1$, we have the parameters of an affine plane, and by Theorem 5.7 we know that a suitable symmetric design (a projective plane) necessarily exists. The corresponding result also holds for $\lambda = 2$, and slightly less can be said for greater values of λ; the proofs are too long to include here.

THEOREM 6.2 Suppose that D is a balanced incomplete block design with parameters (6.1).

(i) If $\lambda = 1$ or $\lambda = 2$, there is a (v, k, λ)-design of which D is the residual.
(ii) There is a number $f(\lambda)$, depending only on λ, such that if $k \geq f(\lambda)$, there is a (v, k, λ)-design of which D is the residual. ∎

A design with parameters (6.1) is called quasi-residual; thus Theorem 6.2(ii) says that "all sufficiently large quasi-residual designs are residual." To give an indication of the meaning of "sufficiently large," $f(3)$ is at most 90.

In order to prove that the theorem cannot be improved significantly, we exhibit a quasi-residual design with $\lambda = 3$. We start with a finite affine geometry $AG(2, 3)$, whose blocks are ordered so that the first three form one parallel class, the next three form another, and so on. Say A is the 9×12 incidence matrix of this design. Define

$$C = \begin{bmatrix} 1 & 1 & 1 & 1 & 1 & 1 & 0 & 0 & 0 & 0 & 0 & 0 \\ 1 & 1 & 1 & 0 & 0 & 0 & 1 & 1 & 1 & 0 & 0 & 0 \\ 1 & 1 & 1 & 0 & 0 & 0 & 0 & 0 & 0 & 1 & 1 & 1 \\ 0 & 0 & 0 & 1 & 1 & 1 & 1 & 1 & 1 & 0 & 0 & 0 \\ 0 & 0 & 0 & 1 & 1 & 1 & 0 & 0 & 0 & 1 & 1 & 1 \\ 0 & 0 & 0 & 0 & 0 & 0 & 1 & 1 & 1 & 1 & 1 & 1 \end{bmatrix},\ B = \begin{bmatrix} 1 & 1 & 1 & 1 & 1 & 1 & 1 & 1 & 1 \\ 1 & 1 & 1 & 0 & 0 & 0 & 0 & 0 & 0 \\ 0 & 0 & 0 & 1 & 1 & 1 & 0 & 0 & 0 \\ 0 & 0 & 0 & 0 & 0 & 0 & 1 & 1 & 1 \end{bmatrix}.$$

6.1 Residual and Derived Designs

TABLE 6.1

1	2	3	4	5	6		7	10	11		2	5	6
2	5	6	7	10	11		7	10	11		2	5	6
1	4	6	7	8	10		7	8	10		1	4	6
2	4	5	7	8	9		7	8	9		2	4	5
3	5	6	8	9	10		8	9	10		3	5	6
3	4	6	7	9	11		7	9	11		3	4	6
1	3	5	7	8	11		7	8	11		1	3	5
1	2	6	8	9	11		8	9	11		1	2	6
1	2	3	7	9	10		7	9	10		1	2	3
2	3	4	8	10	11		8	10	11		2	3	4
1	4	5	9	10	11		9	10	11		1	4	5
		B						R				D	

Then write

$$M = \begin{bmatrix} C^T & A^T & A^T \\ 0 & B & J-B \end{bmatrix}.$$

THEOREM 6.3 M is the incidence matrix of a $(16, 24, 9, 6, 3)$-design which is quasi-residual but not residual.

Proof: It is clear that M has constant row sum 9 and constant column sum 6. To calculate MM^T, we first observe that

$$A^T A = \begin{bmatrix} 3I & J & J & J \\ J & 3I & J & J \\ J & J & 3I & J \\ J & J & J & 3I \end{bmatrix}, \quad C^T C = \begin{bmatrix} 3J & J & J & J \\ J & 3J & J & J \\ J & J & 3J & J \\ J & J & J & 3J \end{bmatrix},$$

$$BB^T = \begin{bmatrix} 9 & 3 & 3 & 3 \\ 3 & 3 & 0 & 0 \\ 3 & 0 & 3 & 0 \\ 3 & 0 & 0 & 3 \end{bmatrix}, \quad (J-B)(J-B)^T = \begin{bmatrix} 0 & 0 & 0 & 0 \\ 0 & 6 & 3 & 3 \\ 0 & 3 & 6 & 3 \\ 0 & 3 & 3 & 6 \end{bmatrix}.$$

Now

$$MM^T = \begin{bmatrix} C^T C + 2A^T A & A^T J \\ JA & BB^T + (J-B)(J-B^T) \end{bmatrix}$$

and since $JA = 3J$, we have $MM^T = 6I + 3J$. So M is the incidence matrix of a design with the desired parameters.

Now consider the seventh and sixteenth blocks of the design. They each contain the four treatments corresponding to the four lines of the AG(2,3) through the first point of the geometry. So these blocks have intersection size 4. If it were possible to add further treatments to this design in order to embed it in a (25,9,3)-design, we would obtain a (25,9,3)-design with a pair of blocks of intersection size at least 4. (In fact, there will be nine such pairs of blocks.) This is impossible, since two blocks of a (25,9,3)-design intersect in three treatments only. ∎

EXERCISES

6.1.1 For the following parameter sets, what are the residual and derived parameters?
  (i) (16,6,2)
  (ii) (22,7,2)
  (iii) (45,12,3)
  (iv) (13,9,6).

6.1.2 Suppose that D is a symmetric balanced incomplete block design with parameters (v, k, λ), where $2k < v$.
  (i) Prove that $2\lambda < k$.
  (ii) Hence show that the residual design of D has no repeated blocks.

6.1.3 Consider the parameters of a finite projective plane: $(n^2 + n + 1, n + 1, 1)$.
  (i) What are the parameters of the complement of this design (as defined in Section 2.2)?
  (ii) What are the parameters of the residual of this complementary design?
  (iii) Prove that a design with these residual parameters always exists.

6.1.4 Suppose that D is a symmetric balanced incomplete block design and E is its complement. Prove that the residual design of D is the complement of the derived design of E.

6.1.5 Let D consist of all the subsets of size n of an (n + 2)-set. Prove that D is a quasi-residual balanced incomplete block design.

6.2 THE MAIN EXISTENCE THEOREM

So far we have seen various families of designs, but know of no parameters (v, b, r, k, λ) that satisfy (1.1) and (1.2) but for which no design exists. In this section we prove that those two conditions are not sufficient. The result is usually called the Bruck-Ryser-Chowla theorem.

6.2 The Main Existence Theorem

THEOREM 6.4 If there exists a symmetric balanced incomplete block design with parameters (v, k, λ), then:

(i) if v is even, $k - \lambda$ must be a perfect square;
(ii) if v is odd, there must exist integers x, y, and z, not all zero, such that

$$x^2 = (k - \lambda)y^2 + (-1)^{(v-1)/2} \lambda z^2. \qquad (6.3)$$

(Note: For those with a knowledge of linear algebra that includes matrix congruence, a shorter proof of this theorem is given in the next section.)

Proof: (i) Suppose that v is even, and that there exists a (v, k, λ)-design with incidence matrix A. Since A is square it has a determinant, $\det(A)$, say, and $\det(A^T) = \det(A)$, so

$$\det(A) = \sqrt{\det(AA^T)}.$$

But from Lemma 2.9,

$$\det(AA^T) = k^2(k - \lambda)^{v-1},$$

so

$$\det(A) = \pm k(k - \lambda)^{(v-1)/2}.$$

Now A is an integer matrix, so $\det(A)$ is an integer. Since k and λ are integers, and v is even, it follows that $\sqrt{k - \lambda}$ is an integer also.

(ii) Now suppose that v is odd. It is important to notice that all the calculations in this proof take place over the rational numbers.

From Theorem 3.22, $k - \lambda$ can be written as the sum of four integer squares; say

$$k - \lambda = a^2 + b^2 + c^2 + d^2,$$

where a, b, c, and d are nonnegative integers.

We consider the matrix

$$B = \begin{bmatrix} a & -b & -c & -d \\ b & a & -d & c \\ c & d & a & -b \\ d & -c & b & a \end{bmatrix}. \qquad (6.4)$$

Since $BB^T = (k - \lambda)I$, $\det(BB^T) = (k - \lambda)^4$ and $\det B = (k - \lambda)^2$, which is nonzero, so B is nonsingular.

Now let W be any rational column vector of length 4, and write

$$U = BW,$$

so that

$$u_1 = aw_1 - bw_2 - cw_3 - dw_4,$$
$$u_2 = bw_1 + aw_2 - dw_3 + cw_4,$$
$$u_3 = cw_1 + dw_2 + aw_3 - bw_4,$$
$$u_4 = dw_1 - cw_2 + bw_3 + aw_4.$$

Then

$$U^T U = (BW)^T \cdot BW = W^T B^T BW;$$

in other words,

$$u_1^2 + u_2^2 + u_3^2 + u_4^2 = (a^2 + b^2 + c^2 + d^2)(w_1^2 + w_2^2 + w_3^2 + w_4^2)$$
$$= (k - \lambda)(w_1^2 + w_2^2 + w_3^2 + w_4^2). \qquad (6.5)$$

Suppose that there is a symmetric balanced incomplete block design with parameters (v, k, λ), whose incidence matrix is $A = (a_{ij})$. Use A to define v homogeneous linear functions $A_1, A_2, \ldots, A_v$ by

$$A_j = A_j(X) = \sum_{i=1}^{v} x_i a_{ij}$$

where X is the row vector of v variables $(x_1, x_2, \ldots, x_j)$. Then

$$XAA^T X^T = \sum_{j=1}^{v} A_j^2. \qquad (6.6)$$

Now (2.11) tells us $AA^T = (k - \lambda)I + \lambda J$, so (6.6) yields

6.2 The Main Existence Theorem

$$\sum_{j=1}^{v} A_j^2 = (k - \lambda)XX^T + \lambda XJX^T$$

$$= (k - \lambda) \sum_{j=1}^{v} (x_j^2) + \lambda \left(\sum_{j=1}^{v} x_j \right)^2 \qquad (6.7)$$

Now assume that $v \equiv 1 \pmod{4}$, so that $(-1)^{(v-1)/2} = 1$. Define a new set of variables $y_1, y_2, \ldots, y_v$ by

$$\begin{bmatrix} y_1 \\ y_2 \\ y_3 \\ y_4 \end{bmatrix} = B \begin{bmatrix} x_1 \\ x_2 \\ x_3 \\ x_4 \end{bmatrix}, \begin{bmatrix} y_5 \\ y_6 \\ y_7 \\ y_8 \end{bmatrix} = B \begin{bmatrix} x_5 \\ x_6 \\ x_7 \\ x_8 \end{bmatrix}, \ldots, \begin{bmatrix} y_{v-4} \\ y_{v-3} \\ y_{v-2} \\ y_{v-1} \end{bmatrix} = B \begin{bmatrix} x_{v-4} \\ x_{v-3} \\ x_{v-2} \\ x_{v-1} \end{bmatrix}, y_v = x_v,$$

and write σ for Σx_j. Then, from (6.5),

$$(k - \lambda) \sum_{j=1}^{v-1} x_j^2 = \sum_{j=1}^{v-1} y_j^2,$$

so (6.7) becomes

$$\sum_{j=1}^{v} A_j^2 = \left(\sum_{j=1}^{v-1} y_j^2 \right) + (k - \lambda)y_v^2 + \lambda \sigma^2. \qquad (6.8)$$

Since B is nonsingular, each x_i is a (rational) linear function of the y_i, so we can write

$$A_1 = \sum_{i=1}^{v} e_i y_i$$

for some rational numbers e_i. Further, the nonsingularity of B implies that by proper choice of the variables x_j, we can give any values we wish to the variables y_j, so that (6.8) remains true whatever values the y_j may take. In particular, it will be true if we restrict ourselves to values such that

$$y_1 = (1 - e_1)^{-1} \sum_{i=2}^{v} e_i y_i$$

or, in the particular case where $e_1 = 1$ and the inverse above does not exist,

$$y_1 = (-1 - e_1)^{-1} \sum_{i=2}^{v} e_i y_i.$$

It is easy to see that this value of y_1 can be specified by the choice of x_1, whatever values of $x_2, x_3, \ldots, x_v$ are chosen.

In both cases we have $A_1^2 = y_1^2$, so equation (6.8) becomes

$$\sum_{j=2}^{v} A_j^2 \left(\sum_{j=2}^{v-1} y_j^2 \right) + (k - \lambda) y_v^2 + \lambda \sigma^2. \qquad (6.9)$$

Each of the A_j, and also σ, are linear homogeneous functions of the variables $y_1, y_2, \ldots, y_v$. But y_1 has been made into a linear homogeneous function of the remaining y_j. So the A_j, and σ, can be considered to be linear functions of $y_2, y_3, \ldots$, and y_v, and they will all have zero constant terms.

One can carry out a similar reduction on (6.9), obtaining an equation in terms of $y_3, y_4, \ldots, y_v$ only, and so on. Eventually,

$$A_v^2 = (k - \lambda) y_v^2 + \lambda \sigma^2. \qquad (6.10)$$

No value for y_v has yet been specified; that choice will be made shortly. The process of eliminating variables has been such that A_v and σ are homogeneous linear rational functions of y_v; for some rational numbers p/q and r/s we have

$$A_v = \frac{p}{q} y_v, \qquad \sigma = \frac{r}{s} y_v.$$

Now we may give y_v any value. We choose to put $y_v = qs$. Then (6.10) becomes

$$\left(\frac{p}{q} qs \right)^2 = (k - \lambda)(qs)^2 + \lambda \left(\frac{r}{s} qs \right)^2$$

which is

$$(ps)^2 = (k - \lambda)(qs)^2 + \lambda (qr)^2.$$

6.2 The Main Existence Theorem

So there are integers x, y, and z such that

$$x^2 = (k - \lambda)y^2 + \lambda z^2,$$

namely, $x = ps$, $y = qs$, and $z = qr$. Since q and s are the denominators of two rational numbers, y is nonzero. So theorem is proved for the case $v \equiv 1 \pmod 4$.

In the case $v \equiv 3 \pmod 4$ we introduce a new variable x_{v+1} and continue the change of variables as far as

$$\begin{bmatrix} y_{v-2} \\ y_{v-1} \\ y_v \\ y_{v+1} \end{bmatrix} = B \begin{bmatrix} x_{v-2} \\ x_{v-1} \\ x_v \\ x_{v+1} \end{bmatrix}.$$

Then (6.7) becomes

$$\left(\sum_{j=1}^{v} A_j^2 \right) + (k - \lambda)x_{v+1}^2 = \left(\sum_{j=1}^{v} y_j^2 \right) + y_{v+1}^2 + \lambda \sigma^2$$

and rather than (6.10) we obtain

$$y_{v+1}^2 = (k - \lambda)x_{v+1}^2 - \lambda \sigma^2.$$

The proof then proceeds as in the case $v \equiv 1 \pmod 4$; the details are left as an exercise. ■

The application of part (i) is quite easy, and we see immediately that many sets of parameters are impossible: for example, $(22, 7, 2)$, $(46, 10, 2)$, $(52, 18, 6)$, and so on. Part (ii) requires some number-theoretic results in some cases. However, a number of designs can be excluded quite simply. As an example we consider the parameters $(141, 21, 3)$.

Suppose that a $(141, 21, 3)$-design exists. Then there exist integers x, y, and z, not all zero, such that

$$x^2 = 18y^2 + 3z^2.$$

Without loss of generality we can assume that x, y, and z have no mutual common factor. Clearly, 3 must divide x^2, so 3 divides x: say $x = 3e$. Then

$$9e^2 = 18y^2 + 3z^2,$$

and z must also be divisible by 3; say z = 3f. We have

$$9e^2 = 18y^2 + 27f^2,$$
$$e^2 = 2y^2 + 3f^2.$$

Reducing modulo 3, we find

$$e^2 \equiv 2y^2 \pmod{3}.$$

If y is not divisible by 3, $2y^2$ is congruent to 2, which is not quadratic—a contradiction. So 3 divides y, and x, y, and z have common factor 3—another contradiction. So no (141,21,3)-design can exist.

Part (ii) is especially interesting when $\lambda = 1$; the designs are finite projective planes. In the case of a PG(2,n) we have $v = n^2 + n + 1$, which is always odd, and $(v-1)/2 = n(n-1)/2$, which is even when n or n+1 is divisible by 4 and odd otherwise. So when $n \equiv 0$ or 3 (mod 4), equation (6.5) becomes

$$x^2 = ny^2 + z^2,$$

which always has the solution x = z = 1, y = 0. But when $n \equiv 1$ or 2 (mod 4), (6.5) is

$$x^2 = ny^2 - z^2,$$

or equivalently

$$ny^2 = x^2 + z^2. \tag{6.11}$$

So ny^2 is the sum of two integer squares. By Theorem 3.28, any prime congruent to 3 (modulo 4) that divides ny^2 must appear to an even power in the factorization of ny^2. Since y^2 is obviously a square, such a prime must divide n itself to an even power. Applying Theorem 3.28 again, we see that n itself can be written as the sum of two integer squares, leading to the following result.

COROLLARY 6.4.1 If there is a PG(2,n), and $n \equiv 1$ or 2 (mod 4), then

$$n = a^2 + b^2$$

for some integers a and b. ∎

6.3 Another Proof of the Existence Theorem

The first few integers congruent to 1 or 2 modulo 4 are 1, 2, 5, 6, 9, 10, 13, and 14. Of these each can be expressed as a sum of two integer squares except 6 and 14. So no PG(2,6) or PG(2,14) can exist.

Everything we have said so far applies only to symmetric designs. Very little is known about the nonexistence of nonsymmetric balanced incomplete block designs, except for the following obvious consequence of Theorems 6.2 and 6.4.

COROLLARY 6.4.2 If there is a $(v - k, v - 1, k, k - \lambda, \lambda)$-design where $\lambda = 1$ or 2, then v, k, and λ must satisfy the conditions of Theorem 6.4. ∎

It follows that no (15, 21, 7, 5, 2)-design can exist; nor can AG(2, 6) or AG(2, 14). Theorem 6.2 enables us to say something about quasi-residual designs with $\lambda \geq 3$, but only when k is sufficiently large: for example, it gives us no information about (120, 140, 21, 18, 3)-designs.

EXERCISES

6.2.1 Complete the proof of Theorem 6.4 in the case $v \equiv 3$ (mod 4).

6.2.2 Suppose that there were a symmetric balanced incomplete block design with $\lambda = 3$ and $k = 21 + 27t$ for some positive integer t.
  (i) According to (1.1) and (1.2), what is the corresponding value of v?
  (ii) Prove that no such design exists.

6.2.3 Use Exercise 6.1.3 to prove the existence of a family of nonembeddable quasi-symmetric designs.

6.2.4 Consider the designs of Exercise 6.1.5, in the case where $n \equiv 0$ or 1 (mod 4) and n is not a perfect square. Prove that the designs are nonembeddable.

6.3 ANOTHER PROOF OF THE EXISTENCE THEOREM

In this section we present a shorter proof of Theorem 6.4, using the algebra of matrix congruence.

Suppose that A and B are symmetric matrices over a field F. Then we say that A and B are <u>congruent</u> over F if there is a nonsingular matrix S over F such that

$$A = S^T BS.$$

In the particular case where F is the field of rational numbers we simply say "congruent," without specifying the field, and write

$A \sim B$.

We require some standard results on congruence.

LEMMA 6.5 Every symmetric matrix is congruent to a diagonal matrix. ■

LEMMA 6.6 If $a_1, a_2, \ldots, a_n, b_1, b_2, \ldots, b_m, c_1, c_2, \ldots, c_m$ are any rational numbers, then

$$\mathrm{diag}(a_1, a_2, \ldots, a_n, b_1, b_2, \ldots, b_m) \sim \mathrm{diag}(a_1, a_2, \ldots, a_n, c_1, c_2, \ldots, c_m)$$

holds if and only if

$$\mathrm{diag}(b_1, b_2, \ldots, b_m) \sim \mathrm{diag}(c_1, c_2, \ldots, c_m). \quad \blacksquare$$

(Lemma 6.6 is called the Witt cancellation law.)

LEMMA 6.7 If n is any positive integer, then

$$\mathrm{diag}(n, n, n, n) \sim \mathrm{diag}(1, 1, 1, 1).$$

Proof: Since n is a positive integer there exist integers a, b, c, d, such that

$$n = a^2 + b^2 + c^2 + d^2.$$

If B is the matrix defined in (6.4), then

$$B^T I B = B^T B = nI,$$

so $nI \sim I$. Since the matrices are 4×4, we have the result. ■

LEMMA 6.8 Suppose that A is a nonsingular symmetric matrix, and that B is a symmetric matrix derived from A by adding a row and column. Then

$$\begin{bmatrix} A & 0 \\ 0 & c \end{bmatrix} \sim B, \quad c = \left[\frac{\det B}{\det A}\right] t^2,$$

for any $t \neq 0$. ■

We can now prove Theorem 6.4. The proof of part (i) proceeds as before. Suppose that v is odd and that there is a (v, k, λ)-design with incidence matrix A. Then

6.3 Another Proof of the Existence Theorem

$$AA^T = (k - \lambda)I + \lambda J.$$

If S is the $(v + 1) \times (v + 1)$ matrix with diagonal entries 1, last row all 1 and other entries zero,

$$S^T[\text{diag}(k - \lambda, k - \lambda, \ldots, k - \lambda, \lambda)]S = \begin{bmatrix} & & & \lambda \\ & AA^T & & \vdots \\ & & & \lambda \\ \lambda & \lambda & \cdots & \lambda \end{bmatrix}$$

so by Lemma 6.8,

$$\text{diag}(k - \lambda, k - \lambda, \ldots, k - \lambda, \lambda) \sim \begin{bmatrix} AA^T & 0 \\ 0 & c \end{bmatrix}, \quad (6.12)$$

where

$$c = \frac{\det(\text{diag}(k - \lambda, k - \lambda, \ldots, k - \lambda, \lambda))}{k^2(k - \lambda)^{v-1}} t^2$$

$$= \frac{(k - \lambda)^v \lambda}{k^2(k - \lambda)^{v-1}} t^2$$

$$= \frac{\lambda(k - \lambda)t^2}{k^2}$$

and putting $t = k$, we see that we can choose $c = \lambda(k - \lambda)$.
Now $AA^T = (A^T)^T I(A^T) \sim I$, so (6.12) yields

$$\text{diag}(k - \lambda, k - \lambda, \ldots, k - \lambda, \lambda) \sim \text{diag}(1, 1, \ldots, 1, \lambda(k - \lambda)). \quad (6.13)$$

Suppose that $v \equiv 1 \pmod 4$. By repeated use of Lemma 6.7 (with $n = k - \lambda$) and Lemma 6.6 we get

$$\text{diag}(k - \lambda, \lambda) \sim \text{diag}(1, \lambda(k - \lambda)),$$

so for some p, q, r, s we have

$$\begin{bmatrix} 1 & 0 \\ 0 & \lambda(k - \lambda) \end{bmatrix} = \begin{bmatrix} r & q \\ r & s \end{bmatrix} \begin{bmatrix} k - \lambda & 0 \\ 0 & \lambda \end{bmatrix} \begin{bmatrix} p & r \\ q & s \end{bmatrix}$$

and comparing (1,1) entries we get

$$1 = (k - \lambda)p^2 + \lambda q^2.$$

Now p and q are rationals, so $p = y/x$ and $q = z/x$ for some integers x, y, and z, whence

$$x^2 = (k - \lambda)y^2 + \lambda z^2$$

and the solution is nontrivial since x cannot be zero.

If $v \equiv 3 \pmod{4}$, we obtain

$$\mathrm{diag}(k - \lambda,\ k - \lambda,\ k - \lambda,\ \lambda) \sim \mathrm{diag}(1,\ 1,\ 1,\ (k - \lambda)\lambda),$$

so by Lemma 6.7,

$$\mathrm{diag}(k - \lambda,\ k - \lambda,\ k - \lambda,\ k - \lambda,\ \lambda) \sim \mathrm{diag}(k - \lambda,\ 1,\ 1,\ 1,\ \lambda(k - \lambda)).$$

On the other hand, applying Lemma 6.7 to Lemma 6.6 yields

$$\mathrm{diag}(k - \lambda,\ k - \lambda,\ k - \lambda,\ k - \lambda,\ \lambda) \sim \mathrm{diag}(1,1,1,1,\lambda),$$

and comparing these results

$$\mathrm{diag}(1,1,1,1,\lambda) \sim \mathrm{diag}(k - \lambda,\ 1,\ 1,\ 1,\ \lambda(k - \lambda))$$

whence

$$\mathrm{diag}(1,\lambda) \sim \mathrm{diag}(k - \lambda,\ \lambda(k - \lambda));$$

for some rational p, q, r, s we have

$$\begin{bmatrix} k - \lambda & 0 \\ 0 & \lambda(k - \lambda) \end{bmatrix} = \begin{bmatrix} p & q \\ r & s \end{bmatrix} \begin{bmatrix} 1 & 0 \\ 0 & \lambda \end{bmatrix} \begin{bmatrix} p & r \\ q & s \end{bmatrix}$$

and comparing (1,1) entries

$$k - \lambda = p^2 + \lambda q^2;$$

putting $p = x/y$ and $q = z/y$ we have

$$x^2 = (k - \lambda)y^2 - \lambda z^2.$$ ∎

6.4 RESOLVABILITY

We now generalize the way in which the lines of an affine plane fall into parallel classes. We define a _parallel class_ in a balanced incomplete block design to be a set of blocks that between them contain every treatment exactly once. A design is called _resolvable_ if one can partition its blocks into parallel classes; such a partition is called a _resolution_.

Clearly, the number of blocks in a parallel class must equal vk^{-1}, and therefore k must divide v in any resolvable design. We write s for the integer vk^{-1}. Another way of putting this is to say that any resolvable design has an additional parameter s, the size of a parallel class, and an additional parameter relation

$$v = sk. \qquad (6.14)$$

Not every design with k a divisor of v is resolvable. In fact, if we take $k = 3$, $\lambda = 2$, and $s = 2$, we obtain the parameters $(6, 10, 5, 3, 2)$. Suppose that there is a resolvable design with these parameters. Without loss of generality we assume that its treatment set is $\{1, 2, 3, 4, 5, 6\}$.

There must be precisely two blocks containing both 1 and 2; after permuting the names of symbols if necessary, these blocks can be taken either as 123, 124 or as 123, 123. If they are 123 and 124, then either 134 is a block, and the set of blocks containing 1 must be

123, 124, 134, 156, 156,

or else 134 is not a block, and the blocks containing 1 must be

123, 124, 135, 146, 156

(or an isomorphic set with 5 and 6 interchanged). If the blocks containing 1 and 2 are 123, 123, we get a set of blocks isomorphic to the first case again.

Since $s = 2$, there are only two blocks in each resolution, so the block in the same resolution as 123 must be its complement, 456. Thus all five remaining blocks are determined. In the first case, they are

456, 356, 256, 234, 234,

and the resulting design is not balanced; in fact, the pair $\{5, 6\}$ occurs in five blocks, instead of two. In the second case, the blocks are

456, 356, 246, 235, 234,

which is again not balanced, as $\{3, 5\}$ and $\{5, 6\}$ occur three times each, while $\{3, 4\}$ occurs only once. So no resolvable balanced incomplete design exists with the parameters stated.

On the other hand, a nonresolvable design with parameters $(6,10,5,3,2)$ is easy to construct.

In 1850, the Reverend T. P. Kirkman proposed the following problem: A schoolmistress has 15 girl pupils and she wishes to take them on a daily walk. The girls are to walk in five rows of three girls each. It is required that no two girls should walk in the same row more than once per week. Can this be done? The problem is generally known as "Kirkman's schoolgirl problem."

First, there are seven days in a week, and every girl will walk in the company of two others each day. As no repetitions are allowed, a girl has 14 companions over the week; since there are only 15 in the class, it follows that every pair of girls must walk together in a row at least once; in view of the requirements, every pair of girls walk together <u>precisely</u> once. Every girl walks on each day of the week. So, if we treat the girls as "treatments" and the rows in which they walk as "blocks," we are required to select 35 blocks of size 3 from a set of 15 treatments, in such a way that every treatment occurs in 7 blocks and every pair of treatments occur together in precisely one block—that is, we require a $(15,35,7,3,1)$-design. But there is a further constraint—since all the girls must walk every day, it is necessary that the blocks be partitioned into seven groups of triples, so that every girl appears once in each group—these will correspond to the seven days. So we require a resolvable solution.

Kirkman's problem has a solution: for example, we may use the schedule:

Monday:	1,2,3	4,5,6	7,8,9	10,11,12	13,14,15
Tuesday:	1,4,11	2,5,10	3,8,13	6,7,14	9,12,15
Wednesday:	1,5,9	2,6,8	3,11,15	4,12,14	7,10,13
Thursday:	1,6,13	2,4,15	3,9,10	5,7,12	8,11,14
Friday:	1,7,15	2,9,14	3,6,12	4,8,10	5,11,13
Saturday:	1,8,12	2,7,11	3,5,14	4,9,13	6,10,15
Sunday:	1,10,14	2,12,13	3,4,7	5,8,15	6,9,11.

More generally, he also asked about resolvable designs with $k = 3$, $\lambda = 1$, and any value of v. We shall discuss this problem in Section 12.5.

The "schoolgirl problem" is a famous resolvable design. If v is even, say $v = 2n$, the $(2n, n(2n - 1), 2n - 1, 2, 1)$-design of all possible unordered pairs is also resolvable. The resolutions—called one-factorizations—were defined in Chapter 1, and will be discussed in more detail in Chapter 11. The third obvious examples are the affine planes.

We define an <u>affine</u> resolvable design (or simply <u>affine design</u>) to be a resolvable design with the property that any two blocks belonging to different parallel classes have m common treatments for some constant m. So the affine planes are affine designs with $m = 1$.

6.4 Resolvability

THEOREM 6.9 An affine design has parameters

$$(s^2 m, \; s^2 m + \lambda m, \; sm + \lambda, \; sm, \; \lambda) \tag{6.15}$$

for some integers s and m, where

$$\lambda = \frac{sm - 1}{s - 1}. \tag{6.16}$$

Proof: Suppose that $B_1, B_2, \ldots, B_s$ are the blocks in one parallel class, and B is a block in another parallel class. Then $B \cap B_i$ has m elements for every i. The sets $B \cap B_i$ are disjoint (since the B_i are disjoint) and they partition B (every treatment belongs to some B_i, so obviously any element of B belongs to some B_i). So

$$k = |B| = \sum_i |B \cap B_i| = sm. \tag{6.17}$$

To establish the value of λ, we count the elements of a set in two ways. Given a fixed block A, we count all the ordered triples (x, y, B) where B is a block other than A and $A \cap B$ contains both x and y. There are $k(k - 1)$ ordered pairs (x, y), and together they belong to λ blocks, one of which is A, so there are $\lambda - 1$ choices for B, and the count is

$$k(k - 1)(\lambda - 1). \tag{6.18}$$

On the other hand, A has nonempty intersection with s blocks in each of the other $r - 1$ parallel classes; since there are m elements in each intersection there are $m(m - 1)$ choices for the ordered pair (x, y), so the count is

$$s(r - 1)m(m - 1). \tag{6.19}$$

Equating (6.18) and (6.19) and applying (6.17), we have

$$(\lambda - 1)(sm - 1) = (r - 1)(m - 1). \tag{6.20}$$

The standard relation (1.2) yields

$$r = \lambda \, \frac{v - 1}{k - 1} = \lambda \, \frac{s^2 m - 1}{sm - 1},$$

and substituting this into (6.20) we have

$$(\lambda - 1)(sm - 1)^2 = [\lambda(s^2m - 1) - (sm - 1)](m - 1),$$

$$\lambda[(sm - 1)^2 - (m - 1)(s^2m - 1)] = (sm - 1)^2 - (m - 1)(sm - 1),$$

$$\lambda(s^2m^2 - 2sm - s^2m^2 + s^2m + m) = (sm - 1)(sm - m),$$

$$\lambda m(s - 1)^2 = (sm - 1)(s - 1)m,$$

$$\lambda = \frac{sm - 1}{s - 1} \qquad (6.16)$$

Using (1.2) and (1.1) we get the parameters (6.15). ∎

COROLLARY 6.9.1 The parameters of an affine design satisfy

$b = rs = v + r - 1$. ∎

The corollary is easy to verify from (6.15) and (6.16); however, observe that $b = rs$ follows directly from (6.14) and (1.1), so that part is true for any resolvable design.

As an application of affine designs we give a construction for symmetric balanced incomplete block designs which uses the existence of an affine design. For convenience we assume that we are given an affine design of parameters (6.15), and we assume the blocks to have been ordered so that the first s blocks constitute one parallel class, the next s constitute another, and so on. Thus the design has incidence matrix

$$A = [A_1 \; A_2 \; \cdots \; A_r],$$

where each A_i is the $s^2m \times s$ incidence matrix of the set of blocks in one parallel class and $r = sm + \lambda$. Clearly, we have

$$A_i J = J \qquad \text{for } i \in (1 \cdots r) \qquad (6.21)$$

$$A_i^T A_i = smI \qquad \text{for } i \in (1 \cdots r) \qquad (6.22)$$

$$A_i^T A_j = mJ \qquad \text{for } i, j \in (1 \cdots r), \; i \neq j \qquad (6.23)$$

$$\sum_{i=1}^{r} A_i A_i^T = (r - \lambda)I + \lambda J \qquad \text{for } i \in (1 \cdots r). \qquad (6.24)$$

(Verifications are left as an exercise.)

6.4 Resolvability

THEOREM 6.10 If there is an affine design with parameters (6.15), there is a symmetric balanced incomplete block design with parameters

$$\left(\frac{s^2 m(s^2 m + s - 2)}{s-1}, \frac{sm(s^2 m - 1)}{s-1}, \frac{sm(sm-1)}{s-1} \right).$$

Proof: For convenience we write r and λ for the relevant parameters of the affine design:

$$r = sm + \frac{sm-1}{s-1}, \quad \lambda = \frac{sm-1}{s-1}.$$

Defining $A_1, A_2, \ldots, A_r$ as above, we consider the square (0,1)-matrix

$$L = \begin{bmatrix} 0 & M_1 & M_2 & \cdots & M_r \\ M_r & 0 & M_1 & & M_{r-1} \\ M_{r-1} & M_r & 0 & & M_{r-2} \\ \vdots & & & & \\ M_1 & M_2 & M_3 & & 0 \end{bmatrix}$$

where $M_i = A_i A_i^T$. This matrix has size $s^2 m(s^2 m + s - 2)/(s-1)$. We shall prove that

$$LL^T = \left(\frac{sm(s^2 m - 1)}{s-1} - \frac{sm(sm-1)}{s-1} \right) I + \frac{sm(sm-1)}{s-1} J,$$

$$JL = \frac{sm(sm^2 - 1)}{s-1} J.$$

By Theorem 2.8 this shows that L is the incidence matrix of the required design. Equation (6.24) tells us that

$$\sum M_i = (r - \lambda)I + \lambda J.$$

So the column sum of L equals $r - \lambda$ plus λ times the number of rows of M_i:

$$r - \lambda + \lambda s^2 m = sm + \frac{(sm-1)s^2 m}{s-1}$$

$$= \frac{s^2 m - sm + s^3 m^2 - s^2 m}{s-1}$$

$$= \frac{sm(s^2 m - 1)}{s-1},$$

so JL has the required value.

If we assume that LL^T is partitioned in the same way as L, then the diagonal blocks of LL^T each equal

$$M_i M_i^T = \sum A_i A_i^T A_i A_i^T$$

$$= sm \sum A_i A_i^T$$

$$= sm[(r-\lambda)I + \lambda J]$$

$$= s^2 m I + \frac{sm(sm-1)}{s-1} J,$$

using (6.22). Off the diagonal, the (i, j) block is the sum of $r - 1$ terms of the form $M_p M_q^T$, where p never equals q, and

$$M_p M_q^T = A_p A_p^T A_q A_q^T$$

$$= A_p (mJ) A_q^T$$

$$= (mJ) A_q^T$$

$$= mJ,$$

using (6.21) and (6.23); so every off-diagonal block equals $(r-1)mJ$. It remains to verify that

$$(r-1)m = \frac{sm(sm-1)}{s-1},$$

but the left-hand side equals

6.4 Resolvability

$$\left(sm + \frac{sm-1}{s-1} - 1\right)m = \frac{(s^2m - sm + sm - 1 - s + 1)t}{s-1}$$

$$= \frac{st(st-1)}{s-1}. \blacksquare$$

In the simplest case, if we use an AG(2,2) as the affine design, we obtain a (16, 6, 2)-design.

One can construct resolvable designs by difference methods. One observation that has been useful in direct searches is the following: If a set of initial blocks contains every element of the treatment set once, the design developed from them is resolvable. If the initial blocks are supplementary difference sets, the result is a resolvable balanced incomplete block design. For example, the designs of Theorem 4.7 are resolvable. Cases with more than two initial blocks can also be constructed; as an example, the blocks

{010 020 101 201}   {011 021 102 202}   {012 022 100 200}
{210 120 221 111}   {211 121 222 122}   {212 122 220 120}
                    {∞   000 001 002}

from a resolvable (28, 63, 9, 4, 1)-design when developed mod (3, 3, -).

Although we have discussed resolvability and affine resolvability only in the case of balanced incomplete block designs, or 2-designs, there is no reason why one should not consider general t-designs in this way: The properties only involve block intersections, and not the degree of balance of the design. We shall discuss affine t-designs in Section 7.4; we see there that the definition forces significant restrictions on t-designs with $t > 2$.

EXERCISES

6.4.1 Prove that if one develops the following blocks modulo 17, one obtains a resolvable (18, 102, 17, 3, 2)-design:

∞, 0, 11   1, 9, 14   2, 5, 7
3, 4, 10   6, 8, 15   12, 13, 17.

6.4.2 Suppose that A is the incidence matrix of a resolvable (v, b, r, k, λ)-design. Prove that A has rank at least $b - r + 1$. Hence prove that $b \geq v + r - 1$.

6.4.3 A <u>PB2-design</u> with parameters $(v; b_1, b_2; r_1, r_2; k_1, k_2; \lambda)$ is defined to be a design on v objects, whose "blocks" are b_1 subsets of size k_1 and b_2 subsets of size k_2, such that every object belongs to r_1 of the blocks of size k_1 and to r_2 of the blocks of size k_2.

(i) Prove that the design formed by deleting all members of one block from a balanced incomplete block design with parameters $(v, b, r, k, 1)$ is a PB2-design $(v - k; b_1, b_2; r_1 r_2; k - 1; 1)$ for some b_1, b_2, r_1, and r_2; and derive expressions for these four parameters in terms of v, b, r, and k. Does a similar result hold if the balanced incomplete block design has $\lambda = 2$ instead of $\lambda = 1$?

(ii) Prove that the existence of a resolvable balanced incomplete block design with parameters (v, b, r, k, λ) and a balanced incomplete block design with parameters $(k, h, s, t, 1)$ implies the existence of a PB2-design $(v; b - vk^{-1}, hvk^{-1}; r - 1, s; k, t; \lambda)$. More generally, show that they imply the existence of PB2-designs

$$(v; b - mvk^{-1}, mhvk^{-1}; r - m, ms; k, t; \lambda) \quad \text{for } m = 1, 2, \ldots, r-1$$

6.4.4 Suppose that one were to insert all the 3-sets out of 3n objects in a square array, so that every object occurs exactly once per row and exactly once per column. How big must the array be? Does such an array exist in the case $n = 2$? For $n = 3$?

6.4.5 Prove equations (6.21), (6.22), (6.23), and (6.24).

6.4.6 Prove the existence of a symmetric balanced incomplete block design with parameters

$$(\lambda^2(\lambda + 2), \lambda(\lambda + 1), \lambda)$$

whenever λ is a prime power.

BIBLIOGRAPHIC REMARKS

The results of Theorem 6.2(i) were proven by Hall and Connor [5]; a proof can also be found in the book by Hall [4]. Part (ii) of Theorem 6.2 was proven by Singhi and Shrikhande [14], [15].

The first example of a quasi-residual design which was not residual was given by Bhattacharya [1]. He exhibited a (16, 24, 9, 6, 3) design, but not the one we construct in the text—that example, whose derivation is much simpler, is due to van Lint [9].

The proof of the "v odd" part of Theorem 6.4 was first presented in two papers: Bruck and Ryser dealt with the case $\lambda = 1$ in [2], and Chowla and Ryser gave the general result in [3]. The "v even" case was proven by Schutzenberger [12]. The new proof in Section 6.3 was discovered independently by Ryser [11] and Lenz [8].

Kirkman introduced his "schoolgirl problem" in 1850 [7]. The general form of the problem—for what values of v is there a resolvable balanced incomplete block design with v treatments, having k = 3 and λ = 1—was solved by Ray-Chaudhuri and Wilson [10] in 1971; they showed that $v \equiv 3$ (mod 6) is both necessary and sufficient. We present a (newer) proof in Chapter 12. The corresponding problem for k = 4 is solvable if and only if $v \equiv 4$ (mod 12); this was proven by the foregoing two authors and Hanani [6]. Theorem 6.10 is from [16].

In 1976 Shrikhande published a comprehensive survey of the theory of affine designs, with an extensive bibliography [13].

1. K. N. Bhattacharya, "A new balanced incomplete block design," Science and Culture 11(1944), 508.
2. R. H. Bruck and H. J. Ryser, "The non-existence of certain finite projective planes," Canadian Journal of Mathematics 1(1949), 88-93.
3. S. Chowla and H. J. Ryser, "Combinatorial problems," Canadian Journal of Mathematics 2(1950), 93-99.
4. M. Hall, Combinatorial Theory, 2nd Ed. (Wiley-Interscience, New York, 1986).
5. M. Hall and W. S. Connor, "An embedding theorem for balanced incomplete block designs," Canadian Journal of Mathematics 6(1954), 35-41.
6. H. Hanani, D. K. Ray-Chaudhuri, and R. M. Wilson, "On resolvable designs," Discrete Mathematics 3(1972), 343-357.
7. T. P. Kirkman, "Query," Lady's and Gentleman's Diary (1850), 48.
8. H. Lenz, "A few simplified proofs in design theory," Expositiones Mathematicae 1(1983), 77-80.
9. J. H. van Lint, "Non-embeddable quasi-residual designs," Indagationes Mathematicae 40(1978), 269-275.
10. D. K. Ray-Chaudhuri and R. M. Wilson, "Solution of Kirkman's school girl problem" Proceedings of Symposia in Pure Mathematics 19, Combinatorics (American Mathematical Society, Providence, R.I., 19 1971), 187-203.
11. H. J. Ryser, "The existence of symmetric block designs," Journal of Combinatorial Theory 32A(1982), 103-105.
12. M. P. Schutzenberger, "A non-existence theorem for an infinite family of symmetrical block designs," Annals of Eugenics 14(1949), 286-287.
13. S. S. Shrikhande, "Affine resolvable balanced incomplete block designs: A survey," Aequationes Mathematicae 14(1976), 251-269.
14. N. M. Singhi and S. S. Shrikhande, "Embedding of quasi-residual designs with λ = 3," Utilitas Mathematica 4(1973), 35-53.
15. N. M. Singhi and S. S. Shrikhande, "Embedding of quasi-residual designs," Geometricae Dedicata 2(1974), 509-517.
16. W. D. Wallis, "Construction of strongly regular graphs using affine designs," Bulletin of the Australian Mathematical Society 4(1971), 41-49.

7
t-Designs

7.1 CONTRACTIONS AND EXTENSIONS

Suppose that E is a t-(v, k, λ) design, as defined in Section 2.3. That is, E consists of v treatments, and a collection of blocks of size k such that every t-set of the treatments is a subset of precisely λ blocks. The number of blocks was seen in Lemma 2.13 to be

$$\lambda \frac{\binom{v}{t}}{\binom{k}{t}}. \tag{7.1}$$

We defined the derivative of a t-design with regard to a set of treatments in Section 2.3. In the special case where the set has only one element x, the derivative is called the <u>contraction</u> at x. More formally, the contraction E_x of E at x is defined as follows. The treatments of E_x are all the treatments of E, other than x. The blocks are all the blocks $B \setminus \{x\}$ where B is a block containing x.

The following result follows from Lemma 2.14.

THEOREM 7.1 Suppose that E is a t-(v, k, λ) design, where $t \geq 2$, and x is a treatment of E. Then E_x is a $(t-1)$-$(v-1, k-1, \lambda)$ design. ∎

We are interested in contractions of t-designs not so much for themselves but rather because of the following question: Is a given t-design the contraction of some $(t+1)$-design or not? For that reason we make the following definition.

Suppose that D is a t-design and E is a $(t+1)$-design such that D is isomorphic to E_x for some treatment x of E. Then we say that E is an <u>extension</u> of D. Given a t-design D, it is not necessarily true that any extension of D

7.1 Contractions and Extensions

exists; we say that D is <u>extendable</u> if it has an extension. There has been some interest in determining which t-designs are extendable. The following easy theorem is useful.

THEOREM 7.2 If D is a t-(v,k,λ) design with b blocks, and E is an extension of D, then E is a $(t+1)$-$(v+1, k+1, \lambda)$ design with

$$b \frac{v+1}{k+1}$$

blocks.

Proof: Suppose that D is a t-(v,k,λ) design and E is an extension of D. By inverting the result of Theorem 7.1 we see that E is a $(t+1)$-$(v+1, k+1, \lambda)$ design.

From (7.1) the number of blocks of D is

$$b = \lambda \frac{\binom{v}{t}}{\binom{k}{t}} = \lambda \frac{v!(k-t)!}{k!(v-t)!} \qquad (7.2)$$

while the number of blocks of E is

$$\lambda \frac{\binom{v+1}{t+1}}{\binom{k+1}{t+1}} = \lambda \frac{(v+1)!(k-t)!}{(k+1)!(v-t)!} = b \frac{v+1}{k+1}. \qquad \blacksquare$$

COROLLARY 7.2.1 If a t-(v,k,λ) design has an extension, then

$$k+1 \text{ divides } b(v+1),$$

where b is given by (7.2). $\blacksquare$

For example, a 2-$(16,6,2)$ design exists; it is a symmetric balanced incomplete block design and has $b = 16$. But $k+1 = 7$ and $b(v+1) = 16 \cdot 17$, so the design cannot be extended.

COROLLARY 7.2.2 No finite projective plane PG$(2,n)$ can have an extension unless $n = 2$, 4, or 10.

Proof: Since a PG$(2,n)$ is a 2-$(n^2+n+1, n+1, 1)$-design with n^2+n+1 blocks, it can have an extension only if $n+2$ divides $(n^2+n+1) \times (n^2+n+2)$. Now

$$n^2 + n + 1 = (n + 2)(n - 1) + 3,$$

$$n^2 + n + 2 = (n + 2)(n - 1) + 4,$$

so $n + 2$ will divide their product if and only if $n + 2$ divides 12. This gives the possible values 1, 2, 3, 4, 6, and 12 for $n + 2$; the first three are too small, and the others give $n = 2, 4$, and 10. ∎

It is easy to show that a $PG(2,2)$ is extendable, since the design is small and is uniquely determined, up to isomorphism, by its parameter. We construct a 3-(8.4.1)-design E, with treatments (0 ·· 7), such that E_0 is the $PG(2,2)$ with blocks

123, 145, 167, 246, 257, 347, 356.

We know that E has 14 blocks, and seven of them are obtained by appending 0 to the triples we just listed. But E_1 must also be a $PG(2,2)$, and three of its blocks must be 023, 045, and 067. To complete E_1 we must use blocks

$\{246, 257, 347, 356\}$ or $\{247, 256, 346, 357\}$.

In the former case, E contains blocks 1246, 1257, 1347, and 1356; $\{2, 4, 6\}$ is a subset of the two blocks 0246 and 1246, which is impossible. So we have blocks

1247, 1256, 1346, 1357

in E. Similarly, one finds that the other three blocks must be

2345, 2367, 4567.

(Consider, for example, E_2 and E_4.) These 14 blocks do make up a 3-(8.4.1)-design, so $PG(2,2)$ is extendable (and the extension is uniquely determined).

This method would be extremely long in the case $n = 4$. However, we show in the next section that $PG(2,4)$ is extendable. The extendability of $PG(2,10)$ is as yet unsolved: we clearly do not know of an extension, since we do not even know whether $PG(2,10)$ exists; on the other hand, no one has shown that a 3-(112, 12, 1)-design is impossible.

EXERCISES

7.1.1 Suppose that E is a t-design and x is a treatment of E. The leave of x, denoted E^x, is the design whose treatments are the treatments of E other than x and whose blocks are the blocks of E that do not contain x.

7.2 Some Examples of 3-Designs

    (i) Prove that E^x is a $(t-1)$-design.
    (ii) If E is a t-(v,k,λ)-design, what are the parameters of E^x?

7.1.2 Recall that a <u>trivial</u> balanced incomplete block design is the set of all k-subsets of a v-set. The designs we now introduce are called trivial t-designs.
    (i) Prove that the set of all k-subsets of a v-set is a t-design provided that $t \leq k \leq v$. What are its parameters?
    (ii) Prove that any trivial t-design is extendable, and that the extension is a trivial $(t+1)$-design.

7.1.3 Prove that no symmetric balanced incomplete block design with $\lambda = 2$ can have an extension unless $k = 3$, 5, or 11.

7.1.4 Prove that $PG(d,n)$ and $AG(d,n)$ do not have extensions when $d > 2$.

7.1.5 Suppose that E is an extension of a nontrivial symmetric balanced incomplete block design D, where D is not trivial and D is not $PG(2,4)$. Show that E cannot be extended.

7.1.6 Prove that there is exactly one 3-(8,4,1)-design (up to isomorphism).

7.2 SOME EXAMPLES OF 3-DESIGNS

We now provide some examples of 3-designs. We construct two families of designs, and also one specific, important example.

Since a 3-design is an extension of a 2-design, it is convenient to work from designs that have already been studied. We find extensions of a family of designs that contain the Hadamard designs (see Section 4.1) and of some affine planes (see Chapter 5).

For our first construction we need a lemma about balanced incomplete block designs.

LEMMA 7.3 Suppose that x, y, and z are any three treatments in a balanced incomplete block design with parameters (v,b,r,k,λ); say there are c blocks which contain none of x, y, and z, and d blocks which contain all three. Then

$$c + d = b - 3r + 3\lambda.$$

Proof: Let us write c_x for the number of blocks which include x but neither y nor z, c_{xy} for the number that includes both x and y but not z, and so on: This notation is consistent with the definition of c, and d might also be called c_{xyz}. The number of blocks that contain x is then

$$r = c_x + c_{xy} + c_{xz} + c_{xyz},$$

and similarly

$$r = c_y + c_{xy} + c_{yz} + c_{xyz},$$
$$r = c_z + c_{xz} + c_{yz} + c_{xyz}.$$

Adding, we obtain

$$3r = c_x + c_y + c_z + 2(c_{xy} + c_{xz} + c_{yz}) + 3d.$$

Since the total number of blocks is

$$b = c + c_x + c_y + c_z + c_{xy} + c_{xz} + c_{yz} + c_{xyz},$$

we have

$$3r = (b - c - d) + (c_{xy} + c_{xz} + c_{yz}) + 3d. \tag{7.3}$$

The number of blocks containing both x and y is $c_{xy} + c_{xyz}$, so

$$c_{xy} + d = \lambda;$$

substituting this and the two similar equations into (7.3), we get

$$3r = b - c - d + 3\lambda,$$

whence $c + d = b - 3r + 3\lambda$, as required. ∎

THEOREM 7.4 Any 2-$(2k + 1, k, \lambda)$-design is extendable.

Proof: A 2-design is a balanced incomplete block design; from (1.1) and (1.2) we see that one with $v = 2k + 1$ has $b = 2(k - 1)^{-1}(2k + 1)\lambda$ and $r = 2(k - 1)^{-1}k\lambda$.

Let D be such a design, with blocks $B_1, B_2, \ldots, B_b$. Write C_i for the complement of B_i in the treatment set; then C_i has $k + 1$ elements. If x, y, and z are any three treatments of D, suppose that there are c of the blocks B_i which contain all three. Then there are $d = c - b + 3r - 3\lambda$ blocks B_i which contain none of x, y, z, so there are d of the C_i that contain all three. So the number of the sets $\{B_1, B_2, \ldots, B_b, C_1, C_2, \ldots, C_b\}$ that contain all three is exactly $b - 3r + 3\lambda$, independently of the choice of x, y, and z; substituting the values of b and r, given above, we have

7.2 Some Examples of 3-Designs

$$b - 3r + 3\lambda = [2(2k+1) - 6k](k-1)^{-1}\lambda + 3\lambda$$

$$= (2 - 2k)(k-1)^{-1}\lambda + 3\lambda$$

$$= \lambda.$$

We construct a design E with 2b blocks as follows. First, we introduce a new treatment, which we label 0. The blocks are the b blocks C_1, C_2, ..., C_b, together with b further blocks C_{b+1}, C_{b+2}, ..., C_{2b}, where $C_{b+i} = B_i \cup \{0\}$. Now every block has size $k + 1$, and every treatment belongs to b blocks (apart from 0, every treatment belongs to exactly one of B_i and C_i for each i). Any triple that does not contain 0 occurs in precisely λ of the blocks; the triple 0xy also occurs in precisely λ blocks, namely those C_{b+i} such that B_i contains both x and y. So E is the required extension, as it is a 3-$(2k+2, k+1, \lambda)$-design and $E_0 = D$. ∎

The 2-designs of Theorem 7.4 have $r = 2(k-1)^{-1}k\lambda$. Since $k - 1$ is prime to k, r will be an integer only if $k - 1$ divides 2λ. Two cases present themselves. If k is even, then $k - 1$ must divide λ; say $\lambda = \alpha(k-1)$. The parameters of the 2-design are

$$(2t + 3,\ 2\alpha(2t+3),\ 2\alpha(t+1),\ t+1,\ \alpha t),$$

where $t = k + 1$. If k is odd, $(k-1)/2$ must divide λ, say $\lambda = [\alpha(k-1)]/2$. The parameters are

$$(4t + 3,\ \alpha(4t+3),\ \alpha(2t+1),\ 2t+1,\ t)$$

where $2t = k - 1$. Recalling the definition of an Hadamard design in Section 4.1, we could restate Theorem 7.4 as follows.

COROLLARY 7.4.1 The following 2-designs are extendable:

(i) the Hadamard designs;
(ii) the designs with parameters

$$(2t + 3,\ 4t + 6,\ 2t + 2,\ t + 1,\ t);$$

(iii) multiples of classes (i) and (ii).

Another important family of 3-designs is derived from the finite affine planes over fields. Every EG(2,n) is extendable. We prove this in the case where $n \equiv 3 \pmod 4$. The cases $n \equiv 1 \pmod 4$ and n even are left as exercises.

Suppose that F is the n-element field, where n is a prime power congruent to 3 (modulo 4), and A is EG(2, n). We define the circle $C(a, b, c)$ to be the set of all points

$$\{(x,y) : (x - a)^2 + (y - b)^2 = c^2\},$$

where x and y range over F. While a and b can be any field elements, c must be nonzero, so there are $n^2(n - 1)$ different circles. They have distinct point sets, and in fact $C(a, b, c)$ and $C(d, e, f)$ have at most two common points if $(a, b, c) \neq (d, e, f)$ (see Exercise 7.2.1).

We now count the points of $C(0, 0, 1)$. For each x we consider the number of solutions y to

$$y^2 = 1 - x^2.$$

If $x = 1$ or -1, there is no solution. If $1 - x^2$ is a nonzero quadratic element, there are two solutions, and otherwise there are none. Say that $1 - x^2$ is a nonzero quadratic element for q values of x. Then it is nonquadratic for the remaining $n - 2 - q$ values. Therefore,

$$\sum_{x \in F} \chi(1 - x^2) = q \cdot 1 + (n - 2 - q) \cdot (-1) + 2 \cdot 0 = 2q + 2 - n.$$

The number of points on $C(0, 0, 1)$ is $2q + 2$, which equals $n + \sum \chi(1 - x^2)$. But if we put $z = x - 1$,

$$\sum_{x \in F} \chi(1 - x^2) = \sum_{x \in F} \chi[(-1)(x - 1)(x + 1)]$$

$$= \chi(-1) \sum_{z \in F} \chi[z(z + 2)]$$

$$= (-1) \cdot (-1)$$

$$= 1,$$

using Theorems 3.10 and 3.9. So $C(0, 0, 1)$ has $n + 1$ points.

LEMMA 7.5 Any circle of EG(2, n) has exactly $n + 1$ points, provided that $n \equiv 3 \pmod{4}$.

Proof: Consider the circle $C(a, b, c)$. Its points can be put into one-to-one correspondence with the points of the circle $C(0, 0, 1)$, via the map

7.2 Some Examples of 3-Designs

$$\varphi: (x,y) \mapsto \left(\frac{x-a}{c}, \frac{y-a}{c}\right).$$

(Since c is nonzero, φ is well-defined and has a well-defined inverse.) So $C(a,b,c)$ and $C(0,0,1)$ are equal in size. ∎

LEMMA 7.6 If $n \equiv 3 \pmod 4$, no circle in $EG(2,n)$ contains three collinear points.

Proof: Consider the circle $C(a,b,c)$ and the straight line $dx + ey = f$. First, suppose that $e \neq 0$. The common points satisfy

$$(x-a)^2 + \left[e^{-1}(f-dx) - b\right]^2 = c^2,$$

which has at most two solutions because it is a quadratic equation over a field; each x-value gives a unique y-value, so point and line have at most two common points. Second, if $e = 0$, the line is $dx = f$, where $d \neq 0$; the common points satisfy

$$(d^{-1}f - a)^2 + (y-b)^2 = c^2,$$

and again there are at most two solutions. ∎

LEMMA 7.7 If $n \equiv 3 \pmod 4$, every point of $EG(2,n)$ belongs to exactly $n^2 - 1$ circles, and every set of three noncollinear points belongs to precisely one circle.

Proof: The circles containing (z,t) are derived from the circles containing (x,y) in a one-to-one fashion; they are

$$\{C(a+z-x, b+z-y, c) : (x,y) \in C(a,b,c)\}.$$

So there is a number r such that every point belong to r circles. Since there are n^2 points in $EG(2,n)$ and $n+1$ points in each of the $n^2(n-1)$ circles, we have

$$r = \frac{n^2(n-1) \cdot (n+1)}{n^2} = n^2 - 1.$$

Since $EG(2,n)$ contains $n^2 + n$ lines and each contains n points, the number of collinear triples of points is

$$(n^2 + n)\binom{n}{3}$$

and the number of noncollinear triples is

$$\binom{n^2}{3} - (n^2+n)\binom{n}{3} = \frac{1}{3!}[n^2(n^2-1)(n^2-2) - (n^2+n)n(n-1)(n-2)]$$

$$= n^2(n-1)\frac{(n+1)(n^2-2) - (n+1)(n-2)}{3!}$$

$$= n^2(n-1)\binom{n+1}{3}.$$

Now each of the $n^2(n-1)$ circles contains $n+1$ points, of which no three are collinear. So if we list all noncollinear triples that arise in circles, we list exactly $n^2(n-1)\binom{n+1}{3}$ objects. But there would be no repetitions—no two circles have three common points—so each noncollinear triple occurs in exactly one circle. ∎

THEOREM 7.8 If $n \equiv 3 \pmod 4$, then $EG(2,n)$ is extendable.

Proof: Suppose that $EG(2,n)$ has lines $B_1, B_2, \ldots, B_{n^2+n}$ and circles $C_1, C_2, \ldots, C_{n^3-n^2}$. We construct a 3-design whose treatments are the points of $EG(2,n)$ together with another point, 0 say, and whose blocks are the sets $C_1, C_2, \ldots, C_{n^3+n}$, defined as follows: If $i \leq n^3 - n^2$, then C_i is the set of points of a circle, as already defined, while

$$C_{n^3-n^2+i} = B_i \cup \{0\}.$$

The design is clearly regular, with $n+1$ treatments per block. Any three treatments other than 0 either constitute a collinear triple in $EG(2,n)$, in which case it is a subset of precisely one C_i, which has $i > n^3 - n^2$; or else it is a noncollinear triple, and is a subset of precisely one C_i, which has $i \leq n^3 - n^2$. Since $\{x,y\}$ belongs to exactly one line, $\{0,x,y\}$ belongs to precisely one C_i (with $i > n^3 - n^2$). So we have a 3-design, E say. Clearly, $E_0 = EG(2,n)$. ∎

These 3-designs are called <u>inversive planes.</u> The results of Theorem 7.8 and Exercises 7.2.2 and 7.2.3 prove that inversive planes exist for all prime power n. It is not clear that every affine plane can be extended—we can only assert this for Euclidean planes. However, every inversive plane <u>is</u> an extension of some affine plane, as is seen in Exercise 7.2.4.

As promised, we now discuss the existence of an extension of $PG(2,4)$. We outline the construction; some details are left for the exercises.

7.2 Some Examples of 3-Designs

Let A be a symmetric balanced incomplete block design with parameters $(11,5,2)$. Such a design exists; for example, it can be generated from the difference set $\{1,3,4,5,9\}$ in Z_{11}. We denote the treatments by $(1 \cdots 11)$ and the blocks by $B_1, B_2, \ldots, B_{11}$.

Suppose that T is any 3-set of elements of $(1 \cdots 11)$. We say that T is a <u>triplet</u> if there is a block of A which contains T, and call T a <u>triangle</u> if no block contains all three members. Since A is symmetric, the intersection of two blocks has λ members, and $\lambda = 2$, so no triplet belongs to two blocks. So the number of triplets in the design equals

$$b \cdot \binom{k}{3} = 11 \cdot \binom{5}{3} = 110,$$

and the number of triangles is therefore

$$\binom{v}{3} - 110 = 165 - 110 = 55.$$

Consider the triangle $T = \{x,y,z\}$. Since $\lambda = 2$, there are two blocks that contain $\{x,y\}$ and two that contain $\{x,z\}$; so there is one further block through x which does not meet T in any other treatment. We call this block the <u>tangent</u> to T at x.

We shall construct a 22-treatment design M with treatments $1, 2, \ldots, 11, b_1, b_2, \ldots, b_{11}$. Treatments 1 to 11 correspond to the treatments of A and b_1 to b_{11} correspond to its blocks (b_i corresponds to B_i). Eleven blocks have the form $\{b_i\} \cup B_i$; 11 more are

$$\{j\} \cup \{b_i : j \in B_i\};$$

the other 55 are derived from the triangles; and if the triangle T has B_i, B_j, and B_k as its tangents, the block is

$$T \cup \{b_i, b_j, b_k\}.$$

It is necessary to prove that every treatment belongs to exactly 21 of these 6-blocks, and that every triple of treatments belongs to exactly one block. These details are left as an exercise. Then:

THEOREM 7.9 There is a 3-$(22,6,1)$-design. ∎

COROLLARY 7.9.1 There is an extension of a $PG(2,4)$. ∎

We shall show in Section 11.5 that only one $PG(2,4)$-design exists (up to isomorphism), so one can say "the $PG(2,4)$ is extendable."

128 7. t-Designs

EXERCISES

7.2.1 Suppose that n is odd prime power, $n \equiv 3 \pmod 4$, and $C(a,b,c)$ and $C(d,e,f)$ are circles in $EG(2,n)$. If $(a,b,c) \neq (d,e,f)$, show that the two circles have at most two common elements.

7.2.2 Suppose that n is a prime power congruent to 1 (mod 4), and r is a nonquadratic element of $GF(n)$. Define an r-<u>ellipse</u> $E(a,b,c)$ in $EG(2,n)$ to be the set of all points (x,y) satisfying

$$(x - a)^2 + r(y - b)^2 = c^2.$$

  (i) Prove that two r-ellipses have at most two common points.
  (ii) Prove that every r-ellipse has $n + 1$ points.
  (iii) Prove analogs of Lemmas 7.6 and 7.7, and hence show that $EG(2,n)$ is extendable.

7.2.3 Prove that $EG(2, 2^k)$ is extendable for any positive integer k.

7.2.4 If E is a $3\text{-}(n^2 + 1, n + 1, 1)$-design, show that E_x is an $AG(2,n)$ for any treatment x.

7.2.5 Prove that any treatment of the design M, constructed in this section, belongs to exactly 21 blocks: 15 of the blocks constructed from triangles, five from one set of 11 blocks, and one from the other set of 11 blocks.

7.2.6 Prove that in the design M constructed in this section, if x and y are points of A and B_z is a block of A, then:
  (i) If $x \in B_z$ and $y \notin B_z$, then there is a unique triangle T of A such that x and y are in T and B_z is a tangent to T.
  (ii) If $x \notin B_z$ and $y \notin B_z$, then there is a unique triangle T such that x and y belong to T and B_z is a tangent to T.
Hence prove that any three treatments of M belong to exactly one block.

7.3 EXTENSIONS OF SYMMETRIC DESIGNS

In some ways symmetric balanced incomplete block designs are the most interesting designs. The finite planes are symmetric designs or their residuals. The Hadamard designs are symmetric. And the main nonexistence theorem applies primarily to symmetric designs. For that reason the following theorem is very interesting.

THEOREM 7.10 If a symmetric balanced incomplete block design D is extendable, then its parameters fit one of the following descriptions:

7.3 Extensions of Symmetric Designs

(i) $v = \lambda + 2$, $k = \lambda + 1$ (trivial designs);
(ii) $v = 4\lambda + 3$, $k = 2\lambda + 1$ (Hadamard designs);
(iii) $v = (\lambda + 2)(\lambda^2 + 4\lambda + 2)$, $k = \lambda^2 + 3\lambda + 1$;
(iv) $v = 111$, $k = 11$, $\lambda = 1$ (PG(2,10));
(v) $v = 495$, $k = 39$, $\lambda = 3$.

Proof: Suppose that the (v, k, λ)-design D is extendable. From Corollary 7.2.1,

$k + 1$ divides $v(v + 1)$.

From (1.2), $\lambda(v - 1) = k(k - 1)$, or

$$v = \frac{k^2 - k + \lambda}{\lambda}, \qquad (7.4)$$

so $k + 1$ divides $(k^2 - k + \lambda)(k^2 - k + 2\lambda)/\lambda^2$, and $k + 1$ therefore divides $(k^2 - k + \lambda)(k^2 - k + 2\lambda)$. But

$$(k^2 - k + \lambda)(k^2 - k + 2\lambda) \equiv 2(\lambda + 1)(\lambda + 2) \pmod{k + 1},$$

so

$k + 1$ divides $2(\lambda + 1)(\lambda + 2)$. (7.5)

Suppose that E is an extension of D. E is a $3\text{-}(v + 1, k + 1, \lambda)$-design. If Y is any block of E, write E^Y for the design whose blocks are those blocks of E that do not meet Y and whose treatments are the treatments of E that do not belong to Y.

If two blocks B and C of E meet in x, then $B\backslash\{x\}$ and $C\backslash\{x\}$ are blocks of the symmetric (v, k, λ)-design E_x, so they have λ common elements. Therefore, two blocks of E either are disjoint or have $\lambda + 1$ common elements. This implies that the blocks of E^Y are precisely those blocks of E (other than Y) that do not meet Y in $\lambda + 1$ treatments.

We now prove that E^Y is a balanced design. Select any two treatments u and v that do not belong to Y. We count the ordered pairs (w, X) where w belongs to Y and X contains both u and v, in two ways. First, suppose that there are m blocks B of E which contain both u and v and which also meet Y. Each of the $\lambda + 1$ treatments common to B and Y is a candidate for w in the list of ordered pairs, so there are $m(\lambda + 1)$ pairs. On the other hand, given w, there are λ blocks that contain u, v, and w (since E is a 3-design with parameter λ) for each of $k + 1$ possible choices of w (the $k + 1$ treatments on Y), so the number of pairs is $\lambda(k + 1)$. Therefore,

$m(\lambda + 1) = \lambda(k + 1)$

and $m = \lambda(k+1)/(\lambda+1)$. Since E_u has constant block size k, there are k blocks of E that contain both u and v, so there are k - m blocks of E^Y that contain both u and v. Therefore, E^Y is balanced. Unless the trivial case v - k = 1 occurs, it is a 2-(v - k, k + 1, k - m)-design. Since m is an integer, $\lambda + 1$ divides $\lambda(k+1)$, so $\lambda + 1$ divides $k + 1$. Say that

$$k + 1 = s(\lambda + 1) \tag{7.6}$$

for some integer s. From (7.5) we have

$$s \text{ divides } 2(\lambda + 2). \tag{7.7}$$

If v - k = 1, we have the parameters (i). If v - k = k + 1, clearly v = 2k + 1 and by (1.2), $k = 2\lambda + 1$, and we have the parameters (ii). Otherwise, E^Y is a balanced incomplete block design, and v - k > k + 1. By Theorem 2.10 the number of blocks of E^Y is therefore at least equal to its number of treatments. From (1.1) and (1.2), E^Y has b blocks, where

$$b(k+1) = (v-k)r$$
$$(k-m)(v-k-1) = kr$$

for some r, so

$$b = (k-m)(v-k)(v-k-1)k^{-1}(k+1)^{-1}$$

and substituting for m yields

$$b = \frac{(k(\lambda+1) - \lambda(k+1))(v-k)(v-k-1)}{k(\lambda+1)(k+1)}$$
$$= \frac{(k-\lambda)(v-k)(v-k-1)}{k(\lambda+1)(k+1)}.$$

This number must be at least v - k, so

$$(k-\lambda)(v-k-1) \geq k(\lambda+1)(k+1);$$

substituting from (7.4), we have

$$(k-\lambda)(k^2 - k - \lambda k) \geq \lambda k(\lambda+1)(k+1),$$
$$(k-\lambda)(k-1-\lambda) \geq \lambda(\lambda+1)(k+1),$$
$$k^2 - k(\lambda+1) - \lambda k + \lambda(\lambda+1) \geq \lambda(\lambda+1)k + \lambda(\lambda+1),$$
$$k(k-2\lambda-1) \geq k\lambda(\lambda+1).$$

7.4 Derived t-Designs and Affine t-Designs

Since $k > 0$ we have

$$k \geq \lambda^2 + 3\lambda + 1,$$

$$k + 1 \geq (\lambda + 1)(\lambda + 2).$$

Substituting from (7.6) we obtain

$$s \geq \lambda + 2.$$

Since (7.5) tells us that s divides $2(\lambda + 2)$, the only possibilities are $s = \lambda + 2$ and $s = 2(\lambda + 2)$.

If $s = \lambda + 2$, then $k = \lambda^2 + 3\lambda + 1$, and we have the parameters (iii). If $s = 2(\lambda + 2)$, then $k = 2\lambda^2 + 6\lambda + 3$. But applying (1.2) to D we have $\lambda(v - 1) = k(k - 1)$, so λ divides $k(k - 1)$, whence

$$\lambda \text{ divides } (2\lambda^2 + 6\lambda + 3)(2\lambda^2 + 6\lambda + 2),$$

and reducing modulo λ we have that λ divides 6. Putting $\lambda = 1$ and 3, we obtain cases (iv) and (v) of the theorem. If $\lambda = 2$, then D had parameters (254, 23, 2), and if $\lambda = 6$, then D had parameters (2036, 111, 6); in both cases v is even and $k - \lambda$ is not a perfect square, so these parameters are outlawed by Theorem 6.4. ∎

The designs of types (i) and (ii) are known to be extendable. In type (iii) the case $\lambda = 1$ gives parameters $(21, 5, 1)$, the PG(2, 4), which is extendable; $\lambda = 2$ gives $(56, 11, 2)$, and while designs of these parameters exist, we know nothing of their extensions; no examples with $\lambda \geq 3$ are known. Neither case (iv) nor case (v) is known to exist.

7.4 DERIVED t-DESIGNS AND AFFINE t-DESIGNS

Suppose that D is a t-(v, k, λ) design. If S is a subset of the treatment set of D, the <u>derivative</u> D_S was defined in Section 2.3 to be the design whose blocks are all the sets $B \backslash S$, where B is a block that contains S, and whose treatments are all those not belonging to S. It was shown in Lemma 2.14 that if $|S| = s$, then D_S is a $(t - s)$-$(v - s, k - s, \lambda)$ design. In particular, if S consists of the single element x, then D_S is denoted D_x and is called the <u>derived design with respect to</u> x (as with 2-designs), as well as the contraction.

We shall use the fact that the derived designs of a t-design are $(t - 1)$-designs, in order to prove the impossibility of certain t-designs. First we consider symmetric t-designs, those with $b = v$. Such a design will have replication number $r_1 = k$, by Theorem 1.1. (The proof of Theorem 1.1

did not depend on the degree of balance.) As we have seen, symmetric 2-designs are very important, but this does not apply for $t > 2$:

THEOREM 7.11 If $t > 2$, the only symmetric t-designs consist of all $(v - 1)$-subsets of a v-set.

Proof: Suppose that D is a symmetric $t-(v,k,\lambda)$ design, with $t > 2$. Select a treatment x of D. Then D_x is a $(t - 1) - (v - 1, k - 1, \lambda)$ design, and $t - 1 \geq 2$. So D_x is also a $2-(v - 1, k - 1, \mu)$ design for some μ, by Corollary 2.15.1. Now D_x has k blocks, since $k = r_1$, so by Theorem 2.10

$$k \geq v - 1.$$

But $k \leq v - 1$ in any balanced incomplete block design, so $k = v - 1$. For balance, D must contain all $(v - 1)$-subsets of its treatment set; for symmetry it will contain each subset just once. ■

At the end of Section 6.4 we pointed out that one can discuss resolvable and affine t-designs for $t > 2$. We can now prove that affine t-designs are very limited when $t > 2$.

THEOREM 7.12 If D is an affine t-design, then the derived designs of D are symmetric.

Proof: Suppose that D is an affine $t-(v,k,\lambda)$ design. Select a treatment x of D and consider the blocks $B_1, B_2, \ldots, B_r$ of D which contain x. They must lie in different parallel classes, and $B_i \cap B_j$ has m elements whenever $i \neq j$ (in the notation of Section 6.4). So the blocks of D_x have common intersection size $m - 1$. From Theorem 2.11 it follows that D_x is symmetric. ■

COROLLARY 7.12.1 If D is an affine 3-design with s blocks in each parallel class, then $s = 2$.

Proof: Suppose that D is an affine $3-(v,k,\lambda)$ design. Then D is also affine when considered as a 2-design; by Corollary 6.9.1, its parameters satisfy

$$sr = b = v + r - 1.$$

But if x is any treatment of D, then from Theorem 7.12, D_x is symmetric. Since D_x has r blocks and $v - 1$ treatments,

$$r = v - 1.$$

So

$sr = b = 2r,$

and $s = 2$. ■

The result above enables us to classify completely the affine 3-designs; see the exercises.

COROLLARY 7.12.2 There are no affine t-designs with $t > 3$.

Proof: Suppose that D is an affine $t-(v,k,\lambda)$ design with $t > 3$, and x is any treatment of D. Then D_x is a symmetric $(t-1) - (v-1, k-1, \lambda)$ design. Since $t - 1 > 2$, Theorem 7.11 tells us that $v - 1 = (k-1) + 1$. So D has $v = k + 1$. But from (6.14), k must divide v, which is impossible. ■

EXERCISES

7.4.1 Prove that any $3-(v,k,\lambda)$ design satisfies

$\lambda(v - 2) \geq (k - 1)(k - 2).$

7.4.2 Prove that the Hadamard 3-designs are affine.

7.4.3 Prove that any affine 3-design is an Hadamard 3-design.

BIBLIOGRAPHIC REMARKS

There is a good discussion of t-designs in [3], and many of the known results can be found there and in [2].

The result of Theorem 7.10, about extending symmetric designs, is from [1] and is often called "Cameron's theorem."

1. P. J. Cameron, "Extending symmetric designs," Journal of Combinatorial Theory 14A(1973), 215-220.
2. P. Dembowski, Finite Geometries (Springer, New York, 1968).
3. D. R. Hughes and F. C. Piper, Design Theory (Cambridge University Press, Cambridge, 1985).

8
Hadamard Matrices

8.1 BASIC IDEAS

A <u>Hadamard matrix</u> of side n is an n × n matrix with every entry either 1 or -1, which satisfies

$$HH^T = nI. \tag{8.1}$$

Suppose that H is an n × n matrix whose (i, r) entry is h_{ir}, where $h_{ir} = \pm 1$. The (i, r) entry of H^T will be h_{ri}, so the (i, j) entry of HH^T is

$$\sum_{r=1}^{n} h_{ir} h_{jr}.$$

In particular, the (i, i) entry is

$$\sum_{r=1}^{n} h_{ir}^2.$$

If every entry of H is 1 or -1, then $h_{ir}^2 = 1$ for every i and r, so the (i, i) entry above equals n. Therefore, an n × n (1,-1)-matrix will be Hadamard if and only if

$$\sum_{r=1}^{n} h_{ir} h_{jr} = 0 \quad \text{when } i \neq j. \tag{8.2}$$

This observation can be expressed by defining a Hadamard matrix to be a square (1,-1)-matrix whose rows are pairwise orthogonal.

8.1 Basic Ideas

Suppose that H is a Hadamard matrix of side n. Then $HH^T = nI$ implies that H has inverse $n^{-1}H^T$. Since $H^{-1}H = I$ we have

$$H^T H = nI,$$

and H^T is also Hadamard. So the transpose of a Hadamard matrix is always Hadamard, and consequently the columns of a Hadamard matrix are orthogonal:

$$\sum_{r=1}^{n} h_{ri} h_{rj} = 0 \quad \text{when } i \neq j. \tag{8.3}$$

LEMMA 8.1 If H is a Hadamard matrix, and if K is obtained from H by negating some or all rows, negating some or all columns, permuting the rows and permuting the columns, then K is Hadamard.

Proof: We consider the four types of transformation separately. In each case K has all entries 1 and -1. We verify equation (8.2).

(i) Suppose that K is obtained by permuting rows—say, rows i and j of K were rows x and y of H. Then

$$\sum_{r=1}^{n} k_{ir} k_{jr} = \sum_{r=1}^{n} h_{xr} h_{yr} = 0.$$

(ii) Suppose that row i of K is the negative of row i of H. Then

$$\sum_{r=1}^{n} k_{ir} k_{jr} = \sum_{r=1}^{n} (-k_{ir}) k_{jr}$$

$$= -\sum_{r=1}^{n} h_{ir} k_{jr}.$$

If row j of K equals row j of H, we have

$$-\sum_{r=1}^{n} h_{ir} k_{jr} = -\sum_{r=1}^{n} h_{ir} h_{jr} = -0 = 0,$$

while if row j of K is the negative of row j of H, then

$$-\sum_{r=1}^{n} h_{ir} k_{jr} = -\sum_{r=1}^{n} h_{ir}(-h_{jr}) = -\left(\sum_{r=1}^{n} h_{ir} h_{jr}\right) = 0.$$

In either case we have (8.2).

(iii) If K is obtained from H by permuting the columns, then $k_{i1}k_{j1} = h_{is}h_{js}$ for some s, $k_{i2}k_{j2} = h_{it}h_{jt}$, and so on. Therefore,

$$\sum_{r=1}^{n} k_{ir} k_{jr} = h_{i1} h_{j1} + h_{i2} h_{j2} + \cdots h_{in} h_{jn} = 0,$$

although the terms may arise in a different order.

(iv) Suppose that K is formed by negating certain columns of H. Since

$$h_{ir} h_{jr} = (-h_{ir})(-h_{jr}),$$

it follows that $k_{ir}k_{jr} = h_{ir}h_{jr}$ whether column r has been negated or not, and

$$\sum_{r=1}^{n} k_{ir} k_{jr} = \sum_{r=1}^{n} h_{ir} h_{jr} = 0.$$

Now suppose that K is obtained from H by a combination of the four operations. Then we can form a sequence of matrices $K_1, K_2, \ldots, K_m$, where K_1 is obtained from H by one operation (possibly a row permutation), K_2 comes from performing one operation on K_1 (for example, column negation), and so on; $K = K_m$. Then

H Hadamard implies K_1 Hadamard;
K_1 Hadamard implies K_2 Hadamard;
. . . .
K_{m-1} Hadamard implies K Hadamard. ∎

We shall say that two Hadamard matrices are <u>equivalent</u> if one can be obtained from the other by a sequence of row and column permutations and negations. Since other equivalence relations on Hadamard matrices are also important, we shall refer to the relation we have just defined as "Hadamard equivalence" whenever any confusion is possible.

Given a Hadamard matrix of side n, suppose that one negates every row in which the first entry is negative, and then negates every column in which the first element is now negative. The resulting matrix has first row and column entries all +1. Such a Hadamard matrix is called <u>standardized.</u> It follows from Lemma 8.1 that every Hadamard matrix is equivalent to a standardized Hadamard matrix. This does not define a unique "standardized"

8.1 Basic Ideas

form for a Hadamard matrix; it is possible to find more than one standardized matrix equivalent to a given Hadamard matrix.

As an example, let us consider standardized Hadamard matrices of side 4. Any such matrix must have the form

$$\begin{bmatrix} 1 & 1 & 1 & 1 \\ 1 & a & b & c \\ 1 & d & e & f \\ 1 & g & h & i \end{bmatrix},$$

where each of a, b, c, ..., i is chosen from $\{1, -1\}$.

Since rows 1 and 2 are orthogonal, we have

$$1 + a + b + c = 0;$$

the only possible solutions have one of $\{a, b, c\}$ equal to +1 and the others equal to -1. Similarly, rows 3 and 4 must contain exactly two entries -1 each. Now consider the orthogonality of the columns. A similar argument to the above proves that each of columns 2, 3, 4, contains exactly two entries -1.

It follows that the matrix

$$A = \begin{bmatrix} a & b & c \\ d & e & f \\ g & h & i \end{bmatrix}$$

has one entry +1 per row and per column. So the entries +1 form a permutation matrix within A. There are six possible permutations, so there are six standardized Hadamard matrices; using the convenient notation that + represents +1 and - represents -1, they are

```
++++    ++++    ++++    ++++    ++++    ++++
++--    ++--    +-+-    +-+-    +--+    +--+
+-+-    +--+    ++--    +--+    ++--    +-+-
+--+    +-+-    +--+    ++--    +-+-    ++--.
```

It is easy to see that these six matrices are equivalent—in fact, each one can be obtained from each other by row permutation alone. So there is exactly one equivalence class of Hadamard matrices of side 4.

It is easy to construct Hadamard matrices of sides 1 and 2:

```
+       ++
        +-.
```

A natural question presents itself: For what sides n do Hadamard matrices exist? We start with a necessary condition.

THEOREM 8.2 If there is a Hadamard matrix of side n, then n = 1, n = 2, or n is a multiple of 4.

Proof: Suppose that n > 2, and that there is a Hadamard matrix of side n. Then there is a Hadamard matrix of side n whose first row has every entry 1. By permuting columns we can make sure that all columns to the left of the matrix have second entry 1 and all columns to the right have second entry -1. Similarly, we can permute columns so that the first three rows of the matrix look like

```
+ + + ...    + + + ...    + + + ...    + + + ...
+ + + ...    + + + ...    - - - ...    - - - ...
+ + + ...    - - - ...    + + + ...    - - - ...
```

Suppose there are a columns which start 1, 1, 1, ..., b columns that start 1, 1, -1, ..., c columns that start 1, -1, 1, ..., and d columns that start 1, -1, -1, ... Then

$$a + b + c + d = n. \tag{8.4}$$

If the matrix has (i, j) entry h_{ij}, then

$$\sum_{r=1}^{n} h_{ir} h_{2r} = a + b - c - d,$$

so equation (8.2) yields

$$a + b - c - d = 0. \tag{8.5}$$

Similarly, we get

$$a - b + c - d = 0, \tag{8.6}$$

$$a - b - c + d = 0, \tag{8.7}$$

by considering the (1,3) and (2,3) entries, respectively, in HH^T. Adding equations (8.5) and (8.6) yields d = a; similarly, c = a and b = a. So (8.4) reduces to

$$4a = n,$$

and n is a multiple of 4. ∎

No other restrictions on the side of a Hadamard matrix are known; in fact, it is conjectured that there is a Hadamard matrix of every side permitted

8.2 Hadamard Matrices and Block Designs

by Theorem 8.2. The smallest side for which the existence of a Hadamard matrix is currently undecided is 412. We shall present a number of methods of construction in later sections.

EXERCISES

8.1.1 Prove that there are exactly 16 (1,-1)-matrices of size 2 × 2, and that exactly 8 of them are Hadamard.

8.1.2 How many (1,-1) matrices are there of size 4 × 4? How many of them are Hadamard?

8.1.3 As a generalization of Hadamard matrices, we define a c-matrix of side n to be an n × n matrix with one entry 0 and all other entries 1 or -1 in each row, whose rows are pairwise orthogonal. Show that the transpose of a c-matrix is necessarily a c-matrix. Find examples for n = 2 and n = 6.

8.1.4 A is a Hadamard matrix of side n, and H is a Hadamard matrix of side m + n, where

$$H = \begin{bmatrix} A & B \\ C & D \end{bmatrix}$$

for some matrices B, C, and D. Prove that $m \geq n$.

8.2 HADAMARD MATRICES AND BLOCK DESIGNS

One of the reasons that Hadamard matrices have been studied is the fact that they are related to balanced incomplete block designs. This has also been the source of several results about Hadamard matrices, including one of the main constructions.

THEOREM 8.3 There exists a Hadamard matrix of side 4m if and only if there exists a (4m - 1, 2m - 1, m - 1)-design.

Proof: Suppose that H is a Hadamard matrix of side 4m. Without loss of generality we may assume that H is normalized. Since H is Hadamard, we know that

$$\sum_{r=1}^{4m} h_{ir} h_{jr} = 0 \quad \text{if } j \neq 1,$$

so that $\sum_{r=1}^{4m} h_{jr} = 0$. Similarly, $\sum_{r=1}^{4m} h_{rj} = 0$ when $j \neq 1$. So the sum of entries in any row or any column, except the first, is zero.

Let A be the $(4m - 1) \times (4m - 1)$ matrix formed by removing the first row and column from H, so that

$$H = \begin{bmatrix} 1 & 1 & \cdots & 1 \\ 1 & & & \\ \vdots & & A & \\ 1 & & & \end{bmatrix}.$$

Then the sum of any row or any column of A must be -1, or equivalently,

$$AJ = JA = -J,$$

where as usual J is the $(4m - 1) \times (4m - 1)$ matrix each of whose entries is $+1$.

Since $HH^T = 4mI$, we see that

$$AA^T = \begin{bmatrix} 4m-1 & -1 & \cdots & -1 \\ -1 & 4m-1 & \cdots & -1 \\ \vdots & \vdots & & \vdots \\ -1 & -1 & & 4m-1 \end{bmatrix} = 4mI - J.$$

Now consider the matrix $B = (A + J)/2$. We have

$$BJ = \frac{1}{2}(AJ + JJ)$$

$$= \frac{1}{2}[-J + (4m-1)J]$$

$$= (2m - 1)J,$$

and similarly

$$JB = (2m - 1)J; \tag{8.8}$$

again

$$BB^T = \frac{1}{4}(A + J)(A^T + J)$$

$$= \frac{1}{4}(AA^T + JA^T + AJ + JJ)$$

$$= \frac{1}{4}\left[(4mI - J) - J - J + (4m-1)J\right]$$

$$= mI + (m-1)J. \tag{8.9}$$

8.2 Hadamard Matrices and Block Designs

From Theorem 2.8, equations (8.8) and (8.9) show that B is the incidence matrix of a (4m - 1, 2m - 1, m - 1)-design.

Conversely, suppose that a (4m - 1, 2m - 1, m - 1)-design exists. Let C be its incidence matrix, and define A to equal 2C - J. Form a matrix H by appending a new first row and column of +1's to A. It is easy to show that H is Hadamard. ∎

The designs with v = 4m - 1, k = 2m - 1, and λ = m - 1 are often called Hadamard 2-designs, as we said in Section 4.1. These are extremal symmetric balanced incomplete block designs, in the sense that they represent the nearest approach to a design with the same parameters as its complement. (The finite projective planes form the opposite extreme.)

In view of Theorems 4.3 and 4.4, we now know:

COROLLARY 8.3.1 If p^r is a prime power congruent to 3 modulo 4, then there is a Hadamard matrix of side $p^r + 1$. If p^r and $p^r + 2$ are both odd prime powers, then there is a Hadamard matrix of side $p^r(p^r + 2) + 1$. ∎

As an example we construct a Hadamard matrix of order 8. The difference set $\{0, 1, 3\}$ can be used to construct a (7.3.1)-design. The incidence matrix of that design, and the corresponding Hadamard matrix, are shown in Figure 8.1.

We saw in Section 7.2 that Hadamard 2-designs are extendable, so they give rise to 3-designs in a natural way. The corresponding 3-(4m, 2m, m - 1) is called a Hadamard 3-design. We now show that Hadamard matrices also give rise to further 2-designs.

THEOREM 8.4 The existence of a Hadamard matrix of side 4m implies the existence of balanced incomplete block designs of parameters:

(i) (2m - 1, 4m - 2, 2m - 2, m - 1, m - 2);
(ii) (2m, 4m - 2, 2m - 1, m, m - 1);
(iii) (2m - 1, 4m - 2, 2m, m, m).

$$\begin{bmatrix} 1 & 0 & 0 & 0 & 1 & 0 & 1 \\ 1 & 1 & 0 & 0 & 0 & 1 & 0 \\ 0 & 1 & 1 & 0 & 0 & 0 & 1 \\ 1 & 0 & 1 & 1 & 0 & 0 & 0 \\ 0 & 1 & 0 & 1 & 1 & 0 & 0 \\ 0 & 0 & 1 & 0 & 1 & 1 & 0 \\ 0 & 0 & 0 & 1 & 0 & 1 & 1 \end{bmatrix} \qquad \begin{bmatrix} + & + & + & + & + & + & + & + \\ + & + & - & - & - & + & - & + \\ + & + & + & - & - & - & + & - \\ + & - & + & + & - & - & - & + \\ + & + & - & + & + & - & - & - \\ + & - & + & - & + & + & - & - \\ + & - & - & + & - & + & + & - \\ + & - & - & - & + & - & + & + \end{bmatrix}$$

Figure 8.1

Proof: Suppose that A is the incidence matrix of the (4m - 1, 2m - 1, m - 1)-design corresponding to the Hadamard matrix. Assume that the rows have been reordered, if necessary, so that all rows with first entry 1 precede all rows with first entry 0. Then

$$A = \begin{bmatrix} e & B \\ 0 & C \end{bmatrix},$$

where e is a column of 2m - 1 entries 1, 0 is a column of 2m entries 0, and B and C are some (0,1)-matrices. Since $AA^T = mI + (m - 1)J$ we have

$$mI + (m - 1)J = \begin{bmatrix} ee^T + BB^T & BC^T \\ CB^T & CC^T \end{bmatrix} = \begin{bmatrix} J + BB^T & BC^T \\ CB^T & CC^T \end{bmatrix}$$

whence

$$BB^T = mI + (m - 2)J, \qquad (8.10)$$

$$CC^T = mI + (m - 1)J. \qquad (8.11)$$

Since A is the incidence matrix of a symmetric design, it also satisfies $A^T A = mI + (m - 1)J$, so

$$mI + (m - 1)J = \begin{bmatrix} e^T & 0 \\ B^T & C^T \end{bmatrix} \begin{bmatrix} e & B \\ 0 & C \end{bmatrix} = \begin{bmatrix} e^T e & e^T B \\ B^T e & B^T B + C^T C \end{bmatrix}.$$

Since $e^T B = (m - 1)J$ we have

$$JB = (m - 1)J. \qquad (8.12)$$

So every column of B has m -1 entries 1. Therefore, $B^T B$ has diagonal (m - 1, m - 1, ..., m - 1). The equation

$$B^T B + C^T C = mI + (m - 1)J$$

therefore implies that $C^T C$ has every diagonal entry $(2m - 1) - (m - 1) = m$, so

$$JC = mJ. \qquad (8.13)$$

By Theorem 2.8, equations (8.10) to (8.13) imply that B and C are the incidence matrices of designs with parameters (i) and (ii). Design (iii) is the complement of (i). ∎

8.3 Kronecker Product Constructions

EXERCISES

8.2.1 Suppose that B and C are the incidence matrices of designs with the parameters $(2m - 1, 4m - 2, 2m - 2, m - 1, m - 2)$ and $(2m, 4m - 2, 2m - 1, m, m - 1)$, respectively. Is

$$\begin{bmatrix} e & B \\ 0 & C \end{bmatrix}$$

necessarily the incidence matrix of a $(4m - 1, 2m - 1, m - 1)$-design?

8.2.2 A balanced incomplete block design is called <u>quasi-symmetric</u> if there are two integers μ and ν such that any two blocks intersect in μ elements, except that given any block there is exactly one other block which intersects it in ν elements.
   (i) Prove that the design formed by taking two copies of a given symmetric balanced incomplete block design is a quasi-symmetric design.
   (ii) Prove that the Hadamard 3-design on $4m$ points, when interpreted as a 2-design, is a

   $(4m, 8m - 2, 4m - 1, 2m, 2m - 1)$-design

   and is quasi-symmetric.
   (iii) Prove that constructions (i) and (ii) yield all quasi-symmetric designs.

8.3 KRONECKER PRODUCT CONSTRUCTIONS

To discuss the easiest of all Hadamard matrix constructions, we introduce the <u>Kronecker product</u> (or <u>direct product</u>) of two matrices: If A is of size $p \times q$ and B is of size $r \times s$, then $A \otimes B$ denotes the $pr \times qs$ matrix

$$A \otimes B = \begin{bmatrix} a_{11}B & a_{12}B & \cdots & a_{1q}B \\ a_{21}B & a_{22}B & \cdots & a_{2q}B \\ & & \cdots & \\ a_{p1}B & a_{p2}B & \cdots & a_{pq}B \end{bmatrix}.$$

We also write $A \otimes B = [a_{ij}B]$ (square brackets denote the general element of a block matrix).

LEMMA 8.5 If A and B are any matrices, then

$$(A \otimes B)^T = A^T \otimes B^T. \tag{8.14}$$

If, further, C and D are any matrices such that the products AC and BD exist, then

$$(A \otimes B)(C \otimes D) = AC \otimes BD. \tag{8.15}$$

Proof: The transpose of any block matrix is formed by using the rule

$$\begin{bmatrix} X & Y \\ Z & W \end{bmatrix}^T = \begin{bmatrix} X^T & Z^T \\ Y^T & W^T \end{bmatrix}.$$

So $(A \otimes B)^T = [(a_{ji}B)^T] = [a_{ji}B^T]$. We have (8.14).

Now $[X_{ij}][Y_{ij}]$ has (i,j) block entry

$$\sum_k X_{ik} Y_{kj}$$

provided that the blocks are conformable and the number of blocks per row in X equals the number of blocks per column in Y. So $(A \otimes B)(C \otimes D)$ has (i,j) block

$$\sum_k (a_{ik}B)(c_{kj}D) = \left(\sum_k a_{ik} c_{kj} \right) BD.$$

But $\Sigma_k\, a_{ik} c_{kj}$ is the (i,j) entry of AC. So

$$(A \otimes B)(C \otimes D) = AC \otimes BD. \qquad \blacksquare$$

THEOREM 8.6 If A and B are Hadamard matrices, then $A \otimes B$ is Hadamard.

Proof: Suppose that A and B are of sides n and r, respectively. Then $AA^T = nI_n$ and $BB^T = rI_r$. So from Lemma 8.5,

$$(A \otimes B)(A \otimes B)^T = (A \otimes B)(A^T \otimes B^T)$$
$$= AA^T \otimes BB^T$$
$$= nI_n \otimes rI_r,$$

which obviously equals nrI_{nr}. The Kronecker product of $(1,-1)$ matrices is clearly a $(1,-1)$ matrix, so $A \otimes B$ is Hadamard. $\blacksquare$

The most important consequence for our work of the discussion above is:

8.3 Kronecker Product Constructions

COROLLARY 8.6.1 If there are Hadamard matrices of sides n and r, then there is a Hadamard matrix of side nr. ■

For example, we can construct a matrix of order 2^k for any k by an inductive process: A is a Hadamard matrix of side 2 and B is a Hadamard matrix of side 2^{k-1}.

Kronecker products are also important in constructions involving two related types of matrices: (symmetric) conference matrices and skew-Hadamard matrices. We introduce the former here and leave the latter to the exercises. In both cases, considerably more results are known than we have room to discuss here.

A conference matrix C of side k is a symmetric k × k (1,-1)-matrix with all diagonal elements +1, which satisfies

$$(C - I)^2 = (k - 1)I. \tag{8.16}$$

A conference matrix is called normalized if every entry in its first row and column is +1. Obviously, row negation and column negation do not alter the validity of (8.16), so every conference matrix can be normalized.

THEOREM 8.7 The order of a conference matrix is congruent to 2 modulo 4.

Proof: Suppose that C is a conference matrix of order k; without loss of generality suppose that C is normalized. We can assume that the first three rows of C - I have the following form (possibly after a simultaneous permutation of rows and columns has been made):

```
0 + +    + ... +    + ... +    + ... +    + ... +
+ 0 +    + ... +    + ... +    - ... -    - ... -
+ + 0    + ... +    - ... -    + ... +    - ... -
          α          β          γ          δ
```

(where "α" means "there are α columns of this type"). Consider the inner products of rows:

(row 1) · (row 2) = $1 + \alpha + \beta - \gamma - \delta$;

(row 1) · (row 3) = $1 + \alpha - \beta + \gamma - \delta$;

(row 2) · (row 3) = $1 + \alpha - \beta - \gamma + \delta$;

and each of these products equals zero by (8.16). Adding gives us

$$3 + 3\alpha - \beta - \gamma - \delta = 0. \tag{8.17}$$

Clearly, $\alpha + \beta + \gamma + \delta$ must equal $k - 3$. Adding $\alpha + \beta + \gamma + \delta$ to both sides of (8.17),

$$3 + 4\alpha = k - 3,$$

so 4 divides $k - 6$ and $k \equiv 2 \pmod{4}$. ∎

Conference matrices are used in the following construction.

THEOREM 8.8 Suppose that there is a Hadamard matrix H of side h, $h > 1$, and there is a conference matrix C of side k. Then there is a Hadamard matrix of side hk.

Proof: By hypothesis h is greater than 1, so h is even. Say that $h = 2g$. Define a signed permutation matrix U by

$$U = I_g \otimes \begin{bmatrix} 0 & 1 \\ -1 & 0 \end{bmatrix},$$

and write

$$M = ((C - I) \otimes H) + (I_k \otimes UH).$$

The matrix M is of size hk. It is a $k \times k$ array of $h \times h$ blocks, and each block is a copy of H, -H, or UH. All of these are (1,-1)-matrices, so M is (1,-1). Now

$$MM^T = [((C - I) \otimes H) + (I \otimes UH)][((C - I) \otimes H^T) + (I \otimes H^T U^T)]$$

$$= ((C - I) \otimes H)(C - I) \otimes H^T) + ((C - I) \otimes H)(I \otimes H^T U^T)$$

$$+ (I \otimes UH)((C - I) \otimes H^T) + (I \otimes UH)(I \otimes H^T U^T)$$

$$= ((C - I)^2 \otimes HH^T) + ((C - I) \otimes HH^T U^T)$$

$$+ ((C - I) \otimes UHH^T) + (I \otimes UHH^T U^T)$$

$$= ((k - 1)I \otimes hI) + ((k - 1)I \otimes hU^T)$$

$$+ ((k - 1)I \otimes hU) + (I \otimes hUU^T).$$

Using the facts that $UU^T = I$ and $U^T = -U$, this reduces to

8.3 Kronecker Product Constructions

$$MM^T = (k-1)hI_{nh} - (k-1)hI \otimes U + (k-1)hI \otimes U + hI_{nh}$$

$$= hkI_{hk},$$

so M is the required Hadamard matrix. ∎

To use this construction, it is necessary to have some conference matrices. Fortunately, they are fairly common, as the next theorem shows.

THEOREM 8.9 If p^r is a prime power congruent to 1 modulo 4, then there is a conference matrix of order $p^r + 1$.

Proof: We use the quadratic elements modulo p^r. Suppose that the elements of $GF(p^r)$ are labeled $\alpha_1, \alpha_2, \ldots$. Define a square matrix P of size p^r by

$p_{ij} = 1$ if $\alpha_i - \alpha_j$ is quadratic;
$p_{ij} = -1$ if $\alpha_i - \alpha_j$ is nonquadratic;
$p_{ii} = 1$.

Then the matrix

$$\begin{bmatrix} 1 & 1 & 1 & \cdots & 1 \\ \hline 1 & & & & \\ 1 & & P & & \\ \vdots & & & & \\ 1 & & & & \end{bmatrix}$$

is found to be a conference matrix. ∎

As an example, consider $p^r = 5$. The quadratic elements are 1 and 4, and

$$P = \begin{bmatrix} + & + & - & - & + \\ + & + & + & - & - \\ - & + & + & + & - \\ - & - & + & + & + \\ + & - & - & + & + \end{bmatrix}.$$

The corresponding conference matrix is

$$\begin{bmatrix} \dot{+} & + & + & + & + & + \\ + & + & + & - & - & + \\ + & + & + & + & - & - \\ + & - & + & + & + & - \\ + & - & - & + & + & + \\ + & + & - & - & + & + \end{bmatrix}.$$

We take

$$H = \begin{bmatrix} + & + \\ + & - \end{bmatrix}, \quad UH = \begin{bmatrix} + & - \\ - & - \end{bmatrix},$$

and obtain the Hadamard matrix

$$\begin{bmatrix}
+ & - & + & + & + & + & + & + & + & + & + & + \\
- & - & + & - & + & - & + & - & + & - & + & - \\
+ & + & + & - & + & + & + & - & + & - & + & + \\
+ & - & - & - & + & - & - & - & - & - & + & - \\
+ & + & + & + & + & - & + & + & + & - & + & - \\
+ & - & + & - & - & - & + & - & - & - & - & - \\
+ & + & + & - & + & + & + & - & + & + & + & - \\
+ & - & - & - & + & - & - & + & - & - & - & - \\
+ & + & + & - & + & - & + & + & + & - & + & + \\
+ & - & - & - & - & - & + & - & - & - & + & - \\
+ & + & + & + & + & - & + & - & + & + & + & - \\
+ & - & + & - & - & - & - & - & + & - & - & -
\end{bmatrix}$$

Not all conference matrices follow the formula "prime power plus 1." For example, a matrix of side 226 is constructed in Exercise 8.3.5. However, not all sides congruent to 2 modulo 4 are possible. To prove this we need the following lemma.

LEMMA 8.10 If B is any $n \times n$ matrix, there exists some diagonal matrix D, with each d_{ii} either 1 or -1, such that $B + D$ has nonzero determinant.

Proof: We proceed by induction on n. The case $n = 1$ is easy. If $n > 1$, assume the lemma to be true for matrices of side $n - 1$, and consider the function of n variables

$$F(x_1, x_2, \ldots, x_n) = \det(B + \text{diag}(x_1, x_2, \ldots, x_n)).$$

Expanding along the first row we see that

$$F(x_1, x_2, \ldots, x_n) = x_1 \det(C + \text{diag}(x_2, x_3, \ldots, x_n)) + d,$$

8.3 Kronecker Product Constructions

where C is obtained from B by deleting the first row and column, and d is a sum of terms that do not involve x_1. By induction there exist values $\epsilon_2, \epsilon_3, \ldots, \epsilon_n$ such that

$$e = \det(C + \mathrm{diag}(\epsilon_2, \epsilon_3, \ldots, \epsilon_n)) \neq 0,$$

and every ϵ_i is 1 or -1. Choose $d_{ii} = \epsilon_i$ for $2 \leq i \leq n$. Now

$$F(x_1, \epsilon_2, \epsilon_3, \ldots, \epsilon_n) = x_1 e + d.$$

If $e + d \neq 0$, choose $d_{11} = 1$. If $e + d = 0$, then $-e + d = -2e \neq 0$, so choose $d_{11} = -1$. In either case these d_{ii} satisfy the lemma. ∎

THEOREM 8.11 Suppose that there is a square rational matrix Q of side q, $q \equiv 2 \pmod{4}$, which satisfies $Q^T Q = mI_q$, m a positive integer. Then m is the sum of two integer squares.

Proof: Let X by the top left 4×4 block of Q: say

$$Q = \begin{bmatrix} X & Y \\ Z & T \end{bmatrix}.$$

By Theorem 3.22, m can be written as the sum of four integer squares, say

$$m = a^2 + b^2 + c^2 + d^2.$$

Define a matrix M by

$$M = \begin{bmatrix} a & -b & -c & -d \\ b & a & -d & c \\ c & d & a & -b \\ d & -c & b & a \end{bmatrix}.$$

Observe that $M^T M = mI$, so M^{-1} exists. From Lemma 8.10, there is a diagonal matrix D, with diagonal entries 1 and -1, such that $XM^{-1} + D$ is nonsingular, so that $X + DM$ has an inverse also. Write $P = T - Z(X + DM)^{-1}Y$. We prove that $P^T P = mI_{q-4}$: If

$$R = \begin{bmatrix} -(X + DM)^{-1} Y \\ I_{q-4} \end{bmatrix},$$

then

$$R^T(Q^TQ)R = R^T(mI)R = mY^T(X^T + M^TD^T)^{-1}(X + DM)^{-1}Y + mI_{q-4},$$

but on the other hand,

$$(R^TQ^T)(QR) = [Y^T(X^T + M^TD^T)^{-1}X^T - Y^T][X(X + DM)^{-1}Y - Y] + P^TP$$

$$= Y^T(X^T + M^TD^T)^{-1}[X^T - (X^T + M^TD^T)]$$

$$\times [X + (X + DM)](X + DM)^{-1}Y + P^TP$$

$$= Y^T(X^T + M^TD^T)^{-1}m(X + DM)^{-1}B + P^TP.$$

These two expressions are equal, so

$$P^TP = mI_{q-4}.$$

If we repeat the process, using P instead of Q, we eventually obtain a rational 2×2 matrix that satisfies

$$\begin{bmatrix} a & b \\ c & d \end{bmatrix}^T \begin{bmatrix} a & b \\ c & d \end{bmatrix} = mI_2,$$

where $m = a^2 + c^2$, a sum of two rational squares. By Corollary 3.18.1, m is the sum of two integer squares. ∎

In particular, if Q is a conference matrix of side k, then (putting $q = k$ and $m = k - 1$) we have

COROLLARY 8.11.1 If there is a conference matrix of side k, then $k - 1$ is the sum of two integer squares. ∎

In particular, there are no conference matrices of sides 22, 34, 58, 70, 78, or 94.

EXERCISES

8.3.1 Prove that the Kronecker product of matrices satisfies the distributive laws

$$(A + B) \otimes C = (A \otimes C) + (B \otimes C)$$
$$A \otimes (B + C) = (A \otimes B) + (A \otimes C)$$

8.3 Kronecker Product Constructions

and the associative law

$$A \otimes (B \otimes C) = (A \otimes B) \otimes C.$$

8.3.2 Verify that the matrix of Theorem 8.9 is in fact a conference matrix.

8.3.3 Suppose that C is a conference matrix of side k. Prove that

$$\begin{bmatrix} C & C - 2I \\ C - 2I & -C \end{bmatrix}$$

is a Hadamard matrix of side 2k. Is this construction related to the case "h = 2" of Theorem 8.8?

8.3.4 A <u>skew-Hadamard</u> matrix is a Hadamard matrix H with the properties that:
  (a) every diagonal entry is +1;
  (b) $H + H^T = 2I$.

  (i) Prove that if there is a skew-Hadamard matrix $H = S + I$ of side n, then

  $$\begin{bmatrix} S + I & S + I \\ S - I & -S + I \end{bmatrix}$$

  is a skew-Hadamard matrix of side 2n.

  (ii) Prove that the Hadamard matrix of side $p^r + 1$ constructed from the difference set of quadratic elements of $GF(p^r)$ is skew-Hadamard.

  (iii) Prove that there is a skew-Hadamard matrix of side 16.

  (iv) A skew-Hadamard matrix is called <u>normalized</u> if its first row and its diagonal have every entry +1 and its first column has every entry -1 except the first. Prove that every skew-Hadamard matrix is equivalent to a normalized skew-Hadamard matrix.

8.3.5 If M is a normalized skew-Hadamard matrix or conference matrix, the <u>core</u> of M is the matrix W obtained from M by deleting the first row and column and setting the diagonal equal to zero.

  (i) Prove that the core of a normalized skew-Hadamard matrix of order k satisfies

  $$WW^T = (k - 1)I - J, \quad WJ = 0, \quad W^T = -W.$$

  (ii) Prove the corresponding results for the core of a normalized conference matrix.

(iii) If W is the core of a normalized skew-Hadamard or conference matrix of side k, prove that

$$(W \otimes W) + (I \otimes J) - (J \otimes I),$$

where I and J are of size $(k-1) \times (k-1)$, is the core of a conference matrix of side $(k-1)^2 + 1$.

(iv) Prove that there is a conference matrix of side 226.

8.3.6 Suppose that H is a normalized skew-Hadamard matrix of side n whose core is W. Write $S = H - I_n$ and $D = W + I_{n-1}$. If

$$K = S \otimes D + I_n \otimes J_{n-1},$$

prove that K is a Hadamard matrix of side $n(n-1)$.

8.4 WILLIAMSON'S METHOD

Corollary 8.6.1 tells us that a Hadamard matrix of side $2^k h$ can be constructed if one of side h is known. So there is particular interest in Hadamard matrices of side $h = 4n$, where n is odd.

To treat this case, John Williamson considered the array

$$H = \begin{bmatrix} A & B & C & D \\ -B & A & -D & C \\ -C & D & A & -B \\ -D & -C & B & A \end{bmatrix} \tag{8.18}$$

which is derived from the quaternions. If A, B, C, and D are replaced by square matrices of order n, H becomes a square matrix of order 4n. One can attempt to choose A, B, C, and D in such a way that H will be Hadamard.

If HH^T is considered as a block matrix with $n \times n$ blocks, the diagonal blocks each equal $AA^T + BB^T + CC^T + DD^T$. This must be 4nI for H to be Hadamard. The (1,2) block is $BA^T - AB^T + DC^T - CD^T$. This will be zero if $AB^T = BA^T$ and $CD^T = DC^T$. Similar results hold for the other off-diagonal elements. So we have

THEOREM 8.12 If there exist square (1,-1) matrices A, B, C, and D of order n that satisfy

$$AA^T + BB^T + CC^T + DD^T = 4nI \tag{8.19}$$

and for every pair X, Y of distinct matrices chosen from A, B, C, D,

8.4 Williamson's Method

$$XY^T = YX^T, \tag{8.20}$$

then the matrix H of (8.18) is a Hadamard matrix of order 4n. ■

As an example, suppose that the submatrices to be inserted were

$$A = \begin{bmatrix} 1 & 1 & 1 \\ 1 & 1 & 1 \\ 1 & 1 & 1 \end{bmatrix}, \quad B = C = D = \begin{bmatrix} + & - & - \\ - & + & - \\ - & - & + \end{bmatrix}$$

Then

$$AB^T = \begin{bmatrix} - & - & - \\ - & - & - \\ - & - & - \end{bmatrix} = BA^T$$

and similarly every other combination satisfies $XY^T = YX^T$. Moreover,

$$AA^T = \begin{bmatrix} 3 & 3 & 3 \\ 3 & 3 & 3 \\ 3 & 3 & 3 \end{bmatrix}, \quad BB^T = CC^T = DD^T = \begin{bmatrix} 3 & -1 & -1 \\ -1 & 3 & -1 \\ -1 & -1 & 3 \end{bmatrix},$$

so $AA^T + BB^T + CC^T + DD^T = 12I = 4nI$. Therefore, these matrices satisfy the theorem. The Hadamard matrix is

$$\begin{bmatrix}
+ & + & + & + & - & - & - & + & - & - & + & - & - \\
+ & + & + & - & + & - & - & - & + & - & - & + & - \\
+ & + & + & - & - & + & - & - & + & - & - & + \\
- & + & + & + & + & + & - & + & + & + & - & - \\
+ & - & + & + & + & + & + & - & + & - & + & - \\
+ & + & - & + & + & + & + & + & - & - & - & + \\
- & + & + & + & - & - & + & + & + & - & + & + \\
+ & - & + & - & + & - & + & + & + & + & - & + \\
+ & + & - & - & - & + & + & + & + & + & + & - \\
- & + & + & - & + & + & + & - & - & + & + & + \\
+ & - & + & + & - & + & - & + & - & + & + & + \\
+ & + & - & + & + & - & - & - & + & + & + & +
\end{bmatrix}$$

Obviously, the main problem in this construction is finding the matrices A, B, C, and D. If they are chosen to be symmetric, the condition (8.20) reduces to a requirement that the four matrices commute. If we further require that each matrix is circulant—its $(i+1, j+1)$ entry equals the (i, j) entry, with the row and column number being reduced modulo n if necessary—the four matrices will be polynomials in the matrix K,

154   8. Hadamard Matrices

$$K = \begin{bmatrix} 0 & 1 & 0 & 0 & \cdots & 0 \\ 0 & 0 & 1 & 0 & \cdots & 0 \\ 0 & 0 & 0 & 1 & \cdots & 0 \\ & & & \cdots & & \\ 0 & 0 & 0 & 0 & \cdots & 1 \\ 1 & 0 & 0 & 0 & \cdots & 0 \end{bmatrix}$$

so they will commute. A further advantage is that the four matrices are completely specified by their first rows.

Figure 8.2 is a table of suitable first rows for the matrices A, B, C, and D for all odd n up to 25.

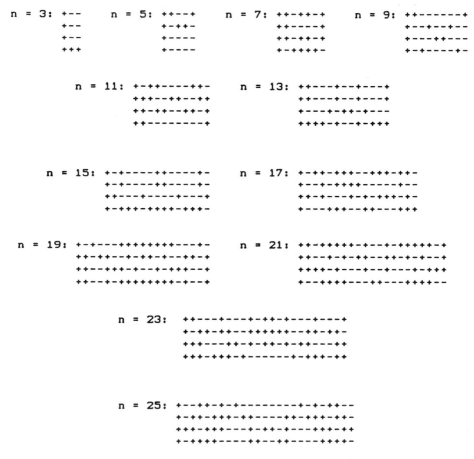

Figure 8.2 Initial rows for matrices used in Williamson's construction.

8.4 Williamson's Method

It is tempting to try to generalize Williamson's technique by changing the size of the array (8.18). The original array has every variable exactly once per row and column. It has been shown by various authors in various ways that such an array exists for sizes 1, 2, 4, and 8 and no other. The arrays of size 1 and 2 are trivial, and the array of size 8 can give rise to no new Hadamard matrices. So attention was directed toward arrays in which the same variable could occur more than once in the same row. We define

$$\begin{bmatrix}
A & A & A & B & -B & C & -C & -D & B & C & -D & -D \\
A & -A & B & -A & -B & -D & D & -C & -B & -D & -C & -C \\
A & -B & -A & A & -D & D & -B & B & -C & -D & C & -C \\
B & A & -A & -A & D & D & D & C & C & -B & -B & -C \\
B & -D & D & D & A & A & A & C & -C & B & -C & B \\
B & C & -D & D & A & -A & C & -A & -D & C & B & -B \\
D & -C & B & -B & A & -C & -A & A & B & C & D & -D \\
-C & -D & -C & -D & C & A & -A & -A & -D & B & -B & -B \\
D & -C & -B & -B & -B & C & C & -D & A & A & A & D \\
-D & -B & C & C & C & B & B & -D & A & -A & D & -A \\
C & -B & -C & C & D & -B & -D & -B & A & -D & -A & A \\
-C & -D & -D & -C & -C & -B & B & B & D & A & -A & -A
\end{bmatrix}$$

$$\begin{bmatrix}
-D & B & -C & -C & -B & C & A & -D & -D & -A & -B & -A & C & -C & -A & A & -B & -D & D & -B \\
-B & -D & B & -C & -C & -A & C & A & -D & -D & -A & -B & -A & C & -C & -B & A & -B & -D & D \\
-C & -B & -D & B & -C & -D & -A & C & A & -D & -C & -A & -B & -A & C & D & -B & A & -B & -D \\
-C & -C & -B & -D & B & -D & -D & -A & C & A & C & -C & -A & -B & -A & -D & D & -B & A & -B \\
B & -C & -C & -B & -D & A & -D & -D & -A & C & -A & C & -C & -A & -B & -B & -D & D & -B & A \\
-C & A & D & D & -A & -D & -B & -C & -C & B & -A & B & -D & D & B & -B & -A & -C & C & -A \\
-A & -C & A & D & D & B & -D & -B & -C & -C & B & -A & B & -D & D & -A & -B & -A & -C & C \\
D & -A & -C & A & D & -C & B & -D & -B & -C & D & B & -A & B & -D & C & -A & -B & -A & -C \\
D & D & -A & -C & A & -C & -C & B & -D & -B & -D & D & B & -A & B & -C & C & -A & -B & -A \\
A & D & D & -A & -C & -B & -C & -C & B & -D & B & -D & D & B & -A & -A & -C & C & -A & -B \\
B & -A & -C & C & -A & A & B & -D & D & B & -D & -B & C & C & B & -C & A & -D & -D & -A \\
-A & B & -A & -C & C & B & A & B & -D & D & B & -D & -B & C & C & -A & -C & A & -D & -D \\
C & -A & B & -A & -C & D & B & A & B & -D & C & B & -D & -B & C & -D & -A & -C & A & -D \\
-C & C & -A & B & -A & -D & D & B & A & B & C & C & B & -D & -B & -D & -D & -A & -C & A \\
-A & -C & C & -A & B & B & -D & D & B & A & -B & C & C & B & -D & A & -D & -D & -A & -C \\
-A & -B & -D & D & -B & B & -A & C & -C & -A & C & A & D & D & -A & -D & B & C & C & -B \\
-B & -A & -B & -D & D & -A & B & -A & C & -C & -A & C & A & D & D & -B & -D & B & C & C \\
D & -B & -A & -B & -D & -C & -A & B & -A & C & D & -A & C & A & D & C & -B & -D & B & C \\
-D & D & -B & -A & -B & C & -C & -A & B & -A & D & D & -A & C & A & C & C & -B & -D & B \\
-B & -D & D & -B & -A & -A & C & -C & -A & B & A & D & D & -A & C & B & C & C & -B & -D
\end{bmatrix}$$

Figure 8.3 Baumert-Hall arrays of orders 12 and 20.

a <u>Baumert-Hall array</u> H of order 4t to be a 4t × 4t array whose elements are ±A, ±B, ±C, and ±D, constructed in such a way that if A, B, C, and D are replaced by (1,-1) matrices obeying (8.20), then HH^T is the 4t × 4t block matrix with diagonal elements $t(AA^T + BB^T + CC^T + DD^T)$ and other elements zero. Clearly, if there is a Baumert-Hall array of order 4t and there are matrices A, B, C, D of order n suitable for Williamson's construction, there is a Hadamard matrix of order 4nt.

There exist Baumert-Hall arrays of many orders; examples of arrays of orders 12 and 20 are shown in Figure 8.3. They may be used together with the matrices of Figure 8.2 to construct Hadamard matrices: For example, a matrix of side 156 can be constructed using the Baumert-Hall array of order 12 and the matrices A, B, C, D for n = 13. (Baumert-Hall arrays were first constructed to deal with this specific case.)

EXERCISES

8.4.1 Find all possible sets of symmetric circulant matrices $\{A,B,C,D\}$ of side 3 for use in Theorem 8.11; and similarly for side 5.

8.4.2 Find an 8 × 8 array based on the symbols ±A, ±B, ±C, ±D, ±E, ±F, ±G, ±H such that each of A, B, C, D, E, F, G, H appear exactly once per row and per column, which would be suitable for use in a generalization of Theorem 8.12.

8.5 REGULAR HADAMARD MATRICES

A Hadamard matrix is called <u>regular</u> if every row contains the same number of elements +1.

THEOREM 8.13 There exists a regular Hadamard matrix of side 4t if and only if there is a symmetric balanced incomplete block design with parameters

$$(4t,\ 2t - t^{1/2},\ t - t^{1/2}).$$

Proof: Suppose that H is a regular Hadamard matrix of side 4t, with h elements +1 per row. Then

$$HH^T = 4tI,$$
$$HJ = (2h - 4t)J.$$

Write $A = (H + J)/2$. A is a square (0,1)-matrix of side 4t. Moreover,

8.5 Regular Hadamard Matrices

$$AA^T = \frac{1}{4}(HH^T + HJ + (HJ)^T + J^2)$$

$$= \frac{1}{4}(4tI + 2(2h - 4t)J + 4tJ)$$

$$= tI + (h - t)J \tag{8.21}$$

$$AJ = \frac{1}{2}(HJ + J^2)$$

$$= hJ. \tag{8.22}$$

By Theorem 2.11, A is the incidence matrix of a symmetric balanced incomplete block design with

$v = b = 4t,$

$r = k = h,$

$\lambda = h - t.$

Then equation (1.2) yields

$h(h - 1) = (h - t)(4t - 1),$

whence

$h^2 - 4th + t(4t - 1) = 0,$

which has solutions

$$h = \frac{1}{2}(4t \pm \sqrt{4t})$$

$$= 2t \pm t^{1/2}.$$

If the smaller root occurs, then A is the incidence matrix of the required design; otherwise, $J - A$ is the incidence matrix.
The converse is left as an exercise. ∎

This proof in fact tells us that a regular Hadamard matrix exists if and only if two designs, with parameters

$(4t, 2t \pm t^{1/2}, t \pm t^{1/2})$

exist; but the two are complements, so either they both exist or neither one does.

COROLLARY 8.13.1 If there is a regular Hadamard matrix of side n, then n is a perfect square.

Proof: If n = 4t, then both $2t - t^{1/2}$ and 4t are required to be integers, so $t^{1/2}$ must be an integer also. It remains to check the two trivial cases: n = 1 is possible, n = 2 is not. ∎

COROLLARY 8.13.2 If H is a regular Hadamard matrix, then each column of H contains the same number of entries +1 as do the rows of H. ∎

This follows from the fact that the dual of a symmetric design is a symmetric design with the same parameters.

Suppose that H and K are regular Hadamard matrices. Then it is easy to see that $H \otimes K$ is regular, and we already know it is Hadamard. So we have

THEOREM 8.14 If there exist regular Hadamard matrices of sides n and r, then there exists a regular Hadamard matrix of side nr. ∎

THEOREM 8.15 Suppose that there is a Hadamard matrix of side n. Then there is a regular Hadamard matrix of side n^2.

Proof: Let H be a Hadamard matrix of side n, and denote by H_j the matrix derived from H by negating columns in such a way that H_j has jth row all +1. Now consider the $n^2 \times n^2$ matrix K, defined to be the n × n array of n × n blocks

$$K = [h_{ij} H_j].$$

Clearly, K is Hadamard. Any row of K contains one block of n consecutive entries either all +1 or all -1, and n - 1 blocks of n entries half of which are +1 and half -1. So any given row contains either $[n(n + 1)]/2$ or $[n(n - 1)]/2$ entries +1. If the row is of the former kind, negate it. The resulting matrix is a regular Hadamard matrix. ∎

These constructions give infinitely many regular Hadamard matrices, but they do not cover any of the cases of side $4t^2$ where t is odd, t > 1. These are difficult to construct. We present one example: a (36,15,6) design (and consequently, a regular Hadamard matrix of side 36) which is constructed using block matrices. The construction of its incidence matrix M is outlined in Figure 8.4; verification is left as an exercise.

In this section we have outlined a few results concerning one special class of Hadamard matrices—regular Hadamard matrices—which have attracted some interest. Other special classes have been studied. Skew-Hadamard matrices have been mentioned earlier, and another case—symmetric Hadamard matrices with constant diagonal, which have sometimes been called graphical Hadamard matrices—are presented in the exercises.

8.5 Regular Hadamard Matrices

$$I = \begin{bmatrix} 1 & 0 & 0 \\ 0 & 1 & 0 \\ 0 & 0 & 1 \end{bmatrix} \quad J = \begin{bmatrix} 1 & 1 & 1 \\ 1 & 1 & 1 \\ 1 & 1 & 1 \end{bmatrix} \quad 0 = \begin{bmatrix} 0 & 0 & 0 \\ 0 & 0 & 0 \\ 0 & 0 & 0 \end{bmatrix}$$

$$K = \begin{bmatrix} 0 & 1 & 0 \\ 0 & 0 & 1 \\ 1 & 0 & 0 \end{bmatrix} \quad L = \begin{bmatrix} 0 & 0 & 1 \\ 1 & 0 & 0 \\ 0 & 1 & 0 \end{bmatrix}$$

$$A = \begin{bmatrix} 0 & J & J \\ J & 0 & J \\ J & J & 0 \end{bmatrix} \quad B = \begin{bmatrix} I & I & I \\ I & I & I \\ I & I & I \end{bmatrix}$$

$$C = \begin{bmatrix} I & I & I \\ K & K & K \\ L & L & L \end{bmatrix} \quad D = \begin{bmatrix} I & I & I \\ L & L & L \\ K & K & K \end{bmatrix} \quad E = \begin{bmatrix} I & L & K \\ L & K & I \\ K & I & L \end{bmatrix}$$

$$M = \begin{bmatrix} A & C & D & B \\ C^T & A & E^T & E \\ D^T & E & A & E^T \\ B & E^T & E & A \end{bmatrix}$$

Figure 8.4

EXERCISES

8.5.1 A regular Hadamard matrix of side $4t^2$ is called <u>subregular</u> if it has $2t^2 - t$ entries +1 per row, and <u>superregular</u> if it has $2t^2 + 1$. Prove that the direct product of two regular Hadamard matrices is superregular if the two constituent matrices are both superregular or both subregular, and is subregular if the two constituents are of different kinds.

8.5.2 (Completion of Theorem 8.13) Suppose that A is the incidence matrix of a symmetric balanced incomplete block design with parameters

$$(4u^2, 2u^2 - u, u^2 - u).$$

Prove that $2A - J$ is a regular Hadamard matrix.

8.5.3 Verify that the matrix M constructed in Figure 8.4 is the incidence matrix of a $(36, 15, 6)$-design.

8.5.4 Prove that there exists a $(15, 7, 3)$-design whose incidence matrix is a 5×5 block matrix based on the matrices I, J, K, L, 0 of Figure 8.4

8.5.5 A Hadamard matrix is called <u>graphical</u> if it is symmetric and has every diagonal entry equal to +1.
   (i) Suppose that H is a symmetric Hadamard matrix of side n. Prove that

$$H^2 = nI$$

   and consequently that the eigenvalues of H are $n^{1/2}$ and $-n^{1/2}$ (with some multiplicities).
   (ii) Prove that there are nonnegative integers a and b auch that $a + b = n$ and $(a - b)n^{1/2}$ equals the trace of H.
   (iii) Suppose further that every diagonal entry in H equals +1. Prove that n is a perfect square. (Thus the side of a graphical Hadamard matrix equals an integer square.)

8.5.6 Suppose that B is a balanced incomplete block design with parameters $(2u^2 - u, 4u^2 - 1, 2u + 1, u, 1)$.
   (i) Prove that any two distinct blocks of B have either no common element or one common element.
   (ii) Order the blocks of B as $B_1, B_2, \ldots, B_{4u^2-1}$ in some way. Define a_{ij} by

$$a_{ij} = |B_i \cap B_j|$$

   when $i = j$, and set $a_{ij} = 0$ for all i. Prove that $A = (a_{ij})$ is the incidence matrix of a symmetric balanced incomplete block design. What are its parameters?
   (iii) Prove that there exists a graphical Hadamard matrix of side $4u^2$.

8.5.7 A <u>cyclic</u> Hadamard matrix of side n is a Hadamard matrix H that satisfies $h_{i+1, j+1} = h_{i, j}$ (with subscripts reduced modulo n when necessary).
   (i) Prove that the side of cyclic Hadamard matrix is necessarily a perfect square.
   (ii) Prove that there is a cyclic Hadamard matrix of side 4, but not one of side 16.

BIBLIOGRAPHIC REMARKS

The study of matrices that are orthogonal to their transposes was instigated by Sylvester [10], who mentioned the case of matrices with all entries 1 and -1, exhibiting Hadamard matrices of orders 2, 4, and 8.

In 1893, Hadamard [4] proved that if a complex n × n matrix has all its entries in the unit circle, then the absolute value of its determinant is at most $\sqrt{n^n}$. If the entries are restricted to be real, so that they lie in the

Bibliographic Remarks

range from −1 to 1, the maximum is attained by A if and only if A is Hadamard (the name was bestowed because of this extremal property). Hadamard constructed the matrices whose side is a power of 2 and proved that the side must be 1, 2, or a multiple of 4. Shortly afterward, Scarpis [8] gave a construction for a Hadamard matrix of side p(p + 1), given that there is a Hadamard matrix of side p + 1, and p is a prime congruent to 3 (modulo 4).

The first major work on Hadamard matrices was the paper of Paley [7], which was generalized by Williamson [13]. It included the use of Kronecker products, and the construction of Hadamard matrices of order $p^r + 1$, where p is a prime power congruent to 3 modulo 4 (Paley constructed the difference sets of Theorem 4.3). Williamson also defined skew-Hadamard matrices and found the construction in Exercise 8.3.6.

Conference matrices were developed by Belevitch [1], [2] because they arose in conference telephony (whence their name). Theorem 8.8 is due to Goethals and Seidel [3], although Williamson [13] had earlier treated the case where the conference matrix is developed from the quadratic elements relative to a prime power congruent to 1 (modulo 4).

Williamson's first paper on the method of Section 8.4 is [12]. The number of developments since then have been too numerous to list here; the reader should consult [11], and also see the references in [9]. Authors who have worked in this area include Baumert, Goethals, Hall, Seberry, Seidel, Turyn, and Whiteman.

For further material on Hadamard matrices, the reader should consult [11]; the textbook of Hall [5] also contains a useful chapter, and the survey [6] indicates some applications and some relationships to other designs.

The best published listing of constructions of Hadamard matrices is probably [9], but new discoveries rapidly make such listings obsolete. The smallest unknown order—412—was reported to us by J. Seberry (private communication).

1. V. Belevitch, "Theory of 2n-terminal networks with applications to conference telephony," Electrical Communications 27(1950), 231-244.
2. V. Belevitch, "Conference networks and Hadamard matrices," Annals de la Societe Scientifique de Bruxelles 82(1968), 13-32.
3. J. M. Goethals and J. J. Seidel, "Orthogonal matrices with zero diagonal," Canadian Journal of Mathematics 19(1967), 1001-1010.
4. J. Hadamard, "Résolution d'une question relative aux déterminants," Bulletin des Sciences Mathematiques (2) 17(1893), 240-246.
5. M. Hall, Combinatorial Theory, 2nd Ed. (Wiley-Interscience, New York, 1986).
6. A. Hedayat and W. D. Wallis, "Hadamard matrices and their applications," Annals of Statistics 6(1978), 1184-1238.
7. R. E. A. C. Paley, "On orthogonal matrices," Journal of Mathematics and Physics 12(1933), 311-320.
8. U. Scarpis, "Sui determinanti di valore massimo," Rendiconti dell'Istituto Lombardo di Scienze e Lettere 31(1898), 1441-1446.

9. J. Seberry, "A computer listing of Hadamard matrices," Combinatorial Mathematics (Lecture Notes in Mathematics 686) (Springer-Verlag, Heidelberg, West Germany, 1978), 275-281.
10. J. J. Sylvester, "Thoughts on inverse orthogonal matrices, simultaneous sign successions, and tessellated pavements in two or more colors, with applications to Newton's rule, ornamental tile-work, and the theory of numbers," Philosophical Magazine (4) $\underline{34}$(1867), 461-475.
11. W. D. Wallis, A. P. Street, and J. S. Wallis, Combinatorics: Room Squares, Sum-Free Sets, Hadamard Matrices (Lecture Notes in Mathematics 292) (Springer-Verlag, Heidelberg, West Germany, 1972).
12. J. Williamson, "Hadamard's determinant theorem and the sum of four squares," Duke Mathematical Journal $\underline{11}$(1944), 65-81.
13. J. Williamson, "Note on Hadamard's determinant theorem," Bulletin of the American Mathematical Society $\underline{53}$(1947), 608-613.

9
The Variability of Hadamard Matrices

9.1 EQUIVALENCE OF HADAMARD MATRICES

As we said in Section 8.1, two Hadamard matrices are called <u>equivalent</u> (or <u>Hadamard equivalent</u>) if one can be obtained from the other by some sequence of row and column permutations and negations. This is the same as saying that two Hadamard matrices are equivalent if one can be obtained from the other by premultiplication and postmultiplication by monomial matrices (signed permutation matrices). So the equivalence class of H is

$\{PHQ : P, Q \text{ monomial}\}$.

Another operation that clearly respects the Hadamard property is transposition. A Hadamard matrix and its transpose may have very different properties as matrices, so this operation is not usually counted as one of the basic equivalence relations. However, it is of special interest, and one speaks of Hadamard matrices H and K as being <u>transpose-equivalent</u> if H is equivalent either to K or to K^T.

Let us denote the number of equivalence classes of Hadamard matrices of side n by h(n). Then it is easy to calculate the value of h(n) for small n: In Section 8.1 we verified that h(4) = 1, and similarly one finds that

h(1) = h(2) = h(4) = h(8) = h(12) = 1.

(Only the last case requires very much calculation.) Exhaustive searches have been carried out which provide

h(16) = 5,   h(20) = 3,   h(24) = 60.

These are the only exact results known. Four inequivalent Hadamard matrices of side 16 are shown in Table 9.1. The matrix H_i^T is equivalent to H_i for i = 1, 2, 3, but H_4^T is a suitable representative for the fifth equivalence

TABLE 9.1

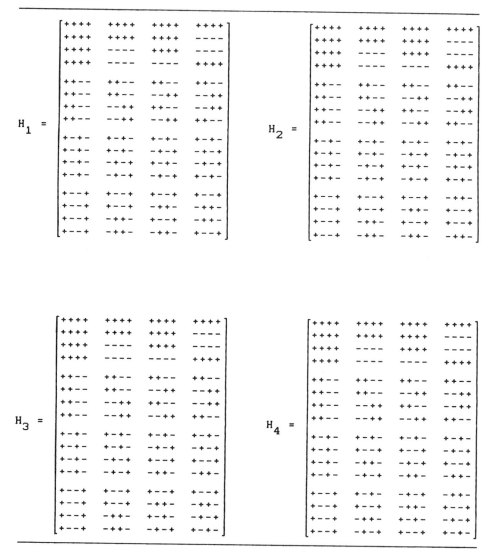

class of Hadamard matrices of side 16. Given our current computing capability, it is most unlikely that h(n) can be determined for n ≥ 32, although it may be possible to calculate h(28).

Precise determination of whether or not two given Hadamard matrices are equivalent is not easy, but it is possible to employ efficient computer

9.2 Integral Equivalence

programs. Perhaps the best approach uses the graph of a Hadamard matrix, which is defined as follows: If H is a Hadamard matrix of side n, then the graph G(H) of H has 4n vertices $\{a_1, a_2, \ldots, a_n, b_1, b_2, \ldots, b_n, c_1, c_2, \ldots, c_n, d_1, d_2, \ldots, d_n\}$. Each of $a_1, a_2, \ldots, a_n, b_1, b_2, \ldots, b_n$ has a loop on it (an edge from the vertex to itself). The other edges are

(a_i, c_j) and (b_i, d_j) if $h_{ij} = 1$;

(a_i, d_j) and (b_i, c_j) if $h_{ij} = -1$.

Then H and K are equivalent Hadamard matrices if and only if the graphs G(H) and G(K) are isomorphic, under the usual definitions of graph isomorphism, and efficient programs are available to test for this.

To investigate Hadamard equivalence, several other equivalence relations and equivalence invariants have been studied. Some of these are investigated in the following sections.

EXERCISES

9.1.1 Verify that $h(8) = 1$.

9.1.2 Prove that if H and K are equivalent Hadamard matrices, then H^T and K^T are equivalent.

9.1.3 The weight of a Hadamard matrix H is defined to be the number of entries +1 in the matrix; w(H) is the maximum weight among all matrices equivalent to H.
   (i) Prove that if H is of side n, then

$$w(H) \geq \frac{1}{2}(n+4)(n-1).$$

   (ii) If, further, $n = 8t + 4$, prove that

$$w(H) \geq \frac{1}{2}(n+6)(n-2).$$

9.2 INTEGRAL EQUIVALENCE

Two integer matrices A and B are called integrally equivalent, or Z-equivalent, if there exist integer matrices P and Q, each of which has an integer inverse, such that $PAQ = B$. It is easy to see that an integer matrix has an integer inverse if and only if it has determinant ± 1. Two matrices are Z-equivalent if and only if one can be obtained from the other by the following row operations: negation, permutation, and adding an integer multiple of one to the other; and the corresponding column operations. So we have:

9. The Variability of Hadamard Matrices

LEMMA 9.1 Two Hadamard matrices which are Hadamard equivalent must be Z-equivalent. ∎

Therefore, one can obtain a lower bound on the number of Hadamard equivalence classes by counting Z-equivalence classes. The significance of this is that Z-equivalence admits of a canonical form which is very useful.

LEMMA 9.2 Given a square integral matrix A, there exists a unique integral diagonal matrix D, which is Z-equivalent to A, say $D = \text{diag}(d_1, d_2, \ldots, d_r, 0, 0, \ldots, 0)$, where r is the rank of A. The d_i are positive integers, and d_i divides d_{i+1} when $1 \leq i < r$. The greatest common divisor of the $k \times k$ subdeterminants of A is $d_1 d_2 d_3 \cdots d_k$. If A is Z-equivalent to the matrix

$$\begin{bmatrix} D_1 & 0 \\ 0 & E \end{bmatrix}$$

where $D_1 = \text{diag}(d_1, d_2, \ldots, d_k)$, then the greatest common divisor of the nonzero elements of E is d_{k+1}. ∎

The proof may be found in books in linear algebra. The matrix D is called the <u>Smith normal form</u> of A.

In particular, consider the Smith normal form of a Hadamard matrix H of order 4m. Since H is invertible, none of the diagonal elements is zero. Let us write $D = \text{diag}(d_1, d_2, \ldots, d_{4m})$ for the Smith form of H.

LEMMA 9.3 For any Hadamard matrix H,

$$d_i d_{4m-i+1} = 4m.$$

Proof: Suppose that P and Q are integer matrices with integer inverses which diagonalize H:

PHQ = D.

Then

$$(PHQ)(Q^{-1}H^T P^{-1}) = PHH^T P^{-1} = 4mPP^{-1} = 4mI,$$

9.2 Integral Equivalence

so

$$Q^{-1}H^T P^{-1} = \text{diag}\left(\frac{4m}{d_1}, \frac{4m}{d_2}, \ldots, \frac{4m}{d_{4m}}\right),$$

and consequently H^T has Smith normal form

$$\text{diag}\left(\frac{4m}{d_{4m}}, \frac{4m}{d_{4m-1}}, \ldots, \frac{4m}{d_1}\right).$$

But (as is obvious from the symmetry of the canonical form) every matrix is Z-equivalent to its transpose, so

$$D = \text{diag}\left(\frac{4m}{d_{4m}}, \ldots, \frac{4m}{d_1}\right).$$

Equating element i in the two forms,

$$\frac{4m}{d_{4m-i+1}} = d_i,$$

so

$$d_i d_{4m-i+1} = 4m. \qquad \blacksquare$$

It is clear that $d_1 = 1$ and easy to see that $d_2 = 2$. (In fact, any 2×2 determinant with all entries 1 and -1 must equal 0, 2, or -2.) Consequently, $d_{4m-1} = 2m$ and $d_{4m} = 4m$. Consider the particular case when m is square-free. We observe that if $i \leq 2m$, then d_i divides d_{4m-i+1}; say $d_{4m-i+1} = \alpha d_i$. Then

$$\alpha d_i^2 = 4m.$$

Consequently, $d_i = 1$ or 2. Unless $i = 1$, d_2 divides d_i, so $d_i = 2$. So:

THEOREM 9.4 If m is square-free, then H has Smith normal form

$$\text{diag}(1, 2^{(2m-1 \text{ times})}, 2m^{(2m-1 \text{ times})}, 4m). \qquad \blacksquare$$

In general, the number of possible Smith normal forms of order 4m is limited by the sort of factors that m can have. For example, when m = 9, the facts that the d_i divide d_{4m} and that $d_{4m} = 4m$ means that the invariants must be divisors of 36, and the fact that $d_2 = 2$ means that they are chosen from $\{1, 2, 4, 6, 12, 18, 36\}$. Since $4m/d_i$ must also be invariant, 4 and 12 are impossible. So the form is

$$\mathrm{diag}(1, 2^{(\alpha \text{ times})}, 6^{(\beta \text{ times})}, 18^{(\alpha \text{ times})}, 36)$$

where $\alpha \geq 0$, $\beta \geq 0$, and $2\alpha + \beta = 34$. A similar analysis applies whenever m is a prime square. (See later in this section.)

We now prove a further restriction on the possible Smith normal forms.

LEMMA 9.5 Any $n \times n$ (0,1)-matrix B with nonzero determinant has at least

$$\lfloor \log_2 n \rfloor + 1$$

entries 1 in its Smith normal form.

Proof: For convenience write t for $\lfloor \log_2 n \rfloor$, so that $2^t \leq n < 2^{t+1}$. We show that B is Z-equivalent to a matrix

$$\begin{bmatrix} I_t & 0 \\ 0 & D \end{bmatrix}$$

where the GCD of the elements of D is 1.

The first part of the process is to reorder the rows and columns of B so that in the reordered matrix, columns 1 and 2 have different entries in the first row, columns 3 and 4 are identical in the first row but have different entries in the second row, and in general columns 2i - 1 and 2i are identical in the first i - 1 rows but differ in row i for i = 1, 2, ..., t. This is done using the following algorithm.

Step 1. Select two columns of B that have different entries in the first row. (If the first row of B has every entry 1, it will first be necessary to subtract some other row from row 1.) Reorder columns so that the two chosen columns become columns 1 and 2 (in either order).

Step 2. Select two columns of the matrix just formed, neither of them being columns 1 or 2, which have identical entries in the first row. Reorder the rows of the matrix other than row 1 so that the two columns chosen have different entries in the new second row. Reorder columns after column 2 so that the new pair become columns 3 and 4.

Step k. In the matrix resulting from step k - 1, select two columns to the right of column 2k - 2 which are identical in rows 1 to k - 1. Reorder

9.2 Integral Equivalence

the rows after row k - 1 and the columns after column 2k - 2 so that the chosen columns become columns 2k - 1 and 2k and differ in their kth row.

It is always possible in step k to find a row in which the two chosen columns differ, since the matrix cannot have two identical columns. Therefore, step k requires only that we can choose two columns from the n - 2k + 2 available ones which are identical in their first k - 1 places. Since there are only 2^{k-1} different (0,1)-vectors of length k - 1, this will be possible provided that

$$2^{k-1} < n - 2k + 2,$$

and this is always true for $1 \leq k \leq t$ except when k = t and n = 4 or n = 8. In the case n = 4, it is easy to check by hand that every (0,1)-matrix of nonzero determinant is Z-equivalent to a matrix on which t steps can be carried out. If n = 8, step 3 will be impossible if the first two rows of the last four columns contain all (0,1)-vectors of length 2; typically, the first two rows would be

```
1 0 1 1 1 1 0 0
* * 1 0 1 0 1 0,
```

and in this case we can proceed with step 3 if we first apply the column permutation (37)(48). Consequently, t steps can always be carried out.

In the second stage, select if possible two columns (columns a and b, say) which are identical in rows 1 to t; reorder the later rows so that columns a and b differ in row t + 1. If the selection was impossible, then $n = 2^t$ and then the first t rows of the matrix constitute the 2^t different column vectors, so one column (column a, say) will start with t zeros; reorder the rows from t + 1 on so that there is a 1 in the (t + 1, a) position. In either case, if column a was in a pair chosen in stage 1, reorder the pair if necessary so that a is even; and similarly for b if two were chosen.

The third stage isolates certain entries ±1 by carrying out t steps.

<u>Step 1.</u> Subtract column 2 from column 1, so that the (1,1) entry becomes ±1. Then add a suitable multiple of the first column to every other column to ensure that row 1 has every entry 0 except the first, and similarly eliminate all entries except the first from column 1 by adding suitable multiples of row 1 to the other rows.

After step 1 the matrix has first row and column (±1, 0, 0, ..., 0); whatever multiple of column 1 was added to column 2k - 1 ($2 \leq k \leq t$), the same multiple was added to column 2k, so that the (k, 2k - 1) and (k, 2k) entries still differ by 1. It will be seen from the description of the general step that after k - 1 steps, the matrix will have its first k - 1 rows zero except for entries ±1 in the (i, 2i - 1) positions, $1 \leq i \leq k - 1$; the first k - 1 odd-numbered columns will be zero except at those positions; and for $k \leq i \leq t$ the (i, 2i - 1) and (i, 2i) entries differ by 1 and the (j, 2i - 1) and (j, 2i) entries are equal when $j < i$. After step k this description can be extended by replacing $k - 1$ by k.

Step k. Subtract column 2k from column $2k - 1$, so that the $(k, 2k - 1)$ entry becomes ± 1. Add a suitable multiple of column $2k - 1$ to every subsequent column, and then add suitable multiples of row k to the later rows, so that row k and column $2k - 1$ become zero except at their intersection.

Observe that if $k < i \leqslant t$, the $(k, 2i - 1)$ and $(k, 2i)$ entries were equal before step k, so the same multiple of column $2k - 1$ was added to both columns $2i - 1$ and $2i$ and the difference between these columns is unchanged. If two columns, a and b, were chosen at the second stage, the difference between those columns is unaltered in the t steps; if only one column was chosen, that column is unaltered in the t steps since it has never had a nonzero entry in its kth row to be eliminated.

Finally, reorder the columns so that the former columns 1, 3, ..., $2t - 1$ become the first t columns. We obtain

$$\begin{bmatrix} I_t & 0 \\ 0 & D \end{bmatrix};$$

D contains either two entries which differ by 1 [corresponding to the former $(a, t + 1)$ and $(b, t + 1)$ entries] or has an entry 1 [the former $(a, t + 1)$ entry] depending on the course followed at stage 2, and in either case the greatest common divisor of entries of D is 1.

Therefore, B has at least $t + 1$ invariants equal to 1. ∎

THEOREM 9.6 Any Hadamard matrix of order 4m has at least $\lfloor \log_2 (4m - 1) \rfloor + 1$ entries 2 in its Smith form.

Proof: We may assume the Hadamard matrix to be normalized. Subtract column 1 from every other column, then subtract row 1 from every other row. Negate all columns except the first. The resulting matrix is

$$\left[\begin{array}{c|c} 1 & 0 \\ \hline 0 & 2B \end{array} \right]$$

for some nonsingular (0,1)-matrix B. Now apply Lemma 9.5 to B. ∎

When m has the form fg^2, where f and g are coprime and square-free, then a Hadamard matrix of side 4m has Smith normal form

$$(1, 2^{(\alpha \text{ times})}, 2g^{(\beta \text{ times})}, 2fg^{(\beta \text{ times})}, 2m^{(\alpha \text{ times})}, 4m),$$

where $\alpha + \beta = 2m - 1$, so α completely determines the Smith normal form (and the Z-equivalence class). For this reason, α is sometimes called the Smith class of H.

9.2 Integral Equivalence

From Theorem 9.6 we see that the Smith class of 16 × 16 Hadamard matrices must be 4, 5, 6, or 7. All of these classes can be realized; the matrices H_1, H_2, H_3, and H_4 of Table 9.1 are of classes 4, 5, 6, and 7, respectively.

For side 32 and for other sides divisible by 8, it is useful to observe the following fact. If H and K are Hadamard matrices of the same side, then

$$M = \begin{bmatrix} H & H \\ K & -K \end{bmatrix}$$

is also Hadamard. If K = H, then the Smith normal form of M is easy to calculate (see Exercise 9.2.1), but if K and H are different, then the situation is not clear. Quite a range of inequivalent matrices can be obtained merely by taking K to be formed from H by column permutation, and in this way one can use the matrices of side 16 to obtain matrices of side 32 of all possible Smith classes from 5 to 15.

Suppose that H is a Hadamard matrix of side 2^r. Then the invariants of the Smith normal form of H must all be powers of 2. Suppose that H has α_i invariants equal to 2^i. Then the sequence $(\alpha_0, \alpha_1, \ldots, \alpha_r)$ determines the Smith normal form of H, and we shall find it useful to define a function

$$f(H, t) = \alpha_0 + \alpha_1 t + \cdots \alpha_r t^r,$$

the generating function of H. If A and B are Hadamard matrices of sides 2^a and 2^b, respectively, the Hadamard matrix $A \otimes B$ of side 2^{a+b} has generating function

$$f(A \otimes B, t) = f(A, t) f(B, t).$$

LEMMA 9.7 Suppose that A and B are Hadamard matrices of side 2^a, and H is a Hadamard matrix of side 2. Then $H \otimes A$ and $H \otimes B$ are integrally equivalent if and only if A and B are integrally equivalent.

Proof: $f(H, t) = 1 + t$. Now $H \otimes A$ and $H \otimes B$ are integrally equivalent if and only if

$$f(H \otimes A, t) = f(H \otimes B, t),$$

that is, if and only if

$$(1 + t) f(A, t) = (1 + t) f(B, t),$$

and this will happen if and only if

$f(A, t) = f(B, t)$,

that is, if and only if A is integrally equivalent to B. ∎

THEOREM 9.8 There are at least $10x + 1$ integral equivalence classes of Hadamard matrices of order 2^{5x}. So the number of integrally inequivalent Hadamard matrices of order 2^m is at least $10 \lfloor m/5 \rfloor + 1$.

Proof: We show by induction that there is a Hadamard matrix of order 2^{5x} with y entries 2 in its Smith normal form whenever $5x \leq y \leq 15x$. This proves the first part of the theorem. The rest then follows from Lemma 9.7.

When $x = 1$ the result follows from the existence of Hadamard matrices of side 32 of all classes from 5 to 15. Assume that the result is true for $x = w$, where $w \geq 1$, and write A_n for a typical Hadamard matrix of order 2^{5x} with n entries in its normal form, so that

$f(A_n, t) = 1 + nt + \cdots$.

Write H_k for a Hadamard matrix of order 32 and class k;

$f(H_k, t) = 1 + kt + \cdots$.

Then

$f(A_n \times H_k, t) = 1 + (n + k)t + \cdots$.

As n ranges from 5w to 15w and k ranges from 5 to 15, $n + k$ will range from $5(w + 1)$ to $15(w + 1)$. So we have the result. ∎

This theorem is certainly not very good. For example, it only guarantees 21 inequivalent matrices of order 2^{10}; but the method of Kronecker products will in fact give 66 different matrices. (This can be checked by looking at α_1 and α_2, instead of only α_1.) But it is sufficient for the following corollary.

COROLLARY 9.8.1 Given any positive integer N, there are infinitely many orders of which there are at least N integrally inequivalent Hadamard matrices. The smallest such order is at most 32^n, where $n = \lfloor (n + 9)/10 \rfloor$. ∎

EXERCISES

9.2.1 (i) Suppose that H is an integer matrix which is Z-equivalent to the diagonal matrix D. Prove that

9.2 Integral Equivalence

$$\begin{bmatrix} H & H \\ H & -H \end{bmatrix}$$

is Z-equivalent to

$$\begin{bmatrix} D & 0 \\ 0 & 2D \end{bmatrix}.$$

(ii) In particular, if H is a Hadamard matrix of side 8m, where m is odd and square-free, what are the invariants of

$$\begin{bmatrix} H & H \\ H & -H \end{bmatrix}?$$

(iii) If H is any Hadamard matrix of side 8m, prove that

$$\begin{bmatrix} H & H \\ H & -H \end{bmatrix}$$

has at least $12m - 1$ invariants divisible by 4.

9.2.2 Prove that if there is a symmetric conference matrix of side n, there is a Hadamard matrix of side 2n with invariants

$$(1, 2^{(n-1)\text{times}}, n^{(n-1)\text{times}}, 2n).$$

9.2.3 (i) Prove that if there is a skew-Hadamard matrix H (see Exercise 8.3.4) of side 4m, the Hadamard matrix

$$\begin{bmatrix} H & H \\ -H^T & H^T \end{bmatrix}$$

has invariants

$$(1, 2^{(4m-1 \text{ times})}, 4m^{(4m-1 \text{ times})}, 8m).$$

(ii) Use Exercise 9.2.1 to prove that whenever there is a skew-Hadamard matrix of side 8m, there are integrally inequivalent Hadamard matrices of side 16m.

9.2.4 Prove that there are at most 351 Z-equivalence classes of Hadamard matrices of side 64.

9.3 PROFILES

Suppose that H is a Hadamard matrix of side n. We can assume that $n \geq 8$. We write $p_{ijk\ell}$ to mean the absolute value of the generalized inner product of rows i, j, k, and ℓ:

$$p_{ijk\ell} = \left| \sum_{x=1}^{n} h_{ix} h_{jx} h_{kx} h_{\ell x} \right|.$$

THEOREM 9.9 $p_{ijk\ell} \equiv n \pmod{8}$.

Proof: For convenience we simply write p instead of $p_{ijk\ell}$. First observe that p is not affected by negations or permutations of the columns of H. So there is no loss of generality in assuming that columns have been negated and arranged such that rows i, j, k, and ℓ look like

```
++... ++... ++... ++... ++... ++... ++... ++...   (i)
++... ++... ++... ++... --... --... --... --...   (j)
++... ++... --... --... ++... ++... --... --...   (k)
++... --... ++... --... ++... --... ++... --...   (ℓ)
```

$$\begin{array}{cccccccc} a & b & c & d & e & f & g & h \\ \text{times} & \text{times} & \text{times} & \text{times} & \text{times} & \text{times} & \text{times} & \text{times} \end{array};$$

and

$$\pm p = a - b - c + d - e + f + g - h.$$

Taking the inner product of row ℓ with rows i, j, and k, we obtain, respectively,

$a + c + e + g = b + d + f + h$,

$a + c + f + h = b + d + e + g$,

$a + d + e + h = b + c + f + g$.

Adding these three equations, we have

$3a - 3b + c - d + e - f - g + h = 0;$

and if we add the left-hand expression to the equation for p, we get

$\pm p = 4a - 4b.$

9.3 Profiles

Now considering the inner products of rows i, j, and k, we see that

$$a + b = c + d = e + f = g + h = \frac{1}{4}n; \qquad (9.1)$$

hence $\pm p = n - 8b$ and $p \equiv n \pmod 8$. ∎

We shall write $\pi(m)$ for the number of sets $\{i, j, k, \ell\}$ of four distinct rows such that $p_{ijk\ell} = m$. From the definition and from the remark above, $\pi(m) = 0$ unless $m \geq 0$ and $m \equiv n \pmod 8$. We call $\pi(m)$ the profile (or 4-profile) of H.

THEOREM 9.10   Equivalent Hadamard matrices have the same profile.

Proof: It is clear that p is unaltered by the column equivalence operations. Row negation does not have any effect, as p is an absolute value. Row permutation has the effect of renaming $p_{ijk\ell}$ as p_{qrst} for some q, r, s, t, but it leaves unchanged the totality of all values p. ∎

The next theorem enables us to say that if there are t Hadamard matrices of side n which all have different profiles, then there will be at least t different profiles at side 2n.

THEOREM 9.11   Suppose that H is a Hadamard matrix of order n with profile π. Then the Hadamard matrix

$$G = \begin{bmatrix} H & H \\ H & -H \end{bmatrix}$$

of order 2n has profile σ, where

$$\sigma(2n) = 8\pi(n) + \binom{n}{2}$$

$$\sigma(m) = 8\pi\left(\frac{1}{2}m\right) \text{ if } m \neq 0 \text{ or } 2n,$$

$$\sigma(0) = 8\pi(0) + \binom{2n}{4} - 8\binom{n}{4} - \binom{n}{2}$$

$$= 8\pi(0) + 2n\left(\frac{1}{2}n - 1\right)(n - 1)(n + 3)/3.$$

Proof: It will be convenient to write $\hat{i}$ for $i - n$, and so on. We let $p_{ijk\ell}$ and $s_{ijk\ell}$ be the absolute values of generalized inner products of rows i, j, k, ℓ of H and G, respectively, so that π and σ count the number of times p and s attain various values. We observe that if $i < j < k < \ell \leq n$, then $s_{ijk\ell} = 2p_{ijk\ell}$; \hfill (9.2)

if $n < i < j < k < \ell$, then $s_{ijk\ell} = 2p_{ij\hat{k}\hat{\ell}}$. \hfill (9.3)

Moreover, if $i < j < k \leq n < \ell$, then

$$s_{ijk\ell} = \sum_{x=1}^{2n} g_{ix}g_{jx}g_{kx}g_{\ell x}$$

$$= \sum_{x=1}^{n} g_{ix}g_{jx}g_{kx}g_{\ell x} + \sum_{x=n+1}^{2n} g_{ix}g_{jx}g_{kx}g_{\ell x}$$

$$= \sum_{x=1}^{n} h_{ix}h_{jx}h_{kx}h_{\hat{\ell}x} + \sum_{x=1}^{n} h_{ix}h_{jx}h_{kx}(-h_{\hat{\ell}x}),$$

$$= 0,$$

and similarly in the case $i \leq n < j < k < \ell$. So

if $i < j < k \leq n < \ell$, then $s_{ijk\ell} = 0$; \hfill (9.4)

if $i \leq n < j < k < \ell$, then $s_{ijk\ell} = 0$. \hfill (9.5)

Finally, suppose that $i < j \leq n < k < \ell$. Clearly, $s_{ijk} = 2p_{ij\hat{k}\hat{\ell}}$. Suppose that $i, j, \hat{k}, \hat{\ell}$ are all distinct; as we range through each of the sets $\{i,j,k,\ell\}$ with all elements distinct and $i < j$, $\hat{k} < \hat{\ell}$, we obtain the sets $\{i,j,\hat{k},\hat{\ell}\}$ with $i < j < k < \ell$ six times each. If $i = k$,

$$p_{ij\hat{k}\hat{\ell}} = \sum_{x=1}^{n} h_{ix}h_{jx}h_{ix}h_{\hat{\ell}x}$$

$$= \sum_{x=1}^{n} h_{jx}h_{\hat{\ell}x},$$

which is $4n$ when $j = \hat{\ell}$ and 0 when $j \neq \hat{\ell}$. Similarly, p will be zero when $j = \hat{\ell}$ but $i \neq k$. So:

If $i < j \leq n < k < \ell$, $s_{ijk\ell}$ attains each nonzero value m on

$6\pi(m/2)$ occasions, except that $2n$ occurs $2\pi(n) + \binom{n}{2}$ times. \hfill (9.6)

Combining (9.2) to (9.6), we get the result. ∎

9.3 Profiles

When n is an odd multiple of 4, a further restriction applies.

THEOREM 9.12 If $n = 8t + 4$, where $t \geq 1$, then $p_{ijk\ell}$ never equals n for any $\{i, j, k, \ell\}$.

Proof: Suppose that $p_{ijk\ell} = n$. If we rearrange rows i, j, k, and ℓ as in the proof of Theorem 9.9, we must have $b = c = e = h = 0$; using (9.2), we obtain $a = d = f = g = n/4$.

Now consider a fifth row, row m. Write x, y, z, and t for the number of entries +1 in the first, second, third, and fourth sets of n/4 columns, respectively, in row m. Taking inner products between row m and rows i, j, k, ℓ, we obtain

$$x + y + z + t = n/2$$
$$x + y - z - t = 0$$
$$x - y + z - t = 0$$
$$x - y - z + t = 0,$$

whence $x = y = z = t$, so $4x = n/2$, and n must be divisible by 8. (The only exception, the case $n = 4$, arises because no fifth row is available to use as row m.) ∎

The Hadamard matrices of sides 8 and 12 have profiles

$$\pi(0) = 56, \quad \pi(8) = 14$$

and

$$\pi(4) = 495, \quad \pi(12) = 0$$

respectively. At side 16, there are four profiles: in terms of the classification of Table 9.1, we have

H_1: $\pi(0) = 1680$, $\pi(8) = 0$, $\pi(16) = 140$,
H_2: $\pi(0) = 1488$, $\pi(8) = 256$, $\pi(16) = 76$,
H_3: $\pi(0) = 1392$, $\pi(8) = 484$, $\pi(16) = 44$,
H_4: $\pi(0) = 1344$, $\pi(8) = 448$, $\pi(16) = 28$.

The fifth equivalence class—the set of transposes of matrices in the fourth class—has the same profile as H_4. The three classes at side 20 all give the same profile:

$$\pi(4) = 4560, \quad \pi(12) = 285, \quad \pi(20) = 0.$$

The only other side for which profiles have been extensively investigated is 36. It has been shown that there are more than 100 different profiles at that side, and that is the best published result on the number of equivalence classes of Hadamard matrices of side 36.

EXERCISE

9.3.1 If H is a Hadamard matrix of side n, where $n \geq k$, define

$$p_{i_1, i_2, \ldots, i_k} = \left| \sum_{x=1}^{n} h_{i_1 x} h_{i_2 x} \cdots h_{i_k x} \right|.$$

The k-<u>profile</u> π_k is defined analogously to the profile: $\pi_k(m)$ equals the number of k-sets for which $p = m$. Prove that if $k \geq 4$ and $n = 8t + 4$, then $\pi_k(n) = 0$ (except when $n = 4$ and $k = 4$).

BIBLIOGRAPHIC REMARKS

The determination of h(16) and h(20) is due to Hall [2], [3]; h(24) is evaluated in [4] and [5]. The idea of using integral equivalence as a clue to Hadamard equivalence was introduced in [8]. Lemma 9.5 and Theorem 9.6 and its consequence appeared first in [6]. Profiles were studied in [1], which is where the results on matrices of side 36 appeared.

Results on the equivalence of Hadamard matrices are surveyed in [7].

1. J. Cooper, J. Milas, and W. D. Wallis, "Hadamard equivalence," Combinatorial Mathematics (Lecture Notes in Mathematics 680) (Springer-Verlag, Heidelberg, West Germany, 1978), 126-135.
2. M. Hall, "Hadamard matrices of order 16," JPL Research Summary 36-10, $\underline{1}$(1961), 21-26.
3. M. Hall, Hadamard Matrices of Order 20 (JPL Technical Report 32-761, Pasadena, 1965).
4. N. Ito, J. S. Leon, and J. Q. Longyear, "Classification of 3-(24,12,5) designs and 24-dimensional Hadamard matrices," Journal of Combinatorial Theory $\underline{31A}$(1981), 66-93.
5. H. Kimura, "New Hadamard matrix of order 24," to appear.
6. W. D. Wallis, "Integral equivalence of Hadamard matrices," Israel Journal of Mathematics $\underline{10}$(1971), 359-368.
7. W. D. Wallis, "Hadamard equivalence," Congressus Numerantium $\underline{28}$ (1980), 15-25.
8. W. D. Wallis and J. Wallis, "Equivalence of Hadamard matrices," Israel Journal of Mathematics $\underline{7}$(1969), 122-128.

10
Latin Squares

10.1 LATIN SQUARES AND SUBSQUARES

A <u>Latin square</u> of side n is an n × n array based on some set S of n symbols (treatments), with the property that every row and every column contains every symbol exactly once. In other words, every row and every column is a permutation of S. Since the arithmetical properties of the symbols are not used, the nature of the elements of S is immaterial; unless otherwise specified, we take them to be (1 ·· n).

As an example, suppose that G is a finite group with n elements g_1, g_2, ..., g_n. Define an array $A = (a_{ij})$ by $a_{ij} = k$ where $g_k = g_i g_j$. We look at column j of A. Consider all the elements $g_i g_j$, for different elements g_i. Recall that g_j has an inverse element in G written g_j^{-1}. If i and ℓ are any two integers from 1 to n, then $g_i g_j = g_k g_j$ implies that $g_i g_j g_j^{-1} = g_k g_j g_j^{-1}$, whence $g_i = g_k$, so i = k. Thus the elements $g_1 g_j$, $g_2 g_j$, ..., $g_n g_j$ are all different. This means that the (1, j), (2, j), ..., and (n, j) entries of the array are all different, so column j contains a permutation of (1 ·· n). This is true of every column. A similar proof applies to rows. So A is a Latin square.

This example gives rise to infinitely many Latin squares. For example, at sides 1, 2, and 3 we have

| 1 |

1	2
2	1

1	2	3
2	3	1
3	1	2

.

There is a Latin square of side 3 that looks quite different from the one just given, namely,

$$\begin{array}{|ccc|}\hline 1 & 2 & 3 \\ 3 & 1 & 2 \\ 2 & 3 & 1 \\ \hline\end{array},$$

but this can be converted into the given 3 × 3 square by exchanging rows 2 and 3.

Suppose that A and B are Latin squares of side n. We shall say that A and B are <u>equivalent</u> if it is possible to reorder the rows of A, reorder the columns of A and/or relabel the treatments of A in such a way as to produce the array B. Two obvious questions are: "Are there Latin squares of the same side which are inequivalent?" and "Is there a Latin square of side n which is not equivalent to the square derived from some group of order n?" In both cases the answer is yes, but the smallest inequivalent squares have side 4 and the smallest square that does not arise from a group has side 5. It is not hard to see that the following Latin squares of side 4 are inequivalent:

$$\begin{array}{|cccc|}\hline 1 & 2 & 3 & 4 \\ 4 & 1 & 2 & 3 \\ 3 & 4 & 1 & 2 \\ 2 & 3 & 4 & 1 \\ \hline\end{array} \quad \begin{array}{|cccc|}\hline 1 & 2 & 3 & 4 \\ 2 & 1 & 4 & 3 \\ 3 & 4 & 1 & 2 \\ 4 & 3 & 2 & 1 \\ \hline\end{array}.$$

Both of these can be derived from groups (from the two nonisomorphic groups of order 4). The following Latin square of side 5 cannot be derived from any group; the proof appears in the exercises.

$$\begin{array}{|ccccc|}\hline 1 & 2 & 3 & 4 & 5 \\ 2 & 1 & 5 & 3 & 4 \\ 3 & 4 & 1 & 5 & 2 \\ 4 & 5 & 2 & 1 & 3 \\ 5 & 3 & 4 & 2 & 1 \\ \hline\end{array}.$$

A <u>Latin rectangle</u> of size k × n is a k × n array with entries from (1 ·· n) such that every row is a permutation and the columns contain no repetitions: for example, the first k rows of an n × n Latin square form a Latin rectangle. Clearly, k can be no larger than n.

THEOREM 10.1 If A is a k × n Latin rectangle, then one can append (n - k) further rows to A so that the resulting array is a Latin square.

10.1 Latin Squares and Subsquares

Proof: If $k = n$, the result is trivial. So assume that $k < n$. Write S for $(1 \cdots n)$ and define S_j to be the set of members of S not occurring in column j of A. Each S_j has k elements. Now consider any element i of S. The k occurrences of i in A are in distinct columns, so i belongs to exactly $(n - k)$ of the sets S_j.

Suppose that there were r of the sets S_j whose union contained less than r members, for some r. If we wrote a list of all members of each of these r sets, we would write out $r(n-k)$ symbols, since each S_j is an $(n-k)$-set. On the other hand, we would write at most $r-1$ different symbols, and each symbol occurs in exactly $n-k$ sets, so there are at most $(r-1)(n-k)$ symbols written. Since $k < n$, $(r-1)(n-k) < r(n-k)$—a contradiction. So no set of r of the S_j contain between them less than r symbols. By Theorem 1.3, there is a system of distinct representatives for the S_j. Say the representative of S_j is i_j. Then $(i_1, i_2, \ldots, i_n)$ is a permutation of S. If we append this to A as row $k+1$, we have a $(k+1) \times n$ Lattin rectangle. The process may be repeated. After $(n-k)$ iterations we have a Latin square with A as its first k rows.

A <u>subsquare</u> of side s in a Latin square of side n is an $s \times s$ subarray which is itself a Latin square of side. s. A necessary and sufficient condition that an $s \times s$ subarray be a subsquare is that it contain only s different symbols.

COROLLARY 10.1.1 If L is a Latin square of side s and $n \geq 2s$, then there is a Latin square of side n with L as a subsquare.

Proof: Suppose that $n = s + t$. Define M to be the $s \times t$ array

$$M = \begin{bmatrix} s+1 & s+2 & \cdots & s+t \\ s+2 & s+3 & \cdots & s+1 \\ \vdots & \vdots & & \vdots \\ s+s & s+s+1 & \cdots & s+s-1 \end{bmatrix}$$

with entries reduced modulo t, where necessary, to ensure that they lie in the range $(s+1, s+2, \ldots, s+t)$. Write A for the array constructed by laying row i of M on the right of row i of L, for $1 \leq i \leq s$:

$$A = \boxed{ L M }.$$

Then A is an $s \times n$ Latin recrangle; if we embed it in an $n \times n$ Latin square, that square has L as a subsquare. ■

We shall prove (in Exercise 10.1.3) that this is the best possible result, provided that one ignores the trivial case $s = n$.

EXERCISES

10.1.1 Prove that there are exactly two different Latin squares of side 4, up to equivalence.

10.1.2 Prove that there exist at least

$$n!(n - 1)! \cdots (n - k + 1)!$$

$k \times n$ Latin rectangles and therefore at least

$$n!(n - 1) \cdots 2!1!$$

Latin squares of side n.

10.1.3 Prove that if a Latin square of side n has a subsquare of side s, where $s < n$, then $n \geq 2s$.

10.1.4 If a Latin square A contains a 2×2 subsquare based on the treatments $\{x,y\}$, we say that x and y form an <u>intercalate</u> in A.
  (i) Suppose that A is derived from the multiplication table of a finite group $G = \{g_1, g_2, \ldots, g_n\}$, where g is the identity, and suppose that 1 and 2 form an intercalate in A. Prove that $g_2^2 = g_1$ and consequently that n is even.
  (ii) Prove that if every symbol is a member of an intercalate in a Latin square A, and if A is derived from the multiplication table of a finite group, then A has an even side.
  (iii) Prove that the following Latin square A is not derived from the multiplication table of any finite group:

$$A = \begin{array}{|ccccc|} \hline 1 & 2 & 3 & 4 & 5 \\ 2 & 1 & 5 & 3 & 4 \\ 3 & 4 & 1 & 5 & 2 \\ 4 & 5 & 2 & 1 & 3 \\ 5 & 3 & 4 & 2 & 1 \\ \hline \end{array}.$$

10.2 ORTHOGONALITY

We now discuss the important concept of <u>orthogonality</u>. Two Latin squares A and B of the same side n are called orthogonal if the n^2 ordered pairs (a_{ij}, b_{ij})—the pairs formed by superimposing one square on the other—are all different. We say "A is orthogonal to B" or "B is orthogonal to A"—clearly, the relation of orthogonality is symmetric. More generally, one can speak

10.2 Orthogonality

of a set of k <u>mutually orthogonal</u> Latin squares: squares $A_1, A_2, \ldots, A_k$ such that A_i is orthogonal to A_j whenever $i \neq j$.

A set of n cells in a Latin square is called a <u>transversal</u> if it contains one cell from each row and one cell from each column and if the n entries include every treatment precisely once. Using this idea we get an alternative definition of orthogonality: A and B are orthogonal if and only if for every treatment x in A, the n cells where A has entry x form a transversal in B. The symmetry of the concept is not so obvious when this definition is used.

Latin squares were first defined by Euler in 1782. He discussed orthogonality, and in particular he considered the following old puzzle. Thirty-six military officers wish to parade in a square formation. They represent six regiments, and the six officers from a regiment hold six different ranks (the same six ranks for each regiment). Is it possible for them to parade so that no two officers of the same rank, or the same regiment, are in the same row or column? If the ranks and regiments are numbered 1, 2, 3, 4, 5, 6, define two 6×6 arrays, A and B, by putting a_{ij} and b_{ij} equal to the regiment number and rank number, respectively, of the officers in the (i, j) position. Clearly, the parade formation has the desired properties if and only if A and B are orthogonal Latin squares of side 6.

Euler could not find a solution and concluded that one was impossible. He further conjectured ("Euler's conjecture") that no pair of orthogonal Latin squares of side n can exist when n is congruent to 2 modulo 4. His intuition about the case n = 6 was proven correct by Tarry, who carried out a complete census of Latin squares of side 6, in 1900. However, the rest of Euler's conjecture is spectacularly wrong; Bose, Parker, and Shrikhande proved in 1959-1960 that there is a pair of orthogonal Latin squares of every side greater than 6. Our main purposes in this chapter are to prove this theorem (in Section 10.4) and to prove a similar result concerning sets of three mutually orthogonal squares (in Section 10.6).

Suppose that A is a Latin square on the symbols $(1 \cdots n)$. There is no loss of generality in assuming that the first row is

$$1 \quad 2 \quad 3 \quad \cdots \quad n \tag{10.1}$$

in ascending order—if not, we simply permute the names of symbols. Similarly, if B is orthogonal to A, we can permute the symbols in B so as to achieve first row (10.1), and this does not affect the orthogonality. We say a Latin square with first row (10.1) is <u>standardized</u> or in <u>standard form</u>.

Now suppose that $A_1, A_2, \ldots, A_k$ are Latin squares of side n, each of which is orthogonal to each other one. Without loss of generality, assume that each has been standardized by symbol permutation, so that each has first row (10.1). Assuming that $n > 1$, write a_i for the (2,1) entry in A_i. No a_i can equal 1 (since the first columns can contain no repetition), and the a_i must be different (if $a_i = a_j$, then the n cells which contain a_i in A_i must contain a repetition in A_j—both the $(1, a_i)$ and $(2,1)$ cells of A_j contain a_i). So $\{a_1, a_2, \ldots, a_k\}$ contains k distinct elements of $(2 \cdots n)$, and $k < n$. We have the following result:

THEOREM 10.2 If there are k mutually orthogonal Latin squares of side n, $n > 1$, then $k < n$. ∎

Let us write $N(n)$ for the number of squares in the largest possible set of mutually orthogonal Latin squares of side n. In this notation,

$N(n) \leq n - 1$ if $n > 1$.

If $n = 1$, we can take $A_1 = A_2 = \cdots$ and it makes sense to write "$N(1) = \infty$" in some situations. Whenever a theorem requires the existence of at least k mutually orthogonal Latin squares of side n, the conditions are satisfied by $n = 1$ for any k.

Theorem 10.2 tells us, for example, that $N(4) \leq 3$. In fact, $N(4) = 3$; one set of three mutually orthogonal Latin squares of side 4 is

1	2	3	4
2	1	4	3
3	4	1	2
4	3	2	1

1	2	3	4
3	4	1	2
4	3	2	1
2	1	4	3

1	2	3	4
4	3	2	1
2	1	4	3
3	4	1	2

.

It is easy to see that the upper bound $n - 1$ is attained whenever n is a prime power. If we write

$$GF(n) = \{f_1, f_2, \ldots, f_n\},$$

where f_n is the zero element, and define

$$a_{ij}^h = f_i + f_h f_j,$$

then $A_h = (a_{ij}^h)$ is a Latin square when $1 \leq h \leq n - 1$, and the $n - 1$ Latin squares are orthogonal. Another proof is a corollary of the following theorem.

THEOREM 10.3 $N(n) = n - 1$ if and only if there is a balanced incomplete block design with parameters $(n^2, n^2 + n, n + 1, n, 1)$.

Proof: Suppose that a block design with the stated parameters exists. From Section 5.1 the set of blocks of the design can be partitioned into $n + 1$

10.2 Orthogonality

parallel classes of n blocks each; blocks in the same parallel class have no common elements, while two blocks in different classes have one element in common. Suppose that the parallel classes have been numbered 0, 1, ..., n in some order, and the blocks in class i have been labeled B_{i1}, B_{i2}, ..., B_{in}. Select two parallel classes for reference purposes, say class 0 and class n. Then construct a square array from class i, where $1 \leq i \leq n - 1$, as follows. Find the point of intersection of B_{0x} with B_{ny}. The point will lie on precisely one line of class i. If it lies in B_{ir}, put r in position (x,y) of the array. Column y of the array will contain all the numbers of the lines in class i which contain a point of B_{ny}. Since the points of B_{ny} lie one on each of the lines in class i, the column will contain the elements of (1 $\cdots$ n) in some order. A similar argument applies to rows. So the array is a Latin square.

Write L_i and L_j for the Latin squares obtained from parallel classes i and j, respectively, where $i \neq j$. Those cells where L_i has entry r all come from points on block B_{ir}. The elements of B_{ir} consist of one element of B_{j1}, one from B_{j2}, ..., and one from B_{jn}. So the cells contain 1, 2, ..., n once each. So L_i is orthogonal to L_j.

Conversely, suppose that a set of n - 1 mutually orthogonal Latin squares exists. The construction above is readily reversed to obtain the design. ∎

COROLLARY 10.3.1 If n is a prime power, then $N(n) = n - 1$. ∎

The theorem in fact tells us more: In view of the Bruck-Chowla-Ryser theorem (Theorem 6.14), we know that $N(n) < n - 1$ for infinitely many values of n: 6, 14, and so on.

It is easy to see that the taking of direct products preserves orthogonality—if A_1 is orthogonal to A_2 and B_1 is orthogonal to B_2, then $A_1 \times B_1$ is orthogonal to $A_2 \times B_2$—so

$$N(nr) \geq \min\{N(n), N(r)\}. \tag{10.2}$$

So we have the following result (MacNeish's theorem):

THEOREM 10.4 Suppose that

$$n = p_1^{a_1} p_2^{a_2} \cdots p_r^{a_r},$$

where the p_i are distinct primes and each $a_i \geq 1$. Then

$$n - 1 \geq N(n) \geq \min_i (p_i^{a_i} - 1). \qquad \blacksquare$$

Suppose that there is a set of k mutually orthogonal Latin squares of side n. It does not follow that every Latin square of side n is a member of such a set. In fact, for every even integer side, there exists a Latin square that is not orthogonal to any other square (or "has no orthogonal mate"). This follows from the next theorem.

THEOREM 10.5 Let A be the Latin square of side 2k defined by

$$a_{ij} \equiv i + j \pmod{2k}, \quad 1 \leq a_{ij} \leq 2k.$$

Then A contains no transversal.

Proof: Suppose that cells $(1, j_1), (2, j_2), \ldots, (2k, j_{2k})$ contain a transversal. Then

$$\{a_{1j_1}, a_{2j_2}, \ldots, a_{2kj_{2k}}\} = (1 \cdots 2k),$$

and summing both sides yields

$$a_{1j_1} + a_{2j_2} + \cdots + a_{2kj_{2k}} = k(2k + 1).$$

Now $a_{ij_i} \equiv i + j_i \pmod{2k}$, so

$$(1 + j_1) + (2 + j_2) + \cdots + (2k + j_{2k}) \equiv k(2k + 1) \pmod{2k}. \tag{10.3}$$

Now $j_1, j_2, \ldots, j_{2k}$ is a reordering of $1, 2, \ldots, 2k$, so

$$(1 + j_1) + (2 + j_2) + \cdots + (2k + j_{2k}) = 2(1 + 2 + \cdots + 2k)$$
$$= 2k(2k + 1).$$

Substituting this into (10.3), we get $0 \equiv k \pmod{2k}$, a contradiction. ∎

One can discuss orthogonality of subsquares; if we say that "A and B are orthogonal Latin squares with orthogonal subsquares S and T," we not only imply that S is a subsquare of A and T a subsquare of B, and that S and T are orthogonal, but also that they occupy the same set of cells in A and B.

EXERCISES

10.2.1 Verify that the two definitions of orthogonality of Latin squares given in the text are equivalent.

10.3 Idempotent Latin Squares

10.2.2 Prove that the Latin square A of Exercise 10.1.4 has no orthogonal mate.

10.2.3 The t mutually orthogonal Latin squares $A_1, A_2, \ldots, A_t$ of side n have mutually orthogonal subsquares $S_1, S_2, \ldots, S_t$ occupying their upper left s × s corners. Prove that $n \geq (t+1)s$.

10.2.4 Suppose that A and B are orthogonal Latin squares whose top left-hand corners are r × r subsquares S and T. Prove that S is orthogonal to T.

10.2.5 A Latin square is called <u>self-orthogonal</u> if it is orthogonal to its own transpose.
   (i) Prove that the diagonal elements of a self-orthogonal Latin square of side n must be $1, 2, \ldots, n$ in some order.
   (ii) Prove that there is no self-orthogonal Latin square of side 3.
   (iii) Find self-orthogonal Latin squares of sides 4 and 5.

10.2.6 For any positive integer n, define two matrices

$$R = \begin{bmatrix} 1 & 1 & \cdots & 1 \\ 2 & 2 & \cdots & 2 \\ \vdots & & & \vdots \\ n & n & \cdots & n \end{bmatrix} \quad \text{and} \quad C = \begin{bmatrix} 1 & 2 & \cdots & n \\ 1 & 2 & \cdots & n \\ \vdots & & & \vdots \\ 1 & 2 & \cdots & n \end{bmatrix}$$

Show that an n × n array, based on (1 ·· n), is a Latin square if and only if it is simultaneously orthogonal to R and C.

10.3 IDEMPOTENT LATIN SQUARES

In this section we use pairwise balanced designs to construct sets of orthogonal Latin squares. We say that an n × n Latin square is <u>idempotent</u> if its main diagonal is

$$(1, 2, \ldots, n). \tag{10.4}$$

An idempotent Latin square is often called a transversal square.

LEMMA 10.6 There exists a set of $N(n) - 1$ mutually orthogonal idempotent Latin squares of side n.

Proof: Suppose that $A_1, A_2, \ldots, A_{N(n)}$ are mutually orthogonal Latin squares of side n. By permuting the columns, transform $A_{N(n)}$ so that the element 1 appears in all the diagonal cells. Carry out the same column permutation on the other squares. The diagonals of the first $N(n) - 1$ new squares will be transversals. Now, in each square, permute the names of the treatments so as to produce the diagonal (10.4). We have the required orthogonal idempotent Latin squares. ∎

THEOREM 10.7 Suppose that there exists a PB(v;K;1), and that for every element k of K, there exists a set of t mutually orthogonal idempotent Latin squares of side k. Then there exists a set of t mutually orthogonal idempotent Latin squares of side v.

Proof: For every k in K, suppose that L_k^o is an idempotent Latin square of side k. If B is any ordered set of k integers, write $L_k(B)$ to mean the array derived from L_k by replacing the number 1 throughout by the first element of B, replacing 2 by the second element of B, and so on.

We construct a Latin square A of side v as follows. Select one block B of the pairwise balanced design; suppose that its order is k, and its elements are $b_1, b_2, \ldots, b_k$, where the b_i have been ordered so that $b_1 < b_2 \cdots < b_k$. Then the entry in position (b_i, b_j) of A is the (i,j) entry of $L_k(B)$. This process is carried out on every block of the design in turn.

If x and y are any two integers in the range from 1 to v, then x and y occur together in precisely one block of the PB(v;K;1), and the (x,y) entry of A is allotted when and only when that block is processed. This means that one and only one entry is placed in position (x,y). When we investigate the diagonal entries, we see that position (x,x) is assigned an entry every time a block containing x is processed, but the entry is always x. So the (x,x) entry is x. Thus the process defines an array A, whose diagonal has the proper form.

To prove that A is a Latin square, we observe that as x and y occur together in one block, they will occur together in one of the squares $L_k(B)$. So y will appear in the row of $L_k(B)$ which has diagonal element x, and consequently it will appear in row x of A. Therefore, each row of A contains each of the integers 1, 2, ..., v at least once; as it has only v elements, it must contain each of 1, 2, ..., v exactly once, and be a permutation of (1 ·· v). A similar argument applies to columns. So A is a Latin square.

Suppose that $L_k^1, L_k^2, \ldots, L_k^t$ are the t mutually orthogonal idempotent Latin squares of side k, for each k in K. Write A^i for the Latin square of side v constructed as above, with L_k replaced by L_k^i for every k. It is easy to see that the A^i are orthogonal. ∎

As an example we shall construct a pair of orthogonal idempotent Latin squares of side 17 using the PB(17;{4,5};1) with blocks

01234	159D	16BG	17CE	18AF
05678	26AE	25CF	28BD	279G
09ABC	37BF	389E	35AG	36CD
0DEFG	48CG	47AD	469F	45BE

10.3 Idempotent Latin Squares

and the pairs of orthogonal Latin squares

$$L_4^1 = \begin{array}{|cccc|} \hline 1 & 3 & 4 & 2 \\ 4 & 2 & 1 & 3 \\ 2 & 4 & 3 & 1 \\ 3 & 1 & 2 & 4 \\ \hline \end{array} \;,\quad L_4^2 = \begin{array}{|cccc|} \hline 1 & 4 & 2 & 3 \\ 3 & 2 & 4 & 1 \\ 4 & 1 & 3 & 2 \\ 2 & 3 & 1 & 4 \\ \hline \end{array}$$

of side 4 and

$$L_5^1 = \begin{array}{|ccccc|} \hline 1 & 4 & 2 & 5 & 3 \\ 4 & 2 & 5 & 3 & 1 \\ 2 & 5 & 3 & 1 & 4 \\ 5 & 3 & 1 & 4 & 2 \\ 3 & 1 & 4 & 2 & 5 \\ \hline \end{array} \;,\quad L_5^2 = \begin{array}{|ccccc|} \hline 1 & 3 & 5 & 2 & 4 \\ 5 & 2 & 4 & 1 & 3 \\ 4 & 1 & 3 & 5 & 2 \\ 3 & 5 & 2 & 4 & 1 \\ 2 & 4 & 1 & 3 & 5 \\ \hline \end{array}$$

of side 5.

For each block of the design we construct a pair of Latin squares. For example, from the block 159D we construct two squares by the substitution $(1, 2, 3, 4) \mapsto (1, 5, 9, D)$, obtaining

$$L_4^1(159D) = \begin{array}{|cccc|} \hline 1 & 9 & D & 5 \\ D & 5 & 1 & 9 \\ 5 & D & 9 & 1 \\ 9 & 1 & 5 & D \\ \hline \end{array} \;,\quad L_4^2(159D) = \begin{array}{|cccc|} \hline 1 & D & 5 & 9 \\ 9 & 5 & D & 1 \\ D & 1 & 9 & 5 \\ 5 & 9 & 1 & D \\ \hline \end{array} .$$

Then these elements are placed in 17×17 squares A^1 and A^2, in all the positions (x, y) with x and y in $\{1, 5, 9, D\}$. Similarly, from the block $\{05678\}$ we construct the two squares

$$L_5^1(05678) = \begin{array}{|ccccc|} \hline 0 & 7 & 5 & 8 & 6 \\ 7 & 5 & 8 & 6 & 0 \\ 5 & 8 & 6 & 0 & 7 \\ 8 & 6 & 0 & 7 & 5 \\ 6 & 0 & 7 & 5 & 8 \\ \hline \end{array} \;,\quad L_5^2(05678) = \begin{array}{|ccccc|} \hline 0 & 6 & 8 & 5 & 7 \\ 8 & 5 & 7 & 0 & 6 \\ 7 & 0 & 6 & 8 & 5 \\ 6 & 8 & 5 & 7 & 0 \\ 5 & 7 & 0 & 6 & 8 \\ \hline \end{array}$$

and from $\{47AD\}$ we get

$$L_4^1(47AD) = \begin{array}{|cccc|} \hline 4 & A & D & 7 \\ D & 7 & 4 & A \\ 7 & D & A & 4 \\ A & 4 & 7 & D \\ \hline \end{array} \;,\quad L_4^2(47AD) = \begin{array}{|cccc|} \hline 4 & D & 7 & A \\ A & 7 & D & 4 \\ D & 4 & A & 7 \\ 7 & A & 4 & D \\ \hline \end{array}.$$

The result of using these squares is shown in Figure 10.1 The other 17 blocks give rise to 34 other squares, and when they are all used we end with the Latin squares in Figure 10.2.

Let us write I(n) for the maximal cardinality of a set of mutually orthogonal idempotent Latin squares of side n. Using Lemma 10.6, we see that for every n,

$$N(n) - 1 \leq I(n) \leq N(n). \tag{10.5}$$

If n is a prime or prime power, equality holds on the left-hand side of (10.5). The only known case where $I(n) = N(n)$ is

$$I(6) = N(6) = 1.$$

[We know that $N(6) = 1$, so $I(6) = 1$ or 0. Proving the existence of an idempotent Latin square of side 6 is left as an exercise.] As easy generalizations of the results of Section 10.2, we have

THEOREM 10.8 $I(n) \leq n - 2$, with equality when n is a prime power. ∎

```
0         7 5 8 6                    0         6 8 5 7
   1      9       D     5               1      D       5       9

          4     A     D     7                  4     D       7       A
7 D       5 8 6 0 1         9        8 9       5 7 0 6 D             1
5         8 6 0 7                    7         0 6 8 5
8         D 6 0 7 5   4     A        6         A 8 5 7 0     D       4
6         0 7 5 8                    5         7 0 6 8
   5      D         9       1          D       1           9         5
          7       D       A     4               D       4       A       7

9         A 1   4   5 7     D        5         7 9   A   1 4         D
```

Figure 10.1

10.3 Idempotent Latin Squares

```
0 3 1 4 2 7 5 8 6 B 9 C A F D G E
3 1 4 2 0 9 B C A D F G E 5 7 8 6
1 4 2 0 3 C A 9 B G E D F 8 6 5 7
4 2 0 3 1 A C B 9 E G F D 6 8 7 5
2 0 3 1 4 B 9 A C F D E G 7 5 6 8
7 D F G E 5 8 6 0 1 3 4 2 9 B C A
5 G E D F 8 6 0 7 4 2 1 3 C A 9 B
8 E G F D 6 0 7 5 2 4 3 1 A C B 9
6 F D E G 0 7 5 8 3 1 2 4 B 9 A C
B 5 7 8 6 D F G E 9 C A 0 1 3 4 2
9 8 6 5 7 G E D F C A 0 B 4 2 1 3
C 6 8 7 5 E G F D A 0 B 9 2 4 3 1
A 7 5 6 8 F D E G 0 B 9 C 3 1 2 4
F 9 B C A 1 3 4 2 5 7 8 6 D G E 0
D C A 9 B 4 2 1 3 8 6 5 7 G E O F
G A C B 9 2 4 3 1 6 8 7 5 E O F D
E B 9 A C 3 1 2 4 7 5 6 8 0 F D G
```

A^1

```
0 2 4 1 3 6 8 5 7 A C 9 B E G D F
4 1 3 0 2 D G E F 5 8 6 7 9 C A B
3 0 2 4 1 F E G D 7 6 8 5 B A C 9
2 4 1 3 0 G D F E 8 5 7 6 3 9 B A
1 3 0 2 4 E F D G 6 7 5 8 A B 9 C
8 9 C A B 5 7 0 6 D G E F 1 4 2 3
7 B A C 9 0 6 8 5 F E G D C 2 4 1
6 C 9 B A 8 5 7 0 G D F E 4 1 3 2
5 A B 9 C 7 0 6 8 E F D G 2 3 1 4
C D G E F 1 4 2 3 9 B 0 A 5 8 6 7
B F E G D 3 2 4 1 0 A C 9 7 6 8 5
A G D F E 4 1 3 2 C 9 B 0 8 5 7 6
9 E F D G 2 3 1 4 B 0 A C 6 7 5 8
G 5 8 6 7 9 C A B 1 4 2 3 D F O E
F 7 6 8 5 B A C 9 3 2 4 1 0 E G D
E 8 5 7 6 C 9 B A 4 1 3 2 G D F O
D 6 7 5 8 A B 9 C 2 3 1 4 F O E G
```

A^2

Figure 10.2

THEOREM 10.9 For every m and n, $I(mn) \geq \min(I(m), I(n))$. If n has prime power decomposition

$$n = p_1^{a_1} p_2^{a_2} \cdots p_k^{a_k},$$

then $I(n) \geq \min(p_i^{a_i} - 1) - 1$. ∎

From Theorem 10.7 we have

COROLLARY 10.7.1 If there exists a PB(v;K;1), then:

(i) if $I(k) \geq t$ for all $k \in K$, then $I(v) \geq t$;
(ii) if $N(k) \geq t$ for all $k \in K$, then $N(v) \geq t - 1$. ∎

EXERCISES

10.3.1 Construct an idempotent Latin square of side 6.

10.3.2 Use the PB(10;{4,3};1) with blocks

$$B_1 = 1348, \quad B_4 = 127, \quad B_7 = 159, \quad B_{10} = 160,$$

$$B_2 = 2568, \quad B_5 = 369, \quad B_8 = 230, \quad B_{11} = 249,$$

$$B_3 = 7890, \quad B_6 = 450, \quad B_9 = 467, \quad B_{12} = 357,$$

to construct a Latin square of side 10 with diagonal
(1, 2, 3, 4, 5, 6, 7, 8, 9, 0).

10.3.3 Self-orthogonal Latin squares were defined in Exercise 10.2.6.
  (i) Modify the proof of Theorem 10.7 to show that if there is a
      PB(v;K;1) and if there is a self-orthogonal Latin square of side
      k for every k ∈ K, there is a self-orthogonal Latin square of
      side v.
  (ii) Construct a self-orthogonal Latin square of side 17.

10.4 TRANSVERSAL DESIGNS

A **simple group divisible design** or SGDD is constructed from a pairwise balanced design with $\lambda = 1$ by selecting a **parallel class**—a set of blocks which between them contain every treatment exactly once—and deleting those blocks. (Of course, not every pairwise balanced design with $\lambda = 1$ contains a parallel class, so they cannot all be used to construct SGDDs.) The deleted blocks are called "groups"; thus two treatments either determine a common block or a common group, but not both. An SGDD whose sets of treatments, groups, and blocks are X, $\mathcal{G}$, and $\mathcal{A}$, respectively, will sometimes simply be denoted by the triple (X, $\mathcal{G}$, $\mathcal{A}$).

By TD(k, n) we denote a **transversal design** or **transversal system**, which is an SGDD with uniform block size k, with uniform group size n, and with k groups. It follows that each block is a **transversal** of the groups: Each block contains precisely one element from each group. The number of blocks is n^2, and there are kn treatments. It will be convenient to accept the existence of a (degenerate) TD(k, 0) with no points, no groups, and k empty blocks.

Suppose that groups G_1 and G_2 are selected from a TD(k, n). Given x belonging to G_1 and y belonging to G_2, there will be exactly one block containing both. Letting y range through G_2, we see that x lies on exactly n blocks (since every block must contain one member of G_2); and as x ranges through G_1, we see that the TD(k, n) has exactly n^2 blocks.

THEOREM 10.10 If $n \geq 2$ and there exists a TD(k, n), then $k \leq n + 1$. If $k = n + 1$, any two blocks have a point of intersection. If $k \leq n$, every block has another block disjoint from it.

Proof: Consider any block A of a TD(k, n), $n \geq 2$. Through each of its k points, there pass n - 1 other blocks, so the number of blocks that meet A is k(n - 1), and there are $n^2 - k(n - 1) - 1$ blocks disjoint from it. If $k > n + 1$, this number is negative, a contradiction; if $k = n + 1$, it is zero; if $k \leq n$, it is positive. ∎

The relationship between transversal designs and Latin squares is very important:

10.4 Transversal Designs

THEOREM 10.11 The existence of a set of $k-2$ mutually orthogonal Latin squares of order n is equivalent to the existence of a TD(k,n).

Proof: Suppose that a TD(k,n) with groups $G_1, G_2, \ldots, G_k$ is given. Relabel the elements so that

$$G_h = \{x_{h1}, x_{h2}, \ldots, x_{hn}\}.$$

Then for every h, i, and j with $1 \leq h \leq k-2$ and $1 \leq i, j \leq n$, define

$$a_{ij}^h = m,$$

where m is the integer such that x_{hm} is the (unique) member of G_h in the (unique) block of the TD(k,n) which contains both $x_{k-1,i}$ and $x_{k,j}$. It is easy to verify that the array A_h with (i,j) element a_{ij}^h is a Latin square of side n, and that $A_1, A_2, \ldots, A_{k-2}$ are orthogonal. The construction may be reversed. ∎

Suppose that a TD(k,n) exists. Then it is easy to see that a PG(k;{k,n};1) exists: one simply takes a pairwise balanced design whose blocks are the blocks and the groups of the transversal design. By adding a common further element to all the blocks of size n, one obtains a PG(kn + 1; {k, n + 1}). So we have:

COROLLARY 10.11.1 If there exist $k-2$ pairwise orthogonal Latin squares of side n, then there exist a PB(kn,{k,n},1) and a PB(kn + 1, {k, n + 1}, 1). ∎

In particular, the existence of a Latin square of side n for every positive integer n proves that there is a PB(3n,{3,n},1) for every n, a fact that was foreshadowed in Section 2.1.

Theorem 10.11 shows that in order to construct orthogonal Latin squares, we may proceed by constructing transversal designs. Our main tool is the following theorem.

THEOREM 10.12 Suppose that $(X, \mathcal{G}, \mathcal{A})$ is a TD(k + ℓ, t) with groups G_1, $G_2, \ldots, G_k, H_1, H_2, \ldots, H_\ell$. Let S be any given subset of $H_1 \cup H_2 \cdots \cup H_\ell$ and m any given nonnegative integer. Suppose that there exist transversal designs of the following kinds:

(i) for each $i = 1, 2, \ldots, \ell$, a TD(k, h_i), where $h_i = |S \cap H_i|$;
(ii) for each block $A \in \mathcal{A}$, a TD(k, m + u_A) that contains a set of u_A pairwise disjoint blocks, where $u_A = |S \cap A|$.

Then there exists a TD(k, mt + u), where u = |S|.

Proof: We first set up some notation. We set $X_0 = G_1 \cup G_2 \cup \cdots \cup G_k$, so that S is a subset of $X \backslash X_0 = H_1 \cup H_2 \cup \cdots \cup H_\ell$; and we write A' for $A \cap S$, write M for some m-set, and write K for $\{1, 2, \ldots, k\}$. We shall construct a transversal design whose treatments are ordered pairs: the elements of $X_0 \cup M$ and $S \times K$.

For each $A \in \mathcal{A}$, there exists a TD(k, $m + u_A$) with u_A disjoint blocks. For each of those u_A blocks allocate an element of A'; in the block that receives the element p, relabel the elements as

$$\{(p,1),\ (p,2),\ \ldots,\ (p,k)\},$$

where (p, i) is the element in group i. Then relabel the remaining elements of the design in such a way that the other m elements of the ith group are $\{(d, s) : s \in M\}$, where d is the member of A in G_i. We denote by $\mathcal{B}(A)$ the set of blocks in the design <u>other than</u> the u_A disjoint blocks.

For each set $S \cap H_i$, $i \in (1 \cdots \ell)$, there is a corresponding TD(k, h_i), where $h_i = |S \cap H_i|$. Take the elements of the first group in this design to be the elements (p, j), where p ranges through $S \cap H_i$.

We now glue together these ingredients. We obtain a design with elements, groups, and blocks $(X', \mathcal{G}', \mathcal{A}')$, where

$$X' = (X_0 \times M) \cup (S \times K);$$

$$\mathcal{G}' = (G_1', G_2', \ldots, G_k'), \quad G_i' = (G_i \times M) \cup (S \times \{i\});$$

$$\mathcal{A}' = \left[\bigcup_{A \in \mathcal{A}} \mathcal{B}(A)\right] \cup \mathcal{C}_1 \cup \mathcal{C}_2 \cup \cdots \cup \mathcal{C}_\ell.$$

We now prove that this design is a TD(k, mt + u). Clearly, $\mathcal{G}'$ is a partition of X' into (mt + u)-sets, and in each block the treatment that belonged to the ith group in the relevant component design is a member of G_i'. This makes it clear that the blocks are transversals of the G_i'. We show that any two members x and y of different blocks lie in a unique common block, which completes the proof of the theorem. We distinguish four cases.

<u>Case 1.</u> x = (a, s) and y = (b, t), where $a \in G_i$, $b \in G_j$, $s \in M$, and $t \in M$. Clearly, neither x nor y is in any block of any $\mathcal{C}_r$. There is a unique block of the TD(k + ℓ, t) which contains both a and b. Call this block A. Then there is a unique block of $\mathcal{B}(A)$ that contains both x and y; if $D \neq A$, no block of $\mathcal{B}(D)$ contains both.

<u>Case 2.</u> x = (a, s) and y = (θ, j), where $a \in G_i$, $s \in M$, $\theta \in S$, and $1 \leq j \leq k$ but $j \neq i$. Again x is not in any $\mathcal{C}_r$. Let A be the unique member of $\mathcal{A}$ that contains both a and θ. Then $\mathcal{B}(A)$ contains a unique block with both x and y; again, if $D \neq A$, no block of $\mathcal{B}(D)$ contains both.

<u>Case 3.</u> x = (θ, i) and y = (φ, j) where $\theta \in S \cap H_p$, $\varphi \in S \cap H_q$ ($q \neq p$), and $1 \leq i < j \leq k$. Since $p \neq q$, no block of any $\mathcal{C}_r$ contains both x and y. As in case 1, there is exactly one block in one $\mathcal{B}(A)$ that contains both.

10.4 Transversal Designs

<u>Case 4.</u> $x = (\theta, i)$ and $y = (\varphi, j)$ where $\theta, \varphi \in S \cap H_p$ and $1 \le i < j \le k$.
No block in any $\mathcal{B}(A)$ contains both x and y, since no A contains both θ and φ.
Since $\theta \notin H_r$ when $r \ne p$, no $\mathcal{C}_r$ except $\mathcal{C}_p$ can contain them; and $\mathcal{C}_p$ has x and y together in precisely one block. ∎

COROLLARY 10.12.1 If $0 \le u \le t$, then

$$N(mt + u) \ge \min\{N(m), N(m+1), N(t) - 1, N(u)\}.$$

Proof: Write $k - 2$ for the minimum on the right-hand side. Then

$N(m) \ge k - 2 \implies TD(k, m)$ exists;
$N(m+1) \ge k - 2 \implies TD(k, m+1)$ exists;
$N(t) - 1 \ge k - 2 \implies TD(k+1, t)$ exists;
$N(u) \ge k - 2 \implies TD(k, u)$ exists.

Consider Theorem 10.12 with $\ell = 1$ and with S any u-subset of the special group H_1. For any block A, $u_A = 0$ or $u_A = 1$. So the required $TD(k + \ell, t)$ is a $TD(k+1, t)$, which exists; the $TD(k, h_i)$ boil down to $TD(k, u)$, which exists; the $TD(k, m + u_A)$ are either $TD(k, m)$ or $TD(k, m+1)$, which exist (the requirement "a set of one disjoint block" is clearly vacuous). Therefore, $TD(k, mt + u)$ exists, and $N(mt + u) \ge k - 2$, as required. ∎

COROLLARY 10.12.2 If $0 \le u, v \le t$, then

$$N(mt + u + v) \ge \min\{N(m), N(m+1), N(m+2), N(t) - 2, N(u), N(v)\}.$$

Proof: Again, let $k - 2$ be the minimum of the right-hand side. We know the following designs exist:

$TD(k, m)$, $TD(k, m+1)$, $TD(k, m+2)$, $TD(k+2, t)$, $TD(k, u)$, $TD(k, v)$.

As in Corollary 10.12.1, we have all the ingredients—this time, for the case $\ell = 2$. However, we require that the $TD(k, m+2)$ have a pair of disjoint blocks. But $k - 2 \le N(m)$, so certainly $k - 2 \le m - 1$, $k \le m + 1$, and Theorem 10.11 assures us of disjoint blocks. ∎

We are now in a position to prove that the Euler conjecture is wrong for all n greater than 6.

THEOREM 10.3 There exists a pair of orthogonal Latin squares of side n for all $n \ge 6$.

Proof: From Theorem 10.4 there will exist a pair of orthogonal Latin squares of side n unless $n \equiv 2 \pmod 4$. We can treat nearly all other cases using Corollary 10.10.1 with $m = 3$. Essentially, we work modulo 36. Table 10.3 shows, for every $v \equiv 2 \pmod 4$, a representation in terms of its residue modulo 36, a value of t and a value of u. In each case, t is prime to 6, so $N(t) \ge 3$. Also, $N(n) \ge 2$. So $N(v) \ge 2$ provided that $t \ge u$. The only cases

Table 10.3

v:	36d+2	36d+6	36d+10	36d+14	36d+18	36d+22	36d+26	36d+30	36d+34
t:	12d−1	12d+1	12d+1	12d+1	12d+5	12d+7	12d+7	12d+7	12d+11
u:	5	3	7	11	3	1	5	9	1

where $t < u$ are $v = 2, 6, 10, 14$, and 30. We know that $N(2) = N(6) = 1$. Latin squares of sides 10 and 14 are presented in Table 10.4; each of these squares is self-orthogonal orthogonal to its own transpose), so $N(10) \geq 2$ and $N(14) \geq 2$. Finally, $30 = 10 \times 3$. So $N(30) \geq 2$, by (10.2). ∎

EXERCISES

10.4.1 We construct a TD(6,46). Write X for $\{x_i : c = 7, 12, 10, 33, 9, 26, 34, 16, 1\}$. (The nine values of c are the quartic residues—perfect fourth powers—modulo 37.) Z is Z_{37}. If Y is any set, Y^i denotes Y with a superscript i appended to each element. (Multiplicative powers are not used in the construction.) The TD(6,46) has $Z^i \cup X^i$ for its ith group. The blocks consist of all the blocks of a TD(6,9) with ith group X^i, together with the blocks

$\{x_c^1, (c+g)^2, (13c+g)^3, (21c+g)^4, (14c+g)^5, (34c+g)^6\}$,

$\{(34c+g)^1, x_c^2, (c+g)^3, (13c+g)^4, (21c+g)^5, (14c+g)^6\}$,

$\{(14c+g)^1, (34c+g)^2, x_c^3, (c+g)^4, (13c+g)^5, (21c+g)^6\}$,

$\{(21c+g)^1, (14c+g)^2, (34c+g)^3, x_c^4, (c+g)^5, (13c+g)^6\}$,

$\{(13c+g)^1, (21c+g)^2, (14c+g)^3, (34c+g)^4, x_c^5, (c+g)^6\}$,

$\{(c+g)^1, (13c+g)^2, (21c+g)^3, (14c+g)^4, (34c+g)^5, x_c^6\}$

for all $g \in Z$ and all $x_c \in X$. Prove that the design is a TD(6,46).

10.4.2 Two tennis clubs, each with k members, wish to play a competition in which doubles matches are played, with two players from one club opposing two players from the other club in each match, subject to the following conditions:

10.4 Transversal Designs

Table 10.4

```
0 3 9 7 8 1 2 5 6 4
1 6 0 9 2 3 5 4 7 8
2 5 1 6 9 4 0 7 8 3
3 4 7 5 1 9 8 6 2 0
4 0 8 2 7 5 9 3 1 6
5 8 6 3 4 2 7 9 0 1
6 7 3 1 0 8 4 2 9 5
9 1 2 0 5 6 3 8 4 7
8 9 5 4 6 7 1 0 3 2
7 2 4 8 3 0 6 1 5 9
```

Side 10

```
 1  9  4 13 10  3  6 11  7 12  2  5 14  8
14  2 10  5  1 11  4  7 12  8 13  3  6  9
 7 14  3 11  6  2 12  5  8 13  9  1  4 10
 5  8 14  4 12  7  3 13  6  9  1 10  2 11
 3  6  9 14  5 13  8  4  1  7 10  2 11 12
12  4  7 10 14  6  1  9  5  2  8 11  3 13
 4 13  5  8 11 14  7  2 10  6  3  9 12  1
13  5  1  6  9 12 14  8  3 11  7  4 10  2
11  1  6  2  7 10 13 14  9  4 12  8  5  3
 6 12  2  7  3  8 11  1 14 10  5 13  9  4
10  7 13  3  8  4  9 12  2 14 11  6  1  5
 2 11  8  1  4  9  5 10 13  3 14 12  7  6
 8  3 12  9  2  5 10  6 11  1  4 14 13  7
 9 10 11 12 13  1  2  3  4  5  6  7  8 14
```

Side 14

198   10. Latin Squares

(a) Every player must play against every member of the other club exactly once.
(b) Two members of the same club can be partners in at most one match.

Prove that k must be even. If k = 2n, prove that there is a solution provided that N(n) ≥ 2.

10.4.3 For the tournament in Exercise 10.4.2, prove that there is no solution when n = 2. Is there a solution when n = 6?

10.4.4 Suppose that a (v, b, r, k, 1)-design exists. One treatment is deleted from the design and the blocks that previously contained it are interpreted as groups.
  (i) Prove that the result is a simple group-divisible design with constant group size and block size.
  (ii) Can the design ever be a transversal design? If so, when? If not, why not?

10.5 SPOUSE-AVOIDING MIXED-DOUBLES TOURNAMENTS

In our discussion of sets of three orthogonal Latin squares we are led to consider another type of design, which has some interest in its own right because it is one of several types of apparently unrelated designs which are related to orthogonal Latin squares.

A spouse-avoiding mixed-doubles tournament is an arrangement for couples to play mixed-doubles tennis so that no player is partnered by, or opposes, his or her spouse; otherwise, every player has each other player as an opponent exactly once and has each other player of the opposite sex as a partner exactly once. The tournament is sharply resolvable if its matches are arranged into rounds such that either every player takes part in every round or else precisely one couple sits out of each round (according as the number of couples is even or odd). A sharply resolvable spouse-avoiding mixed-doubles tournament for n couples is denoted SR(n).

We write [A, B | C, D] to denote a match in which Mr. A and Mrs. B play against Mr. C and Mrs. D. (Observe that [A, B | C, D] and [C, D | A, B] are two ways of denoting the same match.) Given an SR(n) whose couples are labeled 1, 2, ..., n, we construct two n × n arrays, L and S, as follows: If n is odd, reorder the rounds so that couple i sits out of round i. Set $s_{ii} = i$ if n odd and $s_{ii} = n$ if n even. Also, $\ell_{ii} = i$ for all n. If [i, j | k, r] occurs in round m, then set $\ell_{ik} = j$, $\ell_{ki} = r$, and $s_{ik} = s_{ki} = m$. Then L and S are orthogonal Latin squares. Moreover, L is self-orthogonal (orthogonal to its transpose) and S is symmetric. So S, L, and L^T comprise three pairwise orthogonal Latin squares of order n.

As first examples of these designs, we exhibit an SR(4) and an SR(8). The SR(4) has rounds:

10.5 Spouse-Avoiding Mixed-Doubles Tournaments

$[4,1|2,3]$, $[1,4|3,2]$
$[4,2|3,1]$, $[2,4|1,3]$
$[4,3|1,2]$, $[3,4|2,1]$.

The SR(8) has rounds:

$[1,4|2,7]$, $[3,1|4,5]$, $[5,3|6,8]$, $[7,6|8,2]$
$[1,7|3,6]$, $[2,5|7,3]$, $[4,2|5,8]$, $[6,1|8,4]$
$[1,6|4,8]$, $[2,4|8,5]$, $[3,7|5,2]$, $[6,3|7,1]$
$[1,3|5,4]$, $[2,8|6,7]$, $[3,5|8,6]$, $[4,1|7,2]$
$[1,5|6,2]$, $[2,1|3,8]$, $[4,3|8,7]$, $[5,6|7,4]$
$[1,8|7,5]$, $[2,3|4,6]$, $[3,2|6,4]$, $[5,7|8,1]$
$[1,2|8,3]$, $[2,6|5,1]$, $[3,4|7,8]$, $[4,7|6,5]$.

THEOREM 10.14 There exists an SR(n) whenever n is prime to 6.

Proof: Interpret the labels 1, 2, ..., n as integers modulo n. Then the matches

$$\{[3i,\ i+2j|3j,\ 2i+j] : 0 \leq i < j \leq n\}$$

form a spouse-avoiding mixed-doubles tournament, and for every a the matches with $i = a + k$ and $j = a - k$, $0 \leq k \leq n - 1$, that is, the matches

$$\{[3a + 3k,\ 3a - k|3a - 3k,\ 3a + k] : 0 < k \leq (n - 1)/2\},$$

form a round in which couple 3a is omitted. ■

THEOREM 10.15 If there is an SR(n), where n is even, then there is an SR(4n).

Proof: Suppose that the SR(n) has couples labeled 1, 2, ..., n and rounds labeled R(1), R(2), ..., R(n-1). We construct an SR(4n) with couples labeled $1_1, 2_1, \ldots, n_1, 1_2, 2_2, \ldots, n_4$. Define $R(i)_j$ to be the set of all matches $[a_j, b_j | c_j, d_j]$ such that $[a, b | c, d]$ is a match in R(i). If $R^*(i) = R(i)_1 \cup R(i)_2 \cup R(i)_3 \cup R(i)_4$, then $R^*(i)$ contains every player exactly once. We shall take the $R^*(i)$, $1 \leq i \leq n - 1$, as rounds in the tournament of size 4n; these rounds precisely cover the requirements for meetings between players whose labels have the same subscript.

Table 10.5

SR(4) in SR(18):

I_1 = [1, 0 | 3, A_1] J_1 = [0, 8 | A_1, 13]
H_1 = [0, 1 | 7, 8] I_2 = [8, 6 | 11, A_2] J_2 = [4, 9 | A_2, 7]
H_2 = [0, 7 | 1, 4] I_3 = [5, 2 | 9, A_3] J_3 = [6, 10 | A_3, 4]
F_1 = [10, 12 | 2, 11] I_4 = [7, 3 | 12, A_4] J_4 = [13, 5 | A_4, 1]

SR(4) in SR(22):

I_1 = [1, 0 | 3, A_1] J_1 = [7, 9 | A_1, 14]
I_2 = [8, 6 | 11, A_2] J_2 = [0, 3 | A_2, 5]
H_1 = [0, 1 | 9, 10] I_3 = [16, 13 | 2, A_3] J_3 = [12, 4 | A_3, 16]
H_2 = [1, 6 | 0, 9] I_4 = [5, 1 | 10, A_4] J_4 = [17, 11 | A_4, 15]
F_1 = [9, 17 | 15, 10] F_2 = [6, 12 | 13, 2] F_3 = [14, 7 | 4, 8]

SR(4) in SR(26):

H_1 = [0, 1 | 11, 12] I_1 = [1, 0 | 3, A_1] J_1 = [2, 6 | A_1, 9]
H_2 = [0, 9 | 1, 18] I_2 = [13, 11 | 16, A_2] J_2 = [4, 7 | A_2, 14]
F_1 = [15, 8 | 21, 4] I_3 = [6, 3 | 10, A_3] J_3 = [17, 1 | A_3, 15]
F_2 = [5, 17 | 12, 19] I_4 = [9, 5 | 14, A_4] J_4 = [18, 20 | A_4, 21]
F_3 = [0, 10 | 8, 16] F_4 = [11, 2 | 20, 12] F_5 = [19, 13 | 7, 18]

SR(4) in SR(30):

I_1 = [1, 0 | 3, A_1] J_1 = [4, 6 | A_1, 17]
H_1 = [0, 1 | 13, 14] I_2 = [17, 15 | 20, A_2] J_2 = [19, 22 | A_2, 1]
H_2 = [0, 7 | 1, 10] I_3 = [24, 21 | 2, A_3] J_3 = [23, 3 | A_3, 12]
F_1 = [8, 23 | 16, 7] I_4 = [6, 2 | 11, A_4] J_4 = [28, 8 | A_4, 10]
F_2 = [13, 5 | 22, 9] F_3 = [25, 11 | 10, 4] F_4 = [14, 19 | 7, 18]
F_3 = [1, 20 | 17, 12] F_6 = [0, 14 | 12, 20] F_7 = [15, 25 | 9, 13]

10.5 Spouse-Avoiding Mixed-Doubles Tournaments

Table 10.5 (continued)

SR(8) in SR(34):

I_1 = [1, 0 | 2, A_1] J_1 = [0, 13 | A_1, 3]
I_2 = [4, 1 | 6, A_2] J_2 = [7, 14 | A_2, 22]
H_1 = [0, 1 | 13, 14] I_3 = [8, 4 | 11, A_3] J_3 = [13, 17 | A_3, 5]
H_2 = [0, 15 | 7, 16] I_4 = [17, 12 | 21, A_4] J_4 = [14, 16 | A_4, 12]
H_3 = [8, 5 | 9, 20] I_5 = [12, 6 | 20, A_5] J_5 = [16, 19 | A_5, 23]
H_4 = [0, 12 | 11, 9] I_6 = [18, 10 | 23, A_6] J_6 = [19, 25 | A_6, 18]
F_1 = [22, 15 | 10, 24] I_7 = [3, 20 | 9, A_7] J_7 = [24, 8 | A_7, 2]
I_8 = [5, 21 | 15, A_8] J_8 = [25, 7 | A_8, 9]

SR(8) in SR(38):

I_1 = [1, 0 | 2, A_1] J_1 = [26, 28 | A_1, 29]
H_1 = [0, 1 | 15, 16] I_2 = [4, 1 | 6, A_2] J_2 = [17, 20 | A_2, 2]
H_2 = [0, 5 | 9, 24] I_3 = [9, 5 | 12, A_3] J_3 = [11, 15 | A_3, 19]
H_3 = [0, 13 | 11, 20] I_4 = [19, 14 | 23, A_4] J_4 = [27, 3 | A_4, 8]
H_4 = [0, 12 | 23, 9] I_5 = [10, 4 | 16, A_5] J_5 = [18, 25 | A_5, 17]
F_1 = [29, 27 | 13, 6] I_6 = [20, 12 | 25, A_6] J_6 = [15, 23 | A_6, 7]
F_2 = [7, 21 | 24, 11] I_7 = [22, 13 | 0, A_7] J_7 = [14, 24 | A_7, 26]
F_3 = [3, 22 | 21, 9] I_8 = [28, 18 | 8, A_8] J_8 = [5, 16 | A_8, 10]

SR(8) in SR(42):

I_1 = [1, 0 | 2, A_1] J_1 = [16, 21 | A_1, 28]
I_2 = [3, 1 | 6, A_2] J_2 = [0, 8 | A_2, 31]
H_1 = [0, 1 | 17, 18] I_3 = [5, 2 | 10, A_3] J_3 = [18, 25 | A_3, 6]
H_2 = [0, 23 | 15, 10] I_4 = [19, 5 | 11, A_4] J_4 = [24, 26 | A_4, 27]
H_3 = [0, 4 | 21, 13] I_5 = [21, 15 | 28, A_5] J_5 = [12, 22 | A_5, 18]
H_4 = [0, 16 | 25, 11] I_6 = [26, 19 | 30, A_6] J_6 = [19, 30 | A_6, 33]
F_1 = [4, 16 | 15, 24] I_7 = [23, 14 | 29, A_7] J_7 = [31, 10 | A_7, 12]
F_2 = [25, 9 | 7, 29] I_8 = [14, 4 | 22, A_8] J_8 = [8, 11 | A_8, 17]
F_3 = [13, 32 | 33, 20] F_4 = [32, 13 | 20, 3] F_5 = [27, 7 | A , 23]

Now define

$$S^1(i) = \{[a_1,b_2|c_3,d_4], [a_2,b_1|c_4,d_3] : [a,b|c,d] \in R(i)\};$$

$$S^2(i) = \{[a_1,b_3|c_4,d_2], [a_3,b_1|c_2,d_4] : [a,b|c,d] \in R(i)\};$$

$$S^3(i) = \{[a_1,b_4|c_2,d_3], [a_4,b_1|c_3,d_2] : [a,b|c,d] \in R(i)\}.$$

The $3n - 3$ rounds $S^j(i)$, for $j = 1, 2, 3$, and $1 \leq i \leq n - 1$, cover all the remaining requirements except those where both players have symbols k_i and k_j, where $i \neq j$. (In verifying this, bear in mind that each match in $R(i)$ gives rise to <u>four</u> matches in $S^j(i)$; for example, in $S^1(i)$, $[a,b|c,d]$ gives rise not only to $[a_1,b_2|c_3,d_4]$ and $[a_2,b_1|c_4,d_3]$, but also to $[c_1,d_2|a_3,b_4]$ and $[c_2,d_1|a_4,b_3]$, because $[c,d|a,b] = [a,b|c,d]$.)

Finally, write

$$T^1 = \{[i_1,i_2|i_3,i_4], [i_2,i_1|i_4,i_3] : 1 \leq i \leq n\}$$

$$T^2 = \{[i_1,i_3|i_4,i_2], [i_3,i_1|i_2,i_4] : 1 \leq i \leq n\}$$

$$T^3 = \{[i_1,i_4|i_2,i_3], [i_4,i_1|i_3,i_2] : 1 \leq i \leq n\}.$$

The $4n - 1$ rounds

$R^*(i): 1 \leq i \leq n - 1,$

$S^j(i): 1 \leq i \leq n - 1, \; 1 \leq j \leq 3,$

$T^j: 1 \leq j \leq 3$

form the required SR($4n$). ∎

Suppose that some set of k couples play so that all matches which involve more than one of their members have all four chosen from among them, and that these matches are scheduled in the minimum number of rounds (k, if k is odd; k - 1, if k is even). Then the tournament has a <u>subtournament</u> of order k, or sub-SR(k). We present a method for constructing an SR(n) with a sub-SR(k) when n and k are even, and provide the necessary (computer-generated) data to construct an SR(n) with a sub-SR(4) for n = 18, 22, 26, 30 and an SR(n) with a sub-SR(8) for n = 34, 38, 42.

10.5 Spouse-Avoiding Mixed-Doubles Tournaments

Suppose that n and k are even, $n > k$, and that an SR(k) based on the elements $\{A_1, A_2, \ldots, A_k\}$ is given. Write $g = n - k$; let G be the cyclic group of order g, written in multiplicative notation. Write h for $(n/2) - 2k$. The SR(n) will be based on the A_i and the elements of G. The construction uses "starter" matches of the following kinds:

$k/2$ matches $H_1, H_2, \ldots, H_{k/2}$ with all four symbols from G;
h matches $F_1, F_2, \ldots, F_h$ with all four symbols from G;
k matches $I_1, I_2, \ldots, I_k$ of form $[u, v | w, A_j]$ with u, v, $w \in G$;
k matches $J_1, J_2, \ldots, J_k$ of form $[u, v | A_j, w]$ with u, v, $w \in G$.

If P is the match $[p, q | r, s]$ and x belongs to G, then $P + x$ is $[p + x, q + x | r + x, s + x]$, where the convention "$A_i + x = A_i$" is appended to the addition rule of G. The rounds of the SR(n) are:

Round 1: The $g/2$ matches $H_1 + x$, $x = 0, 1, \ldots, g/2 - 1$, together with round 1 of the SR(k);
Round i, $2 \leq i \leq k/2$: the $g/2$ matches $H_i + 2x$, $x = 0, 1, \ldots, g/2 - 1$, together with round i of the SR(k);
Round i, $(k+1)/2 \leq i \leq k - 1$: the $g/2$ matches $H_{1+i-(g/2)} + 2x + 1$, $x = 0, 1, \ldots, g/2 - 1$, together with round i of the SR(k);
Other rounds, one for each $x \in G$: $F_1 + x, F_2 + x, \ldots, F_h + x, I_1 + x, I_2 + x, \ldots, I_k + x, J_1 + x, J_2 + x, \ldots, J_k + x$.

Since h must be nonnegative, the construction can be used only when $n \geq 4k$.

THEOREM 10.16 There exist an SR(n), when n = 18, 22, 26, 30, 34, 38, and 42.

Proof: Suitable "starter" matches are shown in Table 10.5. ■

Not all known constructions for sharply regular spouse-avoiding mixed doubles are listed in this section. Even so, the known methods do not enable us to determine completely those n for which an SR(n) exists. It is known that one exists for all but finitely many values of n, and conjectured that one exists for all n except 2, 3, 6, and possibly 10 and 14.

EXERCISES

10.5.1 An SR(n) is called <u>symmetric</u> if whenever $[a, b | c, d]$ is in a round, $[b, a | d, c]$ is a match in the same round.
  (i) Verify that there is a symmetric SR(4).
  (ii) Is there a symmetric SR(8)?
  (iii) Prove that if there is a symmetric SR(n), where n is even, then there is a symmetric SR(4n).
10.5.2 Prove that there exists an SR(2^k) when $k \geq 2$.

10.5.3 Verify that the construction given before Theorem 10.16 does, in fact, yield an SR(n) provided that suitable "starter" matches exist.

10.5.4 If G is an abelian group with $2k + 1$ elements, a <u>starter sequence</u> in G is a way of ordering the nonzero elements of G as a sequence, say $(a_1, a_2, \ldots, a_{2k})$, so that:
  (a) the 2k differences $(a_2 - a_1)$, $(a_3 - a_2)$, $\ldots$, $(a_{2k} - a_{2k-1})$, $(a_1 - a_{2k})$ are all distinct and nonzero;
  (b) the 2k differences $\pm(a_{k+1} - a_1)$, $\pm(a_{k+2} - a_2)$, $\ldots$, $\pm(a_{2k} - a_k)$ are all distinct and nonzero;
  (c) the 2k differences $(a_{k+2} - a_1)$, $(a_{k+3} - a_2)$, $\ldots$, $(a_{2k} - a_{k-1})$, $(a_{k+1} - a_k)$ and $(a_2 - a_{k-1})$, $(a_3 - a_k)$, $\ldots$, $(a_k - a_{2k-1})$, $(a_1 - a_{2k})$ are all distinct and nonzero.

Prove that if there is a starter sequence in some group of order $2k + 1$, there is an SR($2k + 1$). (<u>Hint</u>: The first round consists of the k matches $[a_i, a_{i+1} | a_{i+k}, a_{i+k+1}]$.)

10.5.5 Using the definition in Exercise 10.5.4, prove that there is a starter sequence in the additive group of the field GF$[2k + 1]$ whenever $2k + 1$ is a prime power greater than 3.

10.6 THREE ORTHOGONAL LATIN SQUARES

None of the results in the preceding sections enable us to construct three orthogonal Latin squares of sides 12, 14, or 15. We discuss them as special cases.

THEOREM 10.17 $N(12) \geq 5$.

Proof: One of the Latin squares is

12	1	2	3	4	5	6	7	8	9	10	11
1	2	3	4	5	12	7	8	9	10	11	6
2	3	4	5	12	1	8	9	10	11	6	7
3	4	5	12	1	2	9	10	11	6	7	8
4	5	12	1	2	3	10	11	6	7	8	9
5	12	1	2	3	4	11	6	7	8	9	10
6	7	8	9	10	11	12	1	2	3	4	5
7	8	9	10	11	6	1	2	3	4	5	12
8	9	10	11	6	7	2	3	4	5	12	1
9	10	11	6	7	8	3	4	5	12	1	2
10	11	6	7	8	9	4	5	12	1	2	3
11	6	7	8	9	10	5	12	1	2	3	4.

Each of the other squares is obtained from this by column permutation: the four permutations give the following four first rows:

10.6 Three Orthogonal Latin Squares

```
12  6   8   2   7   1   9  11   4  10   5   3
12  3   6   1   9  11   2   8   5   4   7  10
12  8   1  11   5   9   3  10   2   7   6   4
12  4  11  10   2   7   8   6   9   1   3   5.
```
∎

THEOREM 10.18 $N(14) \geq 3$.

Proof: We construct three squares L_1, L_2, and L_3 based on the symbols ∞, 0, 1, ..., 12. These are interpreted as the integers modulo 13, with an "infinity element" appended: $\infty + i = \infty$ for all i.

Each square has ∞, 0, 1, 2, 3, 4, 5, 6, 7, 8, 9, 10, 11, 12 as its first column. The first rows are

```
L1:  ∞   0   1   2   3   4   5   6   7   8   9  10  11  12
L2:  ∞   2   3   4   5   6   7   8   9  10  11  12   0   1
L3:  ∞  12   0   1   2   3   4   5   6   7   8   9  10  11.
```

The remainders of the squares satisfy the following rule: The $(i + 1, j + 1)$ entry is obtained by adding 1 to the (i, j) entry (modulo 13); if necessary, row and column numbers are reduced modulo 13 to the range $\{2, 3, \ldots, 14\}$. Thus the squares are determined by their second rows; these are

```
L1:  0   ∞   2  11  10   9   8   5   4   1   7  12   6   3
L2:  0   8  12   9   6   3   1  10   7   5   2  11   ∞   4
L3:  0   1   ∞   7   5  11   2   6  10  12   4   3   9   8.
```
∎

THEOREM 10.19 $N(15) \geq 4$.

Proof: The first rows of the squares are

```
1  15   2  14   3  13   4  12   5  11   6  10   7   9   8
1  14   3  11   6   9   8   7  10   4  13  12   5  15   2
1  10   7  13   4   2  15   6  11   9   8   3  14  12   5
1   6  11  10   7  15   2   5  12  14   3   9   8   4  13.
```

The later rows of each square are obtained by adding 1, 2, ..., 14 to every element (modulo 15). ∎

THEOREM 10.20 There exist three mutually orthogonal Latin squares of every side except 2, 3, 6, and possibly 10.

Proof: From Theorem 10.4, $N(n) \geq 3$ unless either:

2 divides n but 4 does not; or
3 divides n but 9 does not.

So we need only discuss those sides n congruent to 2, 3, 6, 9, 10, 12, 14, 15, 18, 21, 22, or 0 (modulo 24). If t is any integer prime to 6, $N(t) \geq 4$ by Theorem 10.4. So we can interpret Corollary 10.12.1 in the case m = 4, in the following way:

If $0 \leq u \leq t$, where $(t, 6) = 1$, then $N(4t + u) \geq \min \{3, N(u)\}$.

In Table 10.6 we tabulate the 12 residue classes (modulo 24) which are to be considered. In each case we give a value of t and u such that $(t, 6) = 1$ and $N(u) \geq 3$. The column "N.C." shows the necessary condition forced by the demand that $u \leq t$, and the final column lists all sides for which this necessary condition is not satisfied. So $N(n) \geq 3$ except possibly when n is in the last column of Table 10.6.

Cases n = 2 and n = 3 are clearly impossible by Theorem 10.2 and we know that $N(6) = 1$. Theorems 10.16 to 10.19 and Exercise 10.4.1 tell us that $N(n) \geq 3$ for n = 12, 14, 15, 18, 22, 26, 30, 34, 38, 42, and 46. There remain 18 exceptions to be considered:

9, 10, 27, 36, 39, 50, 54, 58, 62, 66, 70, 74, 78, 86, 94, 98, 102, 126.

Six of these numbers are covered by Theorem 10.4, or by (10.2) and the facts that $N(14) \geq 3$ and $N(18) \geq 3$:

$9 = 3^2 \quad 27 = 3^3 \quad 36 = 2^2 \cdot 3^2$
$70 = 5 \cdot 14 \quad 98 = 7 \cdot 14 \quad 126 = 7 \cdot 18.$

Seven more follow from Corollary 10.12.1:

m	t	u	mt+u	m	t	u	mt+u
4	8	7	39	7	11	1	78
7	7	1	50	7	11	9	86
7	7	5	54	18	5	4	94
				7	13	11	102

Finally, we use Corollary 10.12.2 for four values:

m	t	u	v	mt+u+v
7	8	1	1	58
7	8	5	1	62
7	8	5	5	66
7	9	7	4	74

Only the case n = 10 is undecided.

Table 10.6

n	t	u	N.C.	Missing
24r + 2	6r - 5	22	r ⩾ 5	2, 26, 50, 74, 98
24r + 3	6r - 1	7	r ⩾ 2	3, 27
24r + 6	6r - 5	26	r ⩾ 6	6, 30, 54, 78, 102, 126
24r + 9	6r + 1	5	r ⩾ 1	9
24r + 10	6r - 1	14	r ⩾ 3	10, 34, 58
24r + 12	6r + 1	8	r ⩾ 2	12, 36
24r + 14	6r - 1	18	r ⩾ 4	14, 38, 62, 86
24r + 15	6r + 1	11	r ⩾ 2	15, 39
24r + 18	6r + 1	14	r ⩾ 3	18, 42, 66
24r + 21	6r + 5	1	r ⩾ 0	--
24r + 22	6r + 1	18	r ⩾ 4	22, 46, 70, 94
24r	6r - 1	4	r ⩾ 1	--

COROLLARY 10.20 There is a pair of orthogonal idempotent Latin squares of every side except 2, 3, and 6.

Proof: If A, B, and C are orthogonal Latin squares, reorder the columns of A so that it has all its entries 1 on the main diagonal, and reorder the columns of B and C in the same way. The latter squares will have transversals on their main diagonals, and yield orthogonal idempotent Latin squares after symbol permutation. So, by Theorem 10.20, we need only consider side 10. But the square of side 10 given in Table 10.4 and its transpose are orthogonal idempotent Latin squares of that side. ∎

BIBLIOGRAPHIC REMARKS

Latin squares were invented and studied by Euler in 1782 [3]. The Euler conjecture for n = 6 was proven by Tarry [9], and it was disproven for larger n by Bose, Shrikhande, and Parker—see [2]. An interesting article on the disproof of the Euler conjecture has been written by Martin Gardner [4]. MacNeish's theorem was published in 1922, in [7].

The relationship between finite planes and complete sets of orthogonal Latin squares (Theorem 10.3) was discovered by Bose [1] and independently by Levi [6].

We outlined the proof of the existence of three mutually orthogonal Latin squares of all sides greater than 14 in the survey paper [11]. Some important contributions to this theory have included the proof that $N(12) \geq 5$ by Johnson, Dulmage, and Mendelsohn [5], and that $N(15) \geq 4$ by Schellenberg, van Rees, and Vanstone [8]. The list of SR-designs in Table 10.5 comes from Wang's thesis [12]. Our whole approach has been simplified by studying the work of Wilson [13].

After [11] was written—in fact, after the first draft of this chapter was written—the entire situation was changed when Todorov found three mutually orthogonal Latin squares of side 14. His result was published in [10].

1. R. C. Bose, "On the application of the properties of Galois fields to the problem of construction of hyper-Graeco-Latin squares," Sankyha 3 (1938), 323-338.
2. R. C. Bose, S. S. Shrikhande, and E. T. Parker, "Further results on the construction of mutually orthogonal Latin squares and the falsity of Euler's conjecture," Canadian Journal of Mathematics 12(1960), 189-203.
3. L. Euler, "Recherches sur une nouvelle espece de quarrees magiques," Verhandelingen uitgegeven door het Zeeuwsch Genootschap der Wetenschappen te Vlissingen 9(1782), 85-239.
4. M. Gardner, "Euler's spoilers," Martin Gardner's New Mathematical Diversions from Scientific American (Allen & Unwin, London, 1969), 162-172.
5. D. M. Johnson, A. L. Dulmage, and N. S. Mendelsohn, "Orthomorphisms of groups and orthogonal Latin squares I," Canadian Journal of Mathematics 13(1961), 356-372.
6. F. W. Levi, Finite Geometrical Systems (University of Calcutta, Calcutta, India, 1942).
7. H. L. MacNeish, "Euler squares," Annals of Mathematics 23(1922), 221-227.
8. P. J. Schellenberg, G. H. J. van Rees, and S. A. Vanstone, "Four pairwise orthogonal Latin squares of order 15," Ars Combinatoria 6 (1978), 141-150.
9. G. Tarry, "Le probleme des 36 officiers," Comptes Rendus de L'Association Francaise pour L'Avancement de Science 1(1900), 122-123; 2(1901), 170-203.
10. V. Todorov, "Three mutually orthogonal Latin squares of order fourteen," Ars Combinatoria 20(1985), 45-47.
11. W. D. Wallis, "Three orthogonal Latin squares," Congressus Numerantium 42(1984), 69-86.
12. S. P. Wang, On Self-Orthogonal Latin Squares and Partial Transversals in Latin Squares (Ph.D. thesis, Ohio State University, 1978).
13. R. M. Wilson, "Concerning the number of mutually orthogonal Latin squares," Discrete Mathematics 9(1974), 181-198.

11
One-Factors and One-Factorizations

11.1 BASIC IDEAS

Suppose that a balanced incomplete block design has block size $k = 2$. To say "the treatments x and y occur together λ times" is exactly the same as saying "there are λ copies of the block $\{x,y\}$." If there are v treatments, then each of the $v(v-1)/2$ unordered pairs must occur as a block λ times, and the parameters are

$$\left(v, \frac{1}{2}\lambda v(v-1), \lambda(v-1), 2, \lambda\right). \tag{11.1}$$

From this trivial construction we have

THEOREM 11.1 There is a balanced incomplete block design with $k = 2$ for every v and λ. Its parameters are given in (11.1). ∎

It will be seen that the parameters (11.1) are implied by (1.1) and (1.2). So the existence problem for designs with $k = 2$ is completely solved and is not interesting. The designs are not merely determined up to isomorphism—they are completely determined by the parameters.

To impose further structure on these designs, the following definition is used. Given a set S of 2n treatments, a <u>matching</u> of S is a set of n unordered pairs on S which between them contain every member of S precisely once. In the case where v is even, we can discuss matchings among the blocks of the design (11.1).

THEOREM 11.2 If $v = 2n$, the blocks of a balanced incomplete block design with $k = 2$ and $\lambda = 1$ can be partitioned into $2n - 1$ matchings.

Proof: We construct a block design on the treatments $\infty, 0, 1, \ldots, 2n - 2$, which are treated as the integers modulo $2n - 1$ with the additional law "$\infty + x = \infty$." Write F_0 for the set of n unordered pairs

$\{\infty, 0\}, \{1, 2n-2\}, \{2, 2n-3\}, \ldots, \{n-1, n\}$

and define F_i to be the result of adding i to every element of F_0 : F_i is

$\{\infty, i\}, \{1+i, 2n-2+i\}, \{2+i, 2n-3+i\}, \ldots, \{n-1+i, n+i\}$
$= \{\infty, i\}, \{i+1, i-1\}, \{i+2, i-2\}, \ldots, \{i+n-1, i-(n-1)\}$. (11.2)

We show that $\{F_0, F_1, \ldots, F_{2n-2}\}$ form the required partition. It is clear that each F_i is a matching. It remains to show that any given pair $\{x, y\}$ occurs in exactly one of the F_i. But if $x + y = s$ and $x - y = d$, there exist unique integers σ and δ such that $s \equiv 2\sigma \pmod{2n-1}$, $d \equiv 2\delta \pmod{2n-1}$, and $0 \leq \sigma, \delta < 2n-1$. Then $\{x, y\} = \{\sigma + \delta, \sigma - \delta\}$ and occurs in the matching F_σ (and in no other). ∎

COROLLARY 11.2.1 If $v = 2n$, the blocks of a design with parameters (11.1) can be partitioned into $\lambda(2n-1)$ matchings. ∎

Matchings arise naturally in graph theory. A graph $G = G(V, E)$, as defined in Section 1.2, consists of a finite set V of objects called vertices and a set E of unordered pairs of members of V, called edges. The graph $G'(V', E')$ is called a subgraph of $G(V, E)$ if $V' \subseteq V$ and $E' \subseteq E$. In particular, if K_v denotes a complete graph on v vertices, one in which all possible pairs of distinct vertices constitute edges, then any graph on v or less vertices can be interpreted as a subgraph of K_v. A subgraph of G that contains every vertex of G is called a spanning subgraph, or sometimes a factor because one can develop a theory of decomposing graphs into spanning subgraphs which bears some resemblance to the decomposition of integers into factors.

In this terminology, it is clear that a matching F on 2n treatments is a factor of K_{2n} if the treatments are interpreted as vertices and the unordered pairs as edges. Since every vertex lies on precisely one edge, it is called a one-factor or factor of degree 1. (The number of edges containing a given vertex is often called the degree of that vertex.) A set of one-factors of K_{2n} which between them contain every edge precisely once is called a one-factorization of K_{2n}; Theorem 11.2 can be interpreted as saying that K_{2n} has a one-factorization for every n.

If we say "one-factorization" without specifying the graph, we mean "one-factorization of K_{2n}." Obviously, one can discuss the decomposition of general graphs into one-factors. Some graphs have one-factorizations and others do not. Of special interest are the complete bipartite graphs. If S and T are two disjoint sets, the complete bipartite graph on (S, T) has vertex set $S \cup T$; $\{x, y\}$ is an edge if one of x, y is in S and the other is in T, but not otherwise. We write $K_{m,n}$ for a complete bipartite graph whose vertex sets have sizes m and n. In order for $K_{m,n}$ to have a one-factorization—or, indeed, a one-factor—it is necessary that $m = n$ (see Exercise 11.1.4).

11.1 Basic Ideas

THEOREM 11.3 One-factorizations of $K_{n,n}$ are equivalent to Latin squares of side n.

Proof: Suppose that $L = (\ell_{ij})$ is a Latin square of side n. We construct a one-factorization of the $K_{n,n}$ with vertex sets $S = \{1_1, 2_2, \ldots, n_1\}$ and $T = \{1_2, 2_2, \ldots, n_2\}$. Define F_k to consist of the pairs $\{i_1, j_2\}$ such that $\ell_{ij} = k$. Since the symbol k appears once in each row of L, each of the vertices $1_1, 2_1, \ldots, n_1$ appears exactly once in F_k. Similarly, the fact that k appears once per column means that the elements of T appear once each in F_k. Since each cell contains exactly one symbol in L, it follows that every pair with one member in S and one member in T occurs in exactly one F_k.

Conversely, given a $K_{n,n}$, relabel the elements of S and T as $\{1_1, 2_2, \ldots, n_1\}$ and $\{1_2, 2_2, \ldots, n_2\}$, respectively. Then the construction can be reversed to give a Latin square. ∎

Since there are Latin squares of all positive integer sides, $K_{n,n}$ has a one-factorization for all n.

Some special Latin squares are also associated with one-factorizations of complete graphs. Suppose that $F_1, F_2, \ldots, F_{2n-1}$ are the factors in a one-factorization of the complete graph with vertices $\{\infty, 1, 2, \ldots, 2n-1\}$, and suppose—after reordering the factors, if necessary—that F_i contains the pair $\{\infty, i\}$. Given a pair $\{x, y\}$ of symbols, where neither x nor y is ∞, define a_{xy} to be the number of the factor that contains $\{x, y\}$. Then $A = (a_{xy})$ is a symmetric Latin square of side $2n - 1$ with diagonal $(1, 2, \ldots, 2n - 1)$, which is called the <u>Latin square of the one-factorization</u>. These squares are discussed in Exercises 11.1.3 and 11.1.4.

EXERCISES

11.1.1 Prove that the conditions

$$vb = rk, \quad \lambda(v - 1) = r(k - 1), \quad k = 2$$

together imply

$$r = \frac{1}{2}\lambda(v - 1), \quad b = \frac{1}{2}\lambda v(v - 1).$$

11.1.2 Prove that if $m \neq n$, then $K_{m,n}$ has no one-factor.

11.1.3 Verify that the array A, the "Latin square of a one-factorization," <u>is</u> in fact a symmetric Latin square with diagonal $(1, 2, \ldots, 2n - 1)$.

11.1.4 A is a symmetric Latin square of side $2n - 1$.
 (i) Prove that every symbol occurs on the diagonal exactly once.

(ii) Prove that provided the names of the symbols in A are permuted so that the diagonal is $(1, 2, \ldots, 2n-1)$, A is the Latin square of some one-factorization.

11.1.5 Suppose that $F_1, F_2, \ldots, F_{2n-1}$ is a one-factorization of K_{2n}. If F_i consists of the edges $\{a_{1i}, b_{1i}\}, \{a_{2i}, b_{2i}\}, \ldots, \{a_{ni}, b_{ni}\}$, write S_i for the set of the $n(n-1)/2$ 4-sets $\{a_{xi}, b_{xi}, a_{yi}, b_{yi}\}$, for $1 \leq x < y \leq n$. Prove that $S_1 \cup S_2 \cup \cdots \cup S_{2n-1}$ constitute the blocks of a 3-design. What are its parameters?

11.2 STARTERS

The construction of Theorem 11.2 is easy to visualize in graphical terms. Suppose that the vertices of K_{2n} are arranged in a circle with ∞ at the center, as in Figure 11.1. The one-factor F_0 is shown in that figure. F_1 is constructed by rotating the whole one-factor through one-$(2n-1)$th part of a full circle, clockwise, and subsequent factors are obtained in the same way. It is easy to see that the $n-1$ chords separate pairs of vertices which differ (modulo $2n-1$) by different numbers, and that on rotation the one edge which represents distance d will generate all edges with distance d.

To generalize this construction, we make the following definition. A <u>starter</u> in an abelian group G of order $2n-1$ is an ordered partition of the nonzero members of G into 2-sets $\{x_1, y_1\}, \{x_2, y_2\}, \ldots, \{x_{n-1}, y_{n-1}\}$ with the property that the $2n-2$ differences $\pm(x_1 - y_1), \pm(x_2 - y_2), \ldots, \pm(x_{n-1} - y_{n-1})$ are all different and therefore contain every nonzero element of G precisely once.

It is clear that the set F_0 used in the proof of Theorem 11.2 is a starter. On the other hand, suppose that any other starter had been used instead of F_0.

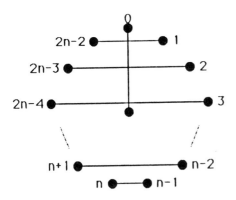

Figure 11.1

11.2 Starters

The pair $\{\infty, i\}$ occurs in the ith factor. Given any unordered pair $\{x, y\}$ where neither symbol is ∞, there will be exactly one factor that contained it: There will be exactly one pair $\{x_j, y_j\}$ in the starter such that $(x_j - y_j) = \pm(x - y)$—assume it happens that $x_j - y_j = x - y$, for convenience—and $\{x, y\}$ will occur in the factor obtained by adding $x - x_j$ to the starter: Since $x_j - y_j = x - y$, y will equal $y_j + x - x_j$, and

$$\{x, y\} = \{x_j + (x - x_j), y_j + (x - x_j)\}.$$

Since all of the $n(2n - 1)$ pairs occur at least once, and there are only $n(2n-1)$ pairs in $2n - 1$ factors, we have each pair precisely once, and we have constructed a one-factorization. So we could use any starter in any group of order $2n - 1$ for Theorem 11.2. The type of starter we used, consisting of all the pairs $\{x, -x\}$ where x ranges through the nonzero elements of some group, is called a **patterned** starter; we used the patterned starter in the cyclic group Z_{2n-1}.

If S is a starter in an abelian group G, the set of pairs $-S$, constructed by replacing each pair $\{x, y\}$ in S by $\{-x, -y\}$, is also a starter. More generally, if the group G has a multiplication operation defined on it—for example, if G is the set of integers modulo $2n - 1$, or if G is the additive group of any integral domain or field—one may define kS by

$$kS = \{(kx, ky) : \{x, y\} \in S\}.$$

The set kS is not always a starter, but in some cases it will be. For example, if S is a starter in the integers modulo $2n - 1$, and if $(k, 2n - 1) = 1$, then kS is a starter (see Exercise 11.2.2).

For small orders, it is possible to enumerate all starters. For example, let us find all starters in the group Z_9 of integers modulo 9. There must be a pair whose difference is ± 1. This must be one of the pairs $\{1, 2\}$, $\{2, 3\}$, $\{3, 4\}$, $\{4, 5\}$, $\{5, 6\}$, $\{6, 7\}$, or $\{7, 8\}$. Any starter, including one of the last three pairs mentioned, will be the negative of one involving one of the first three, so we consider only the first four cases.

Suppose that a starter includes $\{1, 2\}$. Then it must include

one of $\{3, 5\}$, $\{4, 6\}$, $\{5, 7\}$,
one of $\{3, 6\}$, $\{4, 7\}$, $\{5, 8\}$,
one of $\{3, 7\}$, $\{4, 8\}$, $\{8, 3\}$.

As repeated elements are to be avoided, it is easy to show that there are two possibilities:

$$\{1, 2\}, \{4, 6\}, \{5, 8\}, \{3, 7\}, \tag{11.3}$$

$$\{1, 2\}, \{5, 7\}, \{3, 6\}, \{4, 8\}. \tag{11.4}$$

Similar calculations show that there is exactly one starter in each of the next three cases:

$$\{2,3\}, \{6,8\}, \{4,7\}, \{1,5\}, \tag{11.5}$$

$$\{3,4\}, \{5,7\}, \{8,2\}, \{6,1\}, \tag{11.6}$$

$$\{4,5\}, \{8,1\}, \{3,6\}, \{7,2\} \tag{11.7}$$

The negatives of all of these are also starters, but the negative of (11.7) equals (11.7) again. So there are nine starters in Z_9.

EXERCISES

11.2.1 Prove the existence of exactly one starter in the cyclic group of order 3, exactly one starter in the cyclic group of order 5, and exactly three starters in the cyclic group of order 7.

11.2.2 Suppose that S is a starter in the set Z_{2n-1} of integers modulo $2n - 1$.
  (i) Prove that kS is a starter if and only if $(k, 2n - 1) = 1$.
  (ii) Starters S and T are called <u>equivalent</u> if $T = kS$ for some S. Prove that the patterned starter is not equivalent to any other starter.
  (iii) Prove that there are precisely two equivalence classes of starters in Z_7 and three equivalence classes in Z_9.

11.2.3 Find all starters in $Z_3 \times Z_3$.

11.3 ONE-FACTORIZATIONS OF COMPLETE GRAPHS

It is obvious that there is only one-factor in K_2, and it (trivially) constitutes the unique one-factorization. Similarly, K_4 has just three one-factors, and together they form a factorization.

The situation concerning K_6 is more interesting. There are 15 one-factors of K_6, namely

```
01  23  45,    01  24  35,    01  25  34
02  13  45,    02  14  35,    02  15  34
03  12  45,    03  14  25,    03  15  24
04  12  35,    04  13  25,    04  15  23
05  12  34,    05  13  24,    05  14  23.
```

(Brackets and commas are omitted for ease of reading.) Consider the factor

$$F_1 = 01 \quad 23 \quad 45.$$

11.3 One-Factorizations of Complete Graphs

We enumerate the one-factorizations that include F_1. The only factors in the list that contain 02 and have no edge in common with F_1 are

$F_{21} = 02\ \ 14\ \ 35,$

$F_{22} = 02\ \ 15\ \ 34.$

We construct all one-factorizations that contain F_1 and F_{21}; those that contain F_1 and F_{22} can be obtained from them by exchanging the symbols 4 and 5 throughout. But one easily sees that there is exactly one way of completing $\{F_1, F_{21}\}$ to a one-factorization, namely,

03 15 24
04 13 25
05 12 34.

So there are exactly two one-factorizations containing F_1. It follows (by permuting the symbols 0, 1, 2, 3, 4, 5) that there are exactly two one-factorizations containing any given one-factor.

Now let us count all the ordered pairs $(F, \mathcal{F})$, where F is a one-factor of K_6 and $\mathcal{F}$ is a one-factorization containing F. From what we have just said, the total must be $2 \cdot 15$, that is, 30. On the other hand, suppose that there are N one-factorizations. Since there are 5 factors in each, we have

$5N = 30,$

so $N = 6$; there are six one-factorizations of K_6.

The calculations above are simple in theory, but in practice they soon become unwieldy. There are 6240 one-factorizations of K_8 (see Section 11.4) and 1,255,566,720 of K_{10}; no larger numbers are known. In any case, the importance of such numbers is dubious. It is more interesting to know about the existence of "essentially different" one-factorizations. As usual, the first step is to make a sensible definition of isomorphism.

Suppose that G is a graph which has a one-factorization. Two one-factorizations $\mathcal{F}$ and $\mathcal{H}$ of G, say

$\mathcal{F} = \{F_1, F_2, \ldots, F_k\},$

$\mathcal{H} = \{H_1, H_2, \ldots, H_k\},$

are called <u>isomorphic</u> if there exists a map φ from the vertex set of G onto itself such that

$\{F_1\varphi, F_2\varphi, \ldots, F_k\varphi\} = \{H_1, H_2, \ldots, H_k\},$

where $F_i\varphi$ is the set of all the edges $\{x\varphi, y\varphi\}$ where $\{x,y\}$ was an edge in F. In other words, when φ operates on one of the factors in F, it produces one of the factors in $\mathcal{H}$. The map φ is called an <u>isomorphism</u>. In the case where $\mathcal{H}$ equals $\mathcal{F}$ itself, φ is called an <u>automorphism</u>; the automorphisms of a one-factorization $\mathcal{F}$ form a group, the <u>automorphism group</u> of $\mathcal{F}$.

There is a unique one-factorization of K_{2n}, up to isomorphism, for $2n = 2$, 4, or 6. There are exactly six for K_8; we discuss them in Section 11.4. For K_{10} the number is 396, and for K_{12} it is much larger (the exact number is not known).

To discuss isomorphism of factorizations we introduce the ideas of a cycle and of a division. A <u>cycle</u> is a graph consisting of a set of vertices $V = \{v_1, v_2, \ldots, v_k\}$ and the edges $\{v_1, v_2\}$, $\{v_2, v_3\}$, $\ldots$, $\{v_{k-1}, v_k\}$, $\{v_k, v_1\}$. To indicate the size (or length), one sometimes refers to a cycle with k vertices as a <u>k-cycle</u>. The union of two edge-disjoint one-factors will always result in one or more disjoint cycles, always of even length, and one can discuss the way factors fit together in a factorization by considering these cycle lengths. If F_1 and F_2 are two factors in a one-factorization of K_{2n}, their union may be a single 2n-cycle, which is called a Hamiltonian cycle; otherwise, it must decompose into two or more disjoint parts, and we say that F_1 and F_2 constitute a <u>division</u>, or 2-division, in the one-factorization. More generally, a <u>k-division</u> is a set of k one-factors whose union is disconnected. It is useful to observe that each of the connected parts (or <u>components</u>) in a k-division must have at least k + 1 vertices (see Exercise 11.3.3).

It is clear that the image of a d-division, under any permutation of the vertices, remains a d-division. So "having a d-division" is invariant under isomorphism. We shall say a d-division is <u>maximal</u> if it cannot be embedded in a $(d + 1)$-division, and write α_d for the number of maximal d-divisions in the given one-factorization F; then the α_i form a convenient set of invariants for use in discussing one-factorizations.

We shall use the cycle structure to prove the existence of nonisomorphic one-factorizations of K_{2n} when n is an integer greater than 3. It is convenient to work in the integers modulo $2n - 1$. If i and k are any integers, we write d_{ik} for the greatest common divisor of $i - k$ and $2n - 1$,

$$d_{ik} = (i - k, 2n - 1),$$

and we define ν_{ik} by

$$d_{ik}\nu_{ik} = 2n - 1.$$

If $\mathcal{F}_{2n} = \{F_0, F_1, \ldots, F_{2n-2}\}$ is the one-factorization from the patterned starter in the group of the integers modulo $2n - 1$, defined in Section 11.1 by

11.3 One-Factorizations of Complete Graphs

$$F_i = \{\infty, i\}, \{i - 1, i + 1\}, \{i - 2, i + 2\}, \ldots, \{i - n + 1, i + n - 1\},$$

then it is easy to prove the following lemma:

LEMMA 11.4 If F_i and F_k are different factors in $\mathcal{F}_{2n}$, then $F_i \cup F_k$ consists of a cycle of length $\nu_{ik} + 1$ and $(d_{ik} - 1)/2$ cycles each of length $2\nu_{ik}$. ∎

THEOREM 11.5 When $n > 2$, $\mathcal{F}_{2n}$ cannot contain an $(n - 1)$-division.

Proof: An $(n - 1)$-division would have two components of order n. Suppose that $n > 2$, so that $n - 1 \geq 2$, and let F_i and F_k be two different factors in an $(n - 1)$-division. The 2-division $\{F_i, F_k\}$ has one component of size $\nu_{ik} + 1$ and $(d_{ik} - 1)/2$ components of size $2\nu_{ik}$. So one of the components of the $(n - 1)$-division must be a union of disjoint sets of size $2\nu_{ik}$. So ν_{ik} divides n; since ν_{ik} also divides $2n - 1$, we have $\nu_{ik} = 1$ and $d_{ik} = 2n - 1$, which is impossible since $1 \leq i$, $k \leq 2n - 1$ and $i \neq k$. ∎

THEOREM 11.6 If $n \neq 5$, then no 2-division of $\mathcal{F}_{2n}$ has a component of order $2n - 4$.

Proof: Consider the 2-division $\{F_i, F_k\}$. Its component cycles have sizes $2\nu_{ik}$ and $\nu_{ik} + 1$. Since ν_{ik} divides the odd number $2n - 1$, ν_{ik} must be odd; and since ν_{ik} cannot equal 1, $\nu_{ik} \geq 3$. If $\nu_{ik} + 1 = 2n - 4$, we have $\nu_{ik} = 2n - 5$, so ν_{ik} divides $(2n - 1) - (2n - 5)$ and ν_{ik} divides 4, which is a contradiction. So the component of order $\nu_{ik} + 1$ is not a component of order $2n - 4$. If the 2-division is to contain a component of order $2n - 4$, then $2\nu_{ik} = 2n - 4$ and ν_{ik} divides $(2n - 4, 2n - 1)$. So ν_{ik} divides 3, and necessarily $\nu_{ik} = 3$, whence $2n - 4 = 6$ and $n = 5$. Thus $n \neq 5$ implies that the 2-division has no component cycle of order $2n - 4$. ∎

THEOREM 11.7 For all $n \geq 4$, there is a one-factorization of K_{2n} that is not isomorphic to $\mathcal{F}_{2n}$.

Proof: When $n = 4$, the result follows from the existence of six nonisomorphic factorizations, which we prove in Theorem 11.12. To complete the proof, we construct:

(i) a one-factorization $\mathcal{K}_{2n}$ of K_{2n} that contains an $(n - 1)$-division, for every even n;
(ii) a one-factorization $\mathcal{L}_{2n}$ of K_{2n} that contains a 2-division with a component of order $2n - 4$, for every odd n greater than 5;
(iii) a one-factorization of K_{10} not isomorphic to $\mathcal{F}_{10}$.

From Theorem 11.5, $\mathcal{K}_{2n}$ is not isomorphic to $\mathcal{F}_{2n}$; from Theorem 11.6, $\mathcal{L}_{2n}$ is not isomorphic to $\mathcal{F}_{2n}$.

(i) In this case n is even, so K_n is one-factorizable. Label the vertices of K_{2n} as $1_1, 2_1, \ldots, n_1, 1_2, 2_2, \ldots, n_2$, where $1, 2, \ldots, n$ are integers modulo n, and let $F_{\alpha,1}, F_{\alpha,2}, \ldots, F_{\alpha,n-1}$ be the factors in some one-factorization of the K_n with vertices $1_\alpha, 2_\alpha, \ldots, n_\alpha$. For $i = 1, 2, \ldots, n - 1$, define

$$H_i = F_{1,i} \cup F_{2,i}.$$

and for $i = n, n + 1, \ldots, 2n - 1$, define

$$H_i = \{1_1, (1-i)_2\}, \{2_1, (2-i)_2\}, \ldots, \{n_1, (n-i)_2\}. \tag{11.8}$$

Now define

$$\mathcal{K}_{2n} = \{H_1, H_2, \ldots, H_{2n-1}\}.$$

The factors $H_1, H_2, \ldots, H_{n-1}$ contain every edge of the form (x_1, y_1) or (x_2, y_2) exactly once, and the other factors contain every edge of the form (x_1, y_2) exactly once, so $\mathcal{K}_{2n}$ is a one-factorization of K_{2n}. The factors

$$\{H_1, H_2, \ldots, H_{n-1}\}$$

form an n-division.

(ii) Assume that n is odd. Again, write the vertices of K_{2n} as $1_1, 2_1, \ldots, n_1, 1_2, 2_2, \ldots, n_2$. In this case we do not have a one-factorization of K_n available; instead, we use K_{n+1}. Let $F_{\alpha,1}, F_{\alpha,2}, \ldots, F_{\alpha,n}$ be the factors of the one-factorization $\mathcal{F}_n$, from the patterned starter on the symbols $\infty, 1_\alpha, 2_\alpha, \ldots, n_\alpha$, where $1, 2, \ldots, n$ are interpreted as integers modulo n, and write $F^*_{\alpha,i}$ to mean $F_{\alpha,i}$ with the edge (∞, i_α) deleted. Now define

$$L^*_i = F^*_{1,i} \cup F^*_{2,i} \cup \{(i_1, i_2)\}.$$

The factors $L^*_1, L^*_2, \ldots, L^*_n$ contain all the edges of the form (x_1, y_1) or (x_2, y_2) where $x \neq y$, or of form (x_1, x_2), but none of the edges (x_1, y_2) where $x \neq y$. Recall that the factors $H_n, H_{n+1}, \ldots, H_{2n-1}$ as defined in (11.8) contained all the missing edges, and also the edges (x_1, x_2); but the edges (x_1, x_2) are precisely the edges of H_n. So if we write $L_i = H_i$ for $n + 1 \leq i \leq 2n - 1$, the factors $L^*_1, L^*_2, \ldots, L^*_n, L_{n+1}, L_{n+2}, \ldots, L_{2n-1}$ form a one-factorization of K_{2n}.

The factorization we have just found does not have the property we require but is easily transformed into one that does have. We perform the permutation φ,

11.3 One-Factorizations of Complete Graphs

$$(n + 1 - j)_\alpha \longmapsto (2j)_\alpha$$

$$(j + 1)_\alpha \longmapsto (2j + 1)_\alpha$$

$$1_\alpha \longmapsto 1_\alpha$$

for $j = 1, 2, \ldots, (n - 1)/2$ and $\alpha = 1, 2$, on the L_i^*: for $1 \leq i \leq n$, $L_i = L_i^*\varphi$. One readily checks that the factors $L_1, L_2, \ldots, L_n$ contain the same edges as $L_1^*, L_2^*, \ldots, L_n^*$. Now

$$L_1 = \{1_1, 1_2\}, \{2_1, 3_1\}, \{4_1, 5_1\}, \ldots, \{(n-1)_1, n_1\},$$

$$\{2_2, 3_2\}, \ldots, \{(n-1)_2, n_2\},$$

$$L_{n+3} = \{1_1, (n-2)_2\}, \{2_1, (n-1)_2\}, \{3_1, n_2\},$$

$$\{4_1, 1_2\}, \ldots, \{n_1, (n-3)_2\}.$$

So $L_1 \cup L_{n+3}$ contains the $(2n - 4)$-cycle

$$1_1, 1_2, 4_1, 5_1, 2_2, 3_2, 6_1, 7_1, \ldots, (n-2)_2, 1_1.$$

Therefore, $\mathscr{L}_{2n} = \{L_1, L_2, \ldots, L_{2n-1}\}$ is the required factorization.

(iii) Since $\mathscr{F}_{10}$ contains the 3-division $\{F_1, F_4, F_7\}$, it suffices to observe that

```
01 23 45 67 89;   04 13 26 58 79;   07 12 34 59 68
02 14 39 56 78;   05 19 27 38 46;   08 17 25 36 49
03 18 24 57 69;   06 15 29 37 48;   09 16 28 35 47
```

contains no 3-division. ∎

It can, in fact, be shown that the number of nonisomorphic one-factorizations of K_{2n} tends to infinity as n does.

A one-factorization is called <u>perfect</u> if it contains no 2-division. Many perfect factorizations exist, and we know of no order n (greater than 1) for which no perfect factorization exists, but the existence question is not yet settled.

THEOREM 11.8 If p is an odd prime, then K_{p+1} and K_{2p} have perfect factorizations.

Proof: The one-factorization of K_{p+1} from the patterned starter is perfect when p is prime. So we need only consider K_{2p}.

Suppose that p is prime and K_{2p} has vertex set $U \cup V$, where $U = \{u_1, u_2, \ldots, u_p\}$ and $V = \{v_1, v_2, \ldots, v_p\}$. We construct a one-factorization of K_{2p} from two one-factorizations of K_{p+1}: $\mathcal{E}$ is the one-factorization of the K_{p+1} based on the vertex set $U \cup \{u_\infty\}$, with factors

$$E_i = \{u_\infty, u_i\}, \left\{\{u_{i+j}, u_{i-j}\} : 1 \leq j \leq \frac{1}{2}(p-1)\right\};$$

$\mathcal{F}$ is the one-factorization of the K_{p+1} based on vertex set $V \cup \{v_\infty\}$, with factors

$$F_i = \{v_\infty, v_i\}, \left\{\{v_{i+j}, v_{i-j}\} : i \leq j \leq \frac{1}{2}(p-1)\right\}.$$

We first construct p factors $G_1, G_2, \ldots, G_p$ as follows: G_i consists of $E_i \cup F_{p-1-i}$ with the edge (u_i, v_{p-1-i}) added on and the edges (u_i, u_∞) and (v_{p-i-1}, v_∞) deleted. (The subscripts are always taken modulo p, so that G_p is formed from E_p and F_{p-1}.) Then the factors $G_1, G_2, \ldots, G_p$ contain every edge connecting two vertices of U and every edge connecting two vertices of V. The other edges which they contain can be characterized as the edge of $(u_{(p-1)/2}, v_{(p-1)/2})$ and the edges (u_i, v_j), where (u_i, u_j) is an edge of $E_{(p-1)/2}$ and neither i nor j is ∞.

Further factors are labeled $H_1, H_2, \ldots, H_p$, but there is no factor $H_{(p-1)/2}$. To form H_i we take the edge (u_i, v_i) and also all the edges (u_j, v_k) and (u_k, v_j), where (u_j, u_k) is an edge in F_i other than (u_i, u_∞). If H_i and H_j have a common edge, then E_i and E_j had a common edge, which is impossible because E is a one-factorization. If H_i and G_j have a common edge, it must be (u_i, v_i), and we have $(u_i, v_i) = (u_j, v_{p-1-j})$. This occurs only when $i = j = (p-1)/2$, but there is no H_i in that case. So the $2p-1$ edge-disjoint one-factors $G_1, G_2, \ldots, G_p, H_1, H_2, \ldots, H_p$ are edge disjoint, and form a one-factorization.

Now consider the unions of factors in this one-factorization. $E_i \cup E_j$ is a Hamiltonian cycle in the complete graph on vertices $u_1, u_2, \ldots, u_p$, u_∞, and in particular contains the edges (u_i, u_∞) and (u_∞, u_j). So $G_i \cup G_j$ consists of a path from u_i to u_j spanning all the vertices u_k, a path from v_{p-1-i} to v_{p-1-j} spanning all the vertices v_k, and the edges (u_i, v_{p-1-i}) and (u_j, v_{p-1-j}). This constitutes a cycle.

The whole structure of H_i can be expressed by saying that u_x is adjacent to v_{2i-x} for all x. So $H_i \cup H_j$ contains the following sequence Z of vertices, with each vertex being adjacent to the following one:

$$Z = u_x, v_{2i-x}, u_{2j-(2i-x)}, v_{2i-(2j-2i+x)}, \ldots.$$

Every second vertex in this sequence is a member of U; the order in which the members of U arise is

11.3 One-Factorizations of Complete Graphs

$$u_x, u_{2(j-i)+x}, u_{4(j-i)+x}, \ldots.$$

Now, $H_i \cup H_j$ is a union of disjoint cycles, and Z is connected so that Z is a cycle. Therefore, it must eventually happen that u_x is $u_{2k(j-i)+x}$, so that

$$x \equiv 2k(j - i) + x \pmod{p}.$$

But p is a prime; so provided that $j \neq i$, this will not occur until $k = p$. So Z contains 2p vertices—it is all of $H_i \cup H_j$. Therefore, $H_i \cup H_j$ is a Hamiltonian cycle.

Finally, consider $G_j \cup H_i$. This is again necessarily a union of disjoint cycles of even length, including the cycle

$$u_j, v_{p-1-j}, u_{2i+j+1-p}, \ldots.$$

Counting u_j as being in position 1, the vertices of the cycle which are in U lie in the positions 1, 3, 4, 7, 8, ..., 4k - 1, 4k, ... The cycle terminates when u_j arises again. Since its length is even, u_j must arise in position 4k - 1 for some k. The (4k - 1)th vertex is

$$u_{(2k-1)(2i+1)+j-p}$$

and this equals u_j if and only if

$$(2k - 1)(2i + 1) \equiv p \pmod{2p}.$$

Since both expressions on the left-hand side are prime to 2, this equation is equivalent to

$$(2k - 1)(2i + 1) \equiv 0 \pmod{p},$$

which has the unique solution $k \equiv (p + 1)/2 \pmod{p}$. So the cycle has $4[(p + 1)/2] - 2$ vertices, and is Hamiltonian. {This arithmetic is not valid when $i = (p - 1)/2$, but that case does not arise as H_i does not exist for $i = (p - 1)/2$.}

We have now discussed all possible pairs of factors, and shown that their unions are all Hamiltonian cycles. Hence the one-factorization $\{G_1, G_2, \ldots, G_p, H_1, H_2, \ldots, H_{1/2(p-3)}, H_{1/2(p+1)}, \ldots, H_p\}$ is perfect. ∎

We illustrate Theorem 11.8 with the example of K_{10}. The one-factorization $\mathcal{E}$ of K_6 is

$E_1 = (u_\infty, u_1), (u_2, u_5), (u_3, u_4);$

$E_2 = (u_\infty, u_2), (u_3, u_1), (u_4, u_5);$

$E_3 = (u_\infty, u_3), (u_4, u_2), (u_5, u_1);$

$E_4 = (u_\infty, u_4), (u_5, u_3), (u_1, u_2);$

$E_5 = (u_\infty, u_5), (u_1, u_4), (u_2, u_3).$

F is obtained from E by replacing each u by v. The factors G_i are

$G_1 = (u_1, v_3), (u_2, u_5), (u_3, u_4), (v_4, v_2), (v_5, v_1);$

$G_2 = (u_2, v_2), (u_3, u_1), (u_4, u_5), (v_3, v_1), (v_4, v_5);$

$G_3 = (u_3, v_1), (u_4, u_2), (u_5, u_1), (v_2, v_5), (v_3, v_4);$

$G_4 = (u_4, v_5), (u_5, u_3), (u_1, u_2), (v_1, v_4), (v_2, v_3);$

$G_5 = (u_5, v_4), (u_1, u_4), (u_2, u_3), (v_5, v_3), (v_1, v_2);$

the other factors are

$H_1 = (u_1, v_1), (u_2, v_5), (u_5, v_2), (u_3, v_4), (u_4, v_3);$

$H_3 = (u_3, v_3), (u_4, v_2), (u_2, v_4), (u_5, v_1), (u_1, v_5);$

$H_4 = (u_4, v_4), (u_5, v_3), (u_3, v_5), (u_1, v_2), (u_2, v_1);$

$H_5 = (u_5, v_5), (u_1, v_4), (u_4, v_1), (u_2, v_3), (u_3, v_2).$

EXERCISES

11.3.1 Prove that the number of one-factors of K_{2n} is

$(2n-1) \cdot (2n-3) \cdot \cdots \cdot 3 \cdot 1.$

11.3.2 Prove that the number of one-factorizations of K_{2n} is divisible by

$$\frac{(2n-3)!}{(n-2)! 2^{n-2}}$$

11.4 One-Factorizations of K_8

11.3.3 Prove that a one-factorization containing a d-division must have at least $2(d + 1)$ vertices when d is odd, and at least $2(d + 2)$ when d is even.

11.3.4 Prove Lemma 11.4.

11.3.5 Prove that the one-factorization of K_{p+1} from the patterned starter in Z_p is perfect if and only if p is an odd prime.

11.3.6 Construct two perfect factorizations of K_{14}, using the facts that $14 = 13 + 1$ and $14 = 2 \times 7$, respectively. Are the one-factorizations isomorphic?

11.4 ONE-FACTORIZATIONS OF K_8

In this section we derive a complete list of one-factorizations of K_8. The main purpose of this exercise is to provide an example of the "intelligent brute force" methods which are sometimes the best available techniques for listing all designs of a given type.

A one-factorization of K_8 can contain no 4-division. As a first step, we divide our analysis into three parts, according as the one-factorization has a 3-division, has a 2-division but no 3-division, or has no 2-division.

We assume that K_8 has vertices 0, 1, 2, 3, 4, 5, 6, 7, and search for a one-factorization with factors A, B, C, D, E, F, G, where 01 lies in A, 02 in B, and so on.

LEMMA 11.9 There are exactly four nonisomorphic one-factorizations of K_8 that contain a 3-division.

Proof: Without loss of generality we can assume that the 3-division is $\{A, B, C\}$, and

A = 01 23 45 67.
B = 02 12 46 57
C = 03 12 47 56.

There are precisely six one-factors of K_8 that contain 04 and are disjoint from A, B, and C:

D_1 = 04 15 26 37 D_4 = 04 16 27 35
D_2 = 04 15 27 36 D_5 = 04 17 25 36
D_3 = 04 16 25 37 D_6 = 04 17 26 35.

Three of these cases can be eliminated immediately. If we carry out the vertex permutation $1 \mapsto 3 \mapsto 2 \mapsto 1$, $5 \mapsto 7 \mapsto 6 \mapsto 5$ on the one-factor A, we obtain C; the same permutation maps B to A; C goes to B; and D_2 becomes D_3. In symbols

$$(A, B, C, D_2)(132)(576) = (C, A, B, D_3).$$

So any one-factorization that contains $\{A, B, C, D_3\}$ will be isomorphic to one that contains $\{A, B, C, D_2\}$. Similarly,

$$(A, B, C, D_2)(123)(567) = (B, C, A, D_6),$$

$$(A, B, C, D_4)(23)(67) = (B, C, A, D_5).$$

So a complete list of extensions of $\{A, B, C\}$ can be constructed by considering D_1, D_2, and D_4. It is helpful to observe that D_1 forms a 2-division with all of A, B, and C; D_2 forms a 2-division with only one (namely, A), and D_4 forms 2-divisions with none of the three.

(i) Say that $D = D_1$. There are three candidates each for E, F, and G which are disjoint from A, B, C, and D_1, namely,

$$E_1 = 05\ 14\ 27\ 36 \quad F_1 = 06\ 14\ 27\ 35 \quad G_1 = 07\ 14\ 25\ 36$$
$$E_2 = 05\ 16\ 27\ 34 \quad F_2 = 06\ 17\ 24\ 35 \quad G_2 = 07\ 16\ 24\ 35$$
$$E_3 = 05\ 17\ 24\ 36 \quad F_3 = 06\ 17\ 25\ 34 \quad G_3 = 07\ 16\ 25\ 34.$$

Comparing these, we find four compatible sets of three factors:

$$\{E_1, F_2, G_3\}, \{E_1, F_3, G_2\}, \{E_2, F_2, G_1\}, \{E_3, F_1, G_3\}.$$

Now each of E_1, F_2, and G_3 forms a 2-division with each of A, B, C, and D_1, while each of the others forms a 2-division with exactly one of A, B, and C. It follows that the one-factorization which contains all three of these—$\{A, B, C, D_1, E_1, F_2, G_3\}$—cannot be isomorphic to the other three, which contain only one of them. Moreover, if we are to prove—for example—that $\{A, B, C, D_1, E_1, F_3, G_2\}$ and $\{A, B, C, D_1, E_2, F_2, G_1\}$ are isomorphic, we can use the division structure to find a suitable map. The former factorization has the divisions

$$\{A, B, C\}, \{A, D_1, E_1\}, \{A, F_3, G_3\}, \{B, D_1\}, \{B, E_1\}, \{C, D_1\}, \{C, E_1\},$$

while the latter has

11.4 One-Factorizations of K_8

$\{A,B,C\}$, $\{B,D_1,F_2\}$, $\{B,E_2,G_1\}$, $\{A,D_1\}$, $\{A,F_2\}$, $\{C,D_1\}$, $\{C,F_2\}$.

So any isomorphism must map A to B and $\{F_3,G_2\}$ to $\{E_2,G_1\}$, and it must map the two pairs $\{B,C\}$ and $\{D_1,E_1\}$ to the two pairs $\{A,C\}$ and $\{D_1,F_2\}$. One sees that there are 16 possible ways for the factors to be mapped: There are two ways to map F_3 and G_2; there are two ways to select where the sets $\{B,C\}$ and $\{D_1,E_1\}$ should go; and for each of the latter selections there are four choices (e.g., if it has been decided that $\{B,C\}$ goes to $\{D_1,F_2\}$ and therefore that $\{D_1,E_1\}$ goes to $\{B,C\}$, there are two ways to decide where B and C are mapped and two ways to decide where D_1 and E_1 are mapped). If we choose one of these combinations—for example,

$A \mapsto B$, $B \mapsto C$, $D_1 \mapsto D_1$, $F_3 \mapsto G_1$,

$C \mapsto A$, $E_1 \mapsto F_2$, $G_2 \mapsto E_2$,

there will be at most eight permutations that carry out the isomorphism: Once we decide where 0 is to be mapped, we know where every symbol must go. (In the example, suppose that $0 \mapsto 4$. Since A maps to B, the pair 01 must map to some pair in B. The only possible pair is 46, so $1 \mapsto 6$.) One then checks to see if any of the 128 permutations is in fact an isomorphism. This sounds tedious, but it is much shorter than checking all 8 possible permutations. In the actual case we find that the very first permutation tried—the case $0 \mapsto 0$ of our example—is an isomorphism:

$(A,B,C,D_1,E_1,F_3,G_2)(123)(567) = (B,C,A,D_1F_2,G_1E_2)$.

Similarly,

$(A,B,C,D_1,E_1,F_3,G_2)(132)(576) = (C,A,B,D_1,G_2,E_3,F_1)$.

So there are exactly two nonisomorphic factorizations in this case; examples are

$\{A,B,C,D_1,E_1,F_2,G_3\}$. $\{A,B,C,D_1,E_1,F_3,G_2\}$.

We shall call these factorizations $\mathcal{F}_1$ and $\mathcal{F}_2$.

(ii) Say that $D = D_2$. Suppose that we find a factorization in which the E-factor forms a 2-division with all of A, B, and C. Then the permutation (45)(67) leaves A, B, and C unchanged and maps the E-factor into a factor

04, which must mean that any factorization we obtain will be isomorphic to one that starts A, B, C, D_1. Similar arguments apply to the later factors. So we can ignore any factors that form 2-divisions with all of A, B, and C.

There are nine possible factors when the edges of A, B, C, and D_2 have been removed, namely

$$E_4 = 05\ 14\ 26\ 37 \qquad F_4 = 06\ 14\ 25\ 37 \qquad G_4 = 07\ 14\ 26\ 35$$
$$E_5 = 05\ 16\ 24\ 37 \qquad F_5 = 06\ 17\ 24\ 35 \qquad G_5 = 07\ 16\ 24\ 35$$
$$E_6 = 05\ 17\ 26\ 34 \qquad F_6 = 06\ 17\ 25\ 34 \qquad G_6 = 07\ 16\ 25\ 34.$$

F_5 and G_6 each form 2-divisions with A, B, and C, so they can be ignored. There remain three compatible sets:

$$\{E_4, F_6, G_5\},\ \{E_5, F_6, G_4\},\ \{E_6, F_4, G_5\}.$$

The first of the three cases has divisions $\{A, B, C\}$, $\{A, D_2, E_4\}$, $\{A, F_6, G_5\}$, $\{D_2, F_6\}$, $\{D_2, G_5\}$, $\{E_4, F_6\}$, and $\{E_4, G_5\}$. This is the same pattern as $\mathcal{F}_2$, and one suspects that the two factorizations may be isomorphic. This is in fact true:

$$(A, B, C, D_1, E_1, F_3, G_2)(26)(37)(45) = (A, F_6, G_5, E_4, D_2, B, C).$$

The other two factorizations have a new division structure of one 3-division and four maximal 2-divisions: The divisions are

$$\{A, B, C\},\ \{A, D_2\},\ \{A, F_6\},\ \{D_2, F_6\},\ \{E_5, G_4\}$$

and

$$\{A, B, C\},\ \{A, D_2\},\ \{A, G_5\},\ \{D_2, G_5\},\ \{E_6, F_4\},$$

respectively. If the two cases are to be isomorphic, we must find a map that takes

$$A \mapsto A,\quad \{B, C\} \mapsto \{B, C\},$$
$$\{D_2, F_6\} \mapsto \{D_2, G_5\},\quad \{E_5, G_4\} \mapsto \{E_6, F_4\}.$$

There are eight ways in which the factors can be mapped, and in each case there are eight permutations to be tested according as 0 maps to 0, 1, 2, 3, 4, 5, 6, 7: 64 permutations in all. We find that

11.4 One-Factorizations of K_8

$$(A,B,C,D_2,E_5,F_6,G_4)(23)(67) = (A,C,B,D_2,E_6,G_5,F_4).$$

So there is exactly one isomorphism class of one-factorizations here; as a representative we take

$$\mathcal{F}_3 = \{A,B,C,D_2,E_5,F_6,G_4\}.$$

(iii) Say that $D = D_4$. Using the same argument as before, we can assume that none of the E-, F-, or G-factors form a division with any of A, B, or C. The possible factors when A, B, C, and D_4 have been eliminated are nine in number, but six of them form divisions with A, B, or C [see Exercise 11.4.1(ii)]. Those remaining are

$$E_7 = 05\ 17\ 26\ 34, \quad F_7 = 06\ 14\ 25\ 37, \quad G_7 = 07\ 15\ 24\ 36.$$

These are compatible. We write

$$\mathcal{F}_4 = \{A,B,C,D_4,E_7,F_7,G_7\};$$

its maximal divisions are $\{A,B,C\}$, $\{D_4,E_7\}$, $\{D_4,F_7\}$, $\{D_4,G_7\}$, $\{E_7,F_7\}$, $\{E_7,G_7\}$, $\{F_7,G_7\}$. ∎

LEMMA 11.10 There is a unique one-factorization (up to isomorphism) which contains a 2-division but no 3-division.

Proof: We suppose that A and B form a 2-division; without loss of generality we can take

$$A = 01\ 23\ 45\ 67, \quad B = 02\ 13\ 46\ 57.$$

If C forms a division with A, it must contain edges 03 and 12, and we have a 3-division. Therefore, $\{A,C\}$ is not a division. Without loss of generality we can choose

$$C = 03\ 14\ 27\ 56.$$

There are now 16 possible factors:

$D_1 = 04\ 15\ 26\ 37 \qquad E_1 = 05\ 12\ 36\ 47$
$D_2 = 04\ 16\ 25\ 37 \qquad E_2 = 05\ 16\ 24\ 37$
$D_3 = 04\ 17\ 25\ 36 \qquad E_3 = 05\ 17\ 24\ 36$
$D_4 = 04\ 17\ 26\ 35 \qquad E_4 = 05\ 17\ 26\ 34$

$F_1 = 06\ 12\ 35\ 47 \quad G_1 = 07\ 15\ 24\ 36$

$F_2 = 06\ 15\ 24\ 37 \quad G_2 = 07\ 15\ 26\ 34$

$F_3 = 06\ 17\ 24\ 35 \quad G_3 = 07\ 16\ 24\ 35$

$F_4 = 06\ 17\ 25\ 34 \quad G_4 = 07\ 16\ 25\ 34.$

These fall into eight compatible quadruples. However, we can ignore

$\{D_1, E_1, F_3, G_4\}, \{D_1, E_1, F_4, G_2\}, \{D_1, E_3, F_1, G_4\},$

$\{D_2, E_1, F_3, G_2\}, \{D_2, E_3, F_1, G_2\}, \{D_4, E_1, F_2, G_4\},$

because $\{A, F_4, G_3\}, \{B, D_1, F_3\}, \{B, D_4, F_2\}, \{C, E_1, F_3\}$, and $\{C, E_3, F_1\}$ are 3-divisions. We are left with

$\{D_2, E_4, F_1, G_1\}, \{D_3, E_2, F_1, G_2\},$

and we find that

$(A, B, C, D_2, E_4, F_1, G_1)(12)(47) = (B, A, C, G_2, E_2, F_1, D_3).$

The representative we take is

$\mathcal{F}_5 = \{A, B, C, D_2, E_4, F_1, G_1\}.$ ∎

LEMMA 11.11 There is a unique one-factorization of K_8 with no divisions.

Proof: We can take

$A = 01\ 23\ 45\ 67 \quad B = 02\ 14\ 36\ 57.$

Since $\{A, C\}$ is not to be a division, we see that C cannot contain 12; since $\{B, C\}$ is not to be a division we see that C cannot contain 26. Similarly, D cannot contain 15 or 12.

The possible factors containing 03 are

$C_1 = 03\ 15\ 27\ 46 \quad C_3 = 03\ 17\ 24\ 56$

$C_2 = 03\ 16\ 25\ 47 \quad C_4 = 03\ 17\ 25\ 46.$

11.4 One-Factorization of K_8

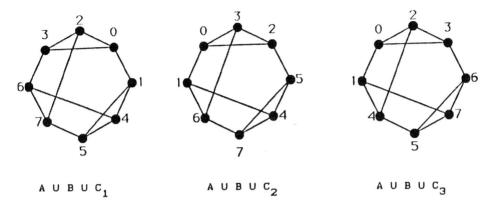

Figure 11.2

In order to see how to reduce the computations, look at Figure 11.2, which shows the graph $A \cup B \cup C_i$ for $i = 1, 2, 3$. These make it clear that $(032)(1675)$ maps $\{A, B, C_2\}$ to $\{A, B, C_1\}$ and $(23)(4657)$ maps $\{A, B, C_3\}$ to $\{A, B, C_1\}$. So only the cases $C = C_1$ and $C = C_4$ need be considered.

(i) Case $C = C_1$. We find that the following factors are compatible with A, B, and C_1 and do not form a division with any of them:

$D_1 = 04\ 16\ 25\ 37 \qquad F_1 = 06\ 12\ 35\ 47$

$D_2 = 04\ 17\ 26\ 35 \qquad F_2 = 06\ 13\ 25\ 47$

$E_1 = 05\ 16\ 24\ 37 \qquad G_1 = 07\ 12\ 34\ 56$

$E_2 = 05\ 17\ 26\ 34 \qquad G_2 = 07\ 13\ 24\ 56.$

The only compatible sets are

$\{C_1, D_1, E_2, F_1, G_2\}, \{C_1, D_2, E_1, F_2, G_1\}.$

(ii) Case $C = C_4$. The compatible factors are

$D_3 = 04\ 13\ 27\ 56 \qquad F_3 = 06\ 12\ 35\ 47$

$D_4 = 04\ 16\ 27\ 35 \qquad F_4 = 06\ 15\ 24\ 37$

$E_3 = 05\ 13\ 26\ 47 \qquad G_3 = 07\ 12\ 34\ 56$

$E_4 = 05\ 16\ 24\ 37 \qquad G_4 = 07\ 15\ 26\ 34.$

The compatible sets are

$$\{C_4, D_3, E_4, F_3, G_4\}, \{C_4, D_4, E_3, F_4, G_3\}.$$

After a little experimentation, we find

$$(A, B, C_1, D_2, E_1, F_2, G_1)(01457632) = (B, A, F_1, C_1, E_2, G_2, D_1),$$

$$(A, B, C_4, D_3, E_4, F_3, G_4)(02367541) = (B, A, E_2, F_1, G_2, C_1, D_1),$$

$$(A, B, C_4, D_4, E_3, F_4, G_3)(04625)(17) = (F_1, A, E_2, C_1, D_1, G_2, B).$$

So the four factorizations are isomorphic. We define

$$\mathcal{F}_6 = \{A, B, C_1, D_1, E_2, F_1, G_2\}. \qquad \blacksquare$$

THEOREM 11.12 There are precisely six nonisomorphic one-factorizations of K_8. $\blacksquare$

The six representative one-factorizations, together with their maximal divisions, are shown in Figure 11.3.

It is instructive to find the automorphism groups of the one-factorizations. We denote the automorphism group of $\mathcal{F}_i$ by G_i. Then the groups are easily calculated—for all but G_6, the division structure is a useful guide; the method for G_6 is outlined in Exercise 11.4.2. We find the following results.

THEOREM 11.13 The groups G_i are as follows:

(i) G_1 has order 1344 and is triply transitive. G_1 is the group $AGL(3,2)$ represented as a permutation group on the points of the 3-space, and may be generated by (0123)(4567), (1245736), and (23)(45);

(ii) G_2 has order 64 and is transitive of rank 4; $G_2(0)$ has orbits $\{0\}$, $\{1\}$, $\{2,3,4,5\}$, and $\{6,7\}$. It is the split extension of the $Z_2 \times Z_2 \times Z_2 \times Z_2$ generated by (01)(67), (06)(17), (23)(45), and (24)(35), by the $Z_2 \times Z_2$ generated by (02)(13)(46)(57) and (45)(67);

(iii) G_3 has order 16 and is transitive of rank 6; $G_3(0)$ has orbits $\{0\}$, $\{1\}$, $\{2\}$, $\{3\}$, $\{46\}$, and $\{57\}$. G_3 is $D_8 \times Z_2$, where D_8 is generated by (01)(23)(45)67);

(iv) G_4 has order 96 and is transitive of rank 3; $G_4(0)$ has orbits $\{0\}$, $\{1,2,3\}$, and $\{4,5,6,7\}$. It is the split extension of the $Z_2 \times Z_2 \times Z_2 \times Z_2$ generated by (01)(23), (02)(13), (45)(67), and (46)(57), by the S_3 generated by (04)(15)(27)(36) and (123)(567);

11.4 One-Factorizations of K_8

$\mathcal{F}_1$: $\alpha_3 = 7$, $\alpha_2 = 0$

A = 01 23 45 67	ABC
B = 02 13 46 57	ADE
C = 03 12 47 56	AFG
D = 04 15 26 37	BDF
E = 05 14 27 36	BEG
F = 06 17 24 35	CDG
G = 07 16 25 34	CEF

$\mathcal{F}_2$: $\alpha_3 = 3$, $\alpha_2 = 4$

A = 01 23 45 67	ABC
B = 02 13 46 57	ADE
C = 03 12 47 56	AFG
D = 04 15 26 37	BD
E = 05 14 27 36	BE
F = 06 17 25 34	CD
G = 07 16 24 35	CE

$\mathcal{F}_3$: $\alpha_3 = 1$, $\alpha_2 = 4$

A = 01 23 45 67	ABC
B = 02 13 46 57	AD
C = 03 12 47 56	AF
D = 04 15 27 36	DF
E = 05 16 24 37	EG
F = 06 17 25 34	
G = 07 14 26 35	

$\mathcal{F}_4$: $\alpha_3 = 1$, $\alpha_2 = 6$

A = 01 23 45 67	ABC
B = 02 13 46 57	DE
C = 03 12 47 56	DF
D = 04 16 27 35	DG
E = 05 17 26 34	EF
F = 06 14 25 37	EG
G = 07 15 24 36	FG

$\mathcal{F}_5$: $\alpha_3 = 0$, $\alpha_2 = 3$

A = 01 23 45 67	AB
B = 02 13 46 57	CF
C = 03 14 27 56	EG
D = 04 16 25 37	
E = 05 17 26 34	
F = 06 12 35 47	
G = 07 15 24 36	

$\mathcal{F}_6$: $\alpha_3 = 0$, $\alpha_2 = 0$

A = 01 23 45 67	
B = 02 14 36 57	
C = 03 15 27 46	
D = 04 16 25 37	
E = 05 17 26 34	
F = 06 12 35 47	
G = 07 13 24 56	

Figure 11.3 One-factorizations of K_8 and their maximal divisions.

(v) G_5 has order 24 and is transitive of rank 4; $G_5(0)$ has orbits $\{0\}$, $\{4\}$, $\{1,3,5\}$, and $\{2,6,7\}$. G_5 is $A_4 \times Z_2$, where A_4 is generated by $(135)(267)$ and $(067)(134)$, and Z_2 is generated by $(04)(16)(25)(37)$;

(vi) G_6 has order 42 and fixes 2; it is the sharply doubly transitive group of degree 7, and is generated by (0643257) and (054637). ∎

Once we know the size of the automorphism group of a one-factorization, we know how many factorizations exist which are isomorphic to it. For example, $\mathcal{F}_1$ has 1344 automorphisms. If we apply all the 8! possible permutations of eight symbols to $\mathcal{F}_1$, we obtain all the one-factorizations isomorphic to it, and each different factorization will occur 1344 times. As the number of distinct factorizations isomorphic to $\mathcal{F}_1$ is

$$\frac{8!}{1344} = \frac{40,320}{1344} = 30.$$

Similar calculations can be carried out for the other factorizations.

COROLLARY 11.13.1 There are 6240 different one-factorizations of K_8.

Proof:

$$\frac{8!}{1344} + \frac{8!}{64} + \frac{8!}{16} + \frac{8!}{96} + \frac{8!}{24} + \frac{8!}{42}$$
$$= 30 + 630 + 2520 + 420 + 1680 + 960$$
$$= 6240. \qquad\blacksquare$$

EXERCISES

11.4.1 In the notation of the proof of Lemma 11.9:
  (i) Find all permutations that map $\{A,B,C,D_2,E_5,F_6,G_4\}$ to $\{A,B,C,D_2,E_6,F_4,G_5\}$;
  (ii) Prove that there are nine factors which extend $\{A,B,C,D_4\}$ and that six of them form divisions with A, B, or C.

11.4.2 Consider the factorization $\mathcal{F}_6$.
  (i) Show that the union of any three factors contains either exactly one or exactly two triangles.
  (ii) If the union of three factors contains exactly one triangle, call the three points of the triangle a <u>triad</u>. Prove that $\mathcal{F}_6$ contains exactly 14 triads, namely,

  025, 027, 034, 036, 047, 056, 236,
  237, 245, 246, 345, 357, 467, 567.

11.5 An Application to Projective Planes

(iii) Observing that the structure of the set of triads is preserved by isomorphism, construct the automorphism group of $\mathcal{F}_6$.

11.5 AN APPLICATION TO PROJECTIVE PLANES

As a somewhat unexpected application, we shall use the one-factorizations of K_6 to prove the uniqueness of the projective plane $PG(2,4)$.

We first return to the study of maximal arcs in projective planes (see Section 5.5). Any 2-arc in a $PG(2,n)$ has at most $n+2$ points; those with $n+1$ and $n+2$ points are called type I and type II ovals, respectively.

LEMMA 11.14 If a maximal 2-arc in a $PG(2,n)$ has m points, then

$$n \leq \binom{m-1}{2} \tag{11.9}$$

Proof: Suppose that c is an m-point 2-arc in a $PG(2,n)$. If $m = n+2$, then (11.9) is true. So let us suppose that $m < n+2$. Then any point p on c lies on at least one tangent to c. Suppose that ℓ is a tangent at p. There are $m-1$ points other than p on c, so there are exactly $\binom{m-1}{2}$ secants to c which do not contain p. If $n > \binom{m-1}{2}$, there must be at least one point of ℓ, other than p, which does not lie on any of these secants; call it q. Clearly, q cannot lie on any secant through p, either. So no line through q contains more than one point of c, and $c \cup \{q\}$ is also a 2-arc. So c was not maximal. It follows that for any maximal 2-arc the assumption that $n > \binom{m-1}{2}$ must be false. ■

Let us apply Lemma 11.14 to the case $n = 4$. Since

$$4 \leq \binom{m-1}{2} \implies m \geq 5$$

it follows that a maximal arc in a $PG(2,4)$ must contain at least 5 points. But if one contained 5 points, it is a type I oval, and can be extended to a type II oval on 6 points, since $n = 4$ which is even (see Theorem 5.11). So any maximal arc in $PG(2,4)$ must contain 6 points (6 is the maximum, since $6 = n+2$). So any $PG(2,4)$ contains a 6-point 2-arc.

Suppose that c is any 6-point 2-arc in a $PG(2,4)$. Select any point p not on c. Every line through p meets c in 0 or 2 points. So the set of lines through p determines a partition of c into three pairs of points. In other words, it determines a one-factor of the K_6 based on the points of c. Call this one-factor $F(p)$.

If p and q are distinct points not on c, then F(p) and F(q) have an edge in common if and only if the line pq is a secant to c; and F(p) and F(q) cannot have more than one common edge. So the 15 points not lying on c correspond to different one-factors of K_6; as we found in Section 11.3 that there are exactly 15 one-factors, each occurs precisely once. Moreover, if ℓ is an external line to c, the five one-factors corresponding to the points on ℓ are disjoint, so they form a one-factorization of K_6. One-factorizations corresponding to the distinct external lines have exactly one common factor, corresponding to the point of intersection of the two lines. Since a 6-point arc has no tangents and 15 secants, it has 6 external lines, corresponding precisely to the six one-factorizations of K_6.

We now observe that these structural facts are sufficient to determine PG(2,4) completely.

THEOREM 11.15 All planes PG(2,4) are isomorphic.

Proof: Any plane PG(2,4) must contain a maximal 2-arc c of 6 points. Label its points 0, 1, 2, 3, 4, 5. There will be 15 secants to c; write ℓ_{ij} for the secant containing i and j. Given two secants ℓ_{ij} and ℓ_{pq}, where $\{i,j,p,q\}$ are all different, label the points of intersection $\ell_{ij} \cap \ell_{pq}$ with the one-factor of K_6 containing the edges ij and pq. The third secant through the point will necessarily be labeled with the third edge of the factor. An elementary counting argument shows that all 15 exterior points will be labeled in this way, and the membership of all the secants is determined.

Finally, there will be six external lines to c. These must correspond to the six one-factorizations; the members of a line are labeled with the factors in one of the factorizations.

The structure is completely determined by these considerations, up to permutation of the point and line labels. So the plane is determined up to isomorphism. ∎

The theorem actually tells us more about PG(2,4): Given any two ovals in the plane, there must be an automorphism that carries one into the other. These considerations apply only to PG(2,4); when eight points are allowed, the variability and number of one-factorizations are already too great for useful application to the isomorphism problem.

BIBLIOGRAPHIC REMARKS

One-factors have been studied for a long time. Many of the standard results appear in Konig's book [5] and other graph theory texts. The discussion of divisions as a clue to isomorphism was introduced in [9], and Theorem 11.7 was proven in that paper. Subsequently, Lindner, Mendelsohn, and Rosa [7] proved that the number of nonisomorphic one-factorizations of K_{2n} tends to infinity with n.

Bibliographic Remarks

Perfect factorizations were mentioned by Kotzig [6]. Anderson reintroduced them in [1] and studied them in a series of papers.

The listing of one-factorizations of K_8 has been reproduced several times in the combinatorial literature. The earliest version which we can find is that of Dickson and Safford [3] in 1906. (Dickson proposed a problem; Safford submitted a partial solution; and Dickson provided most of the remaining parts of the solution—the automorphism group which we called G_3 was not calculated.)

For further references to factorizations, the reader should consult the excellent survey by Mendelsohn and Rosa [8].

The uniqueness of PG(2,4) has been proven in some books on finite geometry—see, for example, Chapter 14 of [4]. Recently, Beutelspacher [2] has published an expository article on the subject.

1. B. A. Anderson, "Finite topologies and Hamiltonian paths," Journal of Combinatorial Theory 14B(1973), 87-93.
2. A Beutelspacher, "21-6=15: A connection between two distinguished geometries," American Mathematical Monthly 93(1986), 29-41.
3. L. E. Dickson and F. H. Safford, "Solution to problem 8 (group theory)," American Mathematical Monthly 13(1906), 150-151.
4. J. W. P. Hirschfeld, Projective Geometries over Finite Fields (Clarendon, Oxford, 1979).
5. D. Konig, Theorie der endlichen und unendlichen Graphen (B. G. Teubner, Leipzig, 1936).
6. A. Kotzig, "Hamilton graphs and Hamilton circuits," Theory of Graphs and Its Applications (Nakl. CSAV, Prague, 1964).
7. C. C. Lindner, E. Mendelsohn, and A. Rosa, "On the number of 1-factorizations of the complete graph," Journal of Combinatorial Theory 20B(1976), 265-282.
8. E. Mendelsohn and A. Rosa, "One-factorizations of the complete graph—A survey," Journal of Graph Theory 9(1985), 43-65.
9. W. D. Wallis, "On one-factorizations of complete graphs," Journal of the Australian Mathematical Society 16(1973), 161-171.

12
Triple Systems

12.1 CONSTRUCTION OF TRIPLE SYSTEMS

Balanced block designs with block size 3 are called <u>triple systems</u>. In particular a <u>Steiner triple system</u> is a triple system with $\lambda = 1$.

Suppose that a triple system on v treatments exists. Then from (1.1) and (1.2) we have

$$3b = vr,$$

$$\lambda(v - 1) = 2r.$$

Therefore, $r = \lambda(v - 1)/2$ and $b = \lambda v(v - 1)/6$. So the parameters λ and v are sufficient to specify all the parameters of the design. We shall denote a triple system with parameters λ and v by $T(\lambda, v)$. [Another common notation for such a system is $S_\lambda(2, 3, v)$.]

Let us consider Steiner triple systems in particular: then $\lambda = 1$, $r = (v - 1)/2$, and $b = v(v - 1)/6$. Since r and b must be integers, v must be odd, and 6 must divide $v(v - 1)$. It follows that $v \equiv 1$ or $3 \pmod 6$. Similar calculations can be carried out for larger λ. It is clear that the conditions depend on the primacy of λ to 6. We easily prove:

LEMMA 12.1 If there is a $T(\lambda, v)$, then:

$(\lambda, 6) = 1$ implies that $v \equiv 1$ or $3 \pmod 6$;
$(\lambda, 6) = 2$ implies that $v \equiv 0$ or $1 \pmod 3$;
$(\lambda, 6) = 3$ implies that $v \equiv 1 \pmod 2$.

Also, v cannot equal 2. ∎

In this section we prove that these necessary conditions are sufficient, giving elementary constructions using Latin squares.

12.1 Construction of Triple Systems

THEOREM 12.2 There is a $T(\lambda, v)$ for all positive integers v and λ not excluded by Lemma 12.1.

Proof: Suppose that we prove the existence of a $T(\lambda, v)$. Then, by taking s copies of each block in that design, we can construct a $T(s\lambda, v)$. So if we can find a $T(1, v)$ for all $v \equiv 1$ or 3 (mod 6), this will prove the existence of a $T(\lambda, v)$ for all λ when $v \equiv 1$ or 3 (mod 6). In particular, we shall have a $T(2, v)$ in those cases; if we also obtain a $T(2, v)$ when $v \equiv 0$ or 4 (mod 6), we shall have a $T(2, v)$ for all $v \equiv 0$ or 1 (mod 3), so we can get a $T(\lambda, v)$ for all those v and for all even λ. By similar arguments we see that it is sufficient to construct the following designs:

$T(1, v)$ for all $v \equiv 1, 3$ (mod 6);
$T(2, v)$ for all $v \equiv 0, 4$ (mod 6);
$T(3, v)$ for all $v \equiv 5$ (mod 6);
$T(6, v)$ for all $v \equiv 2$ (mod 6), $v \neq 2$.

(i) We start with the case $\lambda = 2$. To construct a $T(2, 3n)$ and a $T(2, 3n+1)$, we shall need an idempotent Latin square of side n. This is impossible when $n = 2$, so we first observe that there exist a $T(2, 6)$ and a $T(2, 7)$. Suitable sets of triples are:

$T(2, 6)$: 123, 124, 135, 146, 156, 236, 245, 256, 345, 346
$T(2, 7)$: 123, 124, 135, 146, 157, 167, 236, 247, 256, 257, 345, 347, 367, 456.

Now suppose that $n \neq 2$. Let $A = (a_{ij})$ be an idempotent Latin square of order n. The existence of such a square follows from (10.5), Exercise 10.3.1, and Theorem 10.13 (and a much easier proof is given in Exercise 12.1.2). Consider all the triples

$$\{x^1, y^1, a_{xy}^2\}, \quad 1 \leq x \leq n, \quad 1 \leq y \leq n.$$

If x and z are any two numbers in the range $(1 \cdot \cdot n)$, then $a_{xy} = z$ for exactly one value of y and $a_{yx} = z$ for exactly one value of y. Moreover, if $z \neq x$, the y-values are also not equal to x. So the triples listed will among them contain every unordered pair of the form $\{x^1, z^2\}$ with $1 \leq x \leq n$, $1 \leq z \leq n$, and $x \neq z$, exactly twice each. Moreover, every pair $\{x^1, y^1\}$ with $x \neq y$ occurs exactly twice. To avoid "triples" with repeated elements, we omit the case $x \neq y$. Then the triples

$$\left.\begin{aligned} \{x^1, y^1, a_{xy}^2\} \\ \{x^2, y^2, a_{xy}^3\} \\ \{x^3, y^3, a_{xy}^1\} \end{aligned}\right\} : 1 \leq x \leq n, 1 \leq y \leq n, x \neq y \qquad (12.1)$$

clearly form a T(2, 3n) based on the symbols $1^1, 2^1, \ldots, n^1, 1^2, 2^2, \ldots, n^2, 1^3, 2^3, \ldots, n^3$, except for the fact that no pair $\{x^i, x^j\}$ with $i \neq j$ is represented. If we append

$$\{x^1, x^2, x^3\}, \quad 1 \leq x \leq n, \quad \text{twice each} \tag{12.2}$$

we have a T(2, 3n).

To construct a T(2, 3n + 1) we use the treatments $1^2, 2^2, \ldots, n^3$ again together with a symbol ∞. The triples (12.1) may be used, but they do not contain any copies of the pairs $\{x^i, x^j\}$ with $i \neq j$, nor the pairs $\{x^i, \infty\}$. The triples

$$\left.\begin{matrix} \{x^1, x^2, x^3\} \\ \{x^1, x^2, \infty\} \\ \{x^2, x^3, \infty\} \\ \{x^3, x^1, \infty\} \end{matrix}\right\} : 1 \leq x \leq n, \tag{12.3}$$

together with (12.1), from a T(2, 3n + 1).

(ii) Suppose that the Latin square A used in part (i) was symmetric. Then, for example,

$$\{x^1, y^1, a_{xy}^2\} = \{y^1, x^1, a_{yx}^2\},$$

and the triples in (12.1) and (12.2) consist of a certain set of triples, each one taken twice. So the T(2, 3n) is two copies of a T(1, 3n) joined together. It follows that the existence of a symmetric idempotent Latin square L of side n implies the existence of a T(1, 3n).

If n is odd, say n = 2k - 1, such a square L is easy to construct: One example is $L = (\ell_{ij})$, where

$$\ell_{ij} \equiv k(i + j) \pmod{2k - 1}.$$

(In other words, L is obtained by "back-circulating" its first row; the first row has been chosen in such a way as to give the diagonal elements in the proper order.) So there is a T(1, 3n) whenever n is odd, which is to say that there is a T(1, v) when $v \equiv 3 \pmod 6$.

As one would expect, no such Latin square exists when n is even (see Exercise 12.1.3), so some modification is needed for the case of T(1, v) with $v \equiv 1 \pmod 6$. A slightly different symmetric Latin square must be used.

12.1 Construction of Triple Systems

We let M be a Latin square of side 2k which is symmetric and has diagonal $(1, 2, \ldots, k, 1, 2, \ldots, k)$. Such a square always exists: for example, $M = (m_{ij})$, where

$$\left. \begin{array}{l} m_{ij} \equiv \frac{1}{2}(i + j) \quad \text{when } i + j \text{ is even} \\ m_{ij} \equiv \frac{1}{2}(i + j + 1) \quad \text{when } i + j \text{ is odd} \end{array} \right\} \pmod{2k}.$$

We define a $T(1, 6k + 1)$ based on the symbols $\{x^1\}$, $\{x^2\}$, and $\{x^3\}$ for $1 \leq x \leq 2k$ and a symbol ∞, by the triples

$$\{x^1, x^2, x^3\}, \qquad 1 \leq x \leq k$$

$$\left. \begin{array}{l} \{\infty, x^1, (x-n)^2\} \\ \{\infty, x^2, (x-n)^3\} \\ \{\infty, x^3, (x-n)^1\} \end{array} \right\} : k + 1 \leq x \leq 2k,$$

$$\left. \begin{array}{l} \{x^1, y^1, m_{xy}^2\} \\ \{x^2, y^2, m_{xy}^3\} \\ \{x^3, y^3, m_{xy}^1\} \end{array} \right\} : 1 \leq x < y \leq 2n.$$

The verification is straightforward (see Exercise 12.1.4).

(iii) The case of $T(3, v)$ for $v \equiv 5 \pmod 6$ is the easiest of the constructions. One simply takes all the triples

$$\{y, x + y, 2x + y\} : 0 < x < \tfrac{1}{2}v, \quad 0 \leq y < v$$

where all the additions are reduced modulo v.

(iv) Finally, we construct a $T(6, 6k + 2)$ based on the symbols $1^1, 2^1, \ldots, n^1, 1^2, 2^2, \ldots, n^2, 1^3, 2^3, \ldots, n^3, \infty^1, \infty^2$, where $n = 2k$. The general approach is to take three copies of a $T(2, 3n)$ and modify the triples to allow for the symbols ∞^1 and ∞^2.

Suppose that A is an idempotent Latin square of side n, $n \neq 2$. If we take each of the triples (12.1) three times, we have covered all pairs of symbols exactly six times except that the pairs of type $\{x^i, x^j\}$ and $\{x^i, \infty^j\}$ are completely missing. If one were to take all the triples

$$\{\infty^i, x^1, x^2\}, \{\infty^i, x^2, x^3\}, \{\infty^i, x^3, x^1\} : 1 \leq x \leq n, i = 1, 2, \qquad (12.4)$$

three times each, all pairs would be covered except $\{\infty^1, \infty^2\}$. So we take all the triples (12.4) except for those with $x = 1$ and append them to three copies of the list (12.1). Then it is necessary to find a set of triples that contain three copies of all the pairs in the triples

$$\{\infty^1, 1^1, 1^2\}, \{\infty^1, 1^2, 1^3\}, \{\infty^1, 1^3, 1^1\}, \{\infty^2, 1^1, 1^2\},$$
$$\{\infty^2, 1^2, 1^3\}, \{\infty^2, 1^3, 1^1\}$$

as well as six copies of the pair $\{\infty^1, \infty^2\}$. What is required is in fact a $T(6, 5)$ on the symbols $1^1, 1^2, 1^3, \infty^1,$ and ∞^2. But such a system exists, from part (iii).

To summarize, the following triples are used:

$$\left.\begin{array}{l}\{x^1, y^1, a_{xy}^2\} \\ \{x^2, y^2, a_{xy}^3\} \\ \{x^3, y^3, a_{xy}^1\}\end{array}\right\} : 1 \leq x \leq n,\ 1 \leq y \leq n,\ x \neq y,\ \text{each triple taken three times}$$

$$\left.\begin{array}{l}\{\infty^1, x^1, x^2\}, \{\infty^2, x^1, x^2\} \\ \{\infty^1, x^2, x^3\}, \{\infty^2, x^2, x^3\} \\ \{\infty^1, x^3, x^1\}, \{\infty^2, x^3, x^1\}\end{array}\right\} : 2 \leq x \leq n,\ \text{each triple taken three times}$$

$$\{\infty^1, \infty^2, 1^1\}, \{\infty^1, \infty^2, 1^2\}, \{\infty^1, \infty^2, 1^3\},$$
$$\{\infty^1, 1^1, 1^2\}, \{\infty^1, 1^2, 1^3\}, \{\infty^1, 1^3, 1^1\},$$
$$\{\infty^2, 1^1, 1^2\}, \{\infty^2, 1^2, 1^3\}, \{\infty^2, 1^3, 1^1\},$$
$$\{1^1, 1^2, 1^3\}.$$

This covers all the cases with $v \equiv 2 \pmod 6$ except for $v = 8$. A $T(6, 8)$ is easily constructed. Suppose that B is the set of all blocks of a $T(5, 7)$—for example, one can take five copies of the well-known $T(1, 7)$—based on $S = \{1, 2, 3, 4, 5, 6, 7\}$. Append to B the 21 blocks formed by taking the 21 unordered pairs of members of S and adding the element ∞ to each. These blocks form a $T(6, 8)$.

12.2 Subsystems

The assumption that n is even was not used in the last part of Theorem 12.2, so our proof of part (iv) provides a construction for a T(6, v) when $v \equiv 5 \pmod{6}$. This was not needed, since one can double the T(3, v) constructed in part (iii). However, the proof cannot be simplified by omitting this duplication. Moreover, the two T(6, v)'s constructed are not isomorphic. (The proof is left as an exercise.)

EXERCISES

12.1.1 Verify that the construction given for a T(1, 6n + 1), given in the proof of Theorem 12.2, actually works.

12.1.2 If n is odd, define $L_n = (\ell_{ij})$ by $\ell_{ij} = 2i - j \pmod{n}$. If n is even, n > 2, define T_n to be the Latin square of order n derived from L_{n-1} by replacing the (1,2), (2,3), ..., (n - 2, n - 1) and (n - 1, 1) elements by n and then appending a last row and column which make the square Latin; the last column is (n - 1, 1, 2, ..., n - 2, n) and the last column (n - 2, n - 1, 1, ..., n - 3, n). Prove that the L_n and T_n are idempotent Latin squares.

12.1.3 Suppose that L is a symmetric Latin square of side n, and that the symbol x occurs k times on the diagonal.
 (i) How many times does x occur above the diagonal?
 (ii) Prove that $k \equiv n \pmod{2}$.
 (iii) If L is also idempotent, prove that n is odd.

12.1.4 Show that the triples listed in part (iii) of the proof of Theorem 12.2 actually form a T(1, 6k + 1).

12.1.5 Prove that there are nonisomorphic triple systems T(6, 6n + 5) for all n.

12.1.6 Suppose that a T(v - 3, v - 1) exists. Show that there is a T(v - 2, v) with the blocks of the T(v - 3, v - 1) as some of its blocks.

12.2 SUBSYSTEMS

By a <u>subsystem</u> of a triple system we mean a subset of the set of blocks which forms a triple system on some subset of the set of treatments. Although these can be discussed for any λ, we shall consider only the case $\lambda = 1$, which is the most interesting in many ways.

THEOREM 12.3 Suppose that there exist a Steiner triple system A with v_1 treatments, and a Steiner triple system B with v_2 treatments which contains a subsystem C on v_3 treatments. Then there exists a Steiner triple system D on $v = v_3 + v_1(v_2 - v_3)$ treatments, which contains v_1 subsystems on v_2 treatments each, one subsystem on v_1 treatments and one subsystem on v_3 treatments.

Proof: Write $s = v_2 - v_3$. We construct a $T(1,v)$ based on the v-set

$$S_0 \cup S_1 \cup \cdots \cup S_{v_1},$$

where

$$S_0 = \{m_1, m_2, \ldots, m_{v_3}\},$$
$$S_1 = \{n_{11}, n_{12}, \ldots, n_{1s}\},$$
$$S_2 = \{n_{21}, n_{22}, \ldots, n_{2s}\},$$
$$\vdots$$
$$S_{v_1} = \{n_{v_1 1}, n_{v_1 2}, \ldots, n_{v_1 s}\}.$$

The blocks are constructed according to the following rules.

(i) Associate the elements of S_0 with the treatments of C and form the triple $\{m_i, m_j, m_k\}$ if and only if $\{i, j, k\}$ is a triple of C. This gives $v_3(v_3-1)/6$ triples, which we take as blocks of D.

(ii) For each i, associate $S_0 \cup S_i$ with the treatments of B in such a way that the elements of S_0 are always associated with the treatments of C, just as they were in (i), and form the triples on $S_0 \cup S_i$ which correspond to the triples of B. Some of these contain three members of S_0, and they are precisely the triples which were constructed in (i); these can be ignored. There remain

$$\frac{v_2(v_2-1) - v_3(v_3-1)}{6}$$

triples which each contain at most one element of S_0; we take these triples as blocks of D.

(iii) Finally, consider the given system A on the treatments $\{1, \ldots, v_1\}$; if $\{j, k, \ell\}$ is a triple of this system, adjoin to D all triples $\{n_{jx}, n_{ky}, n_{\ell z}\}$, where x, y, and z satisfy

$$x + y + z \equiv 0 \pmod{s}, \quad 1 \leq x, y, z \leq s.$$

In solving this congruence, x and y can be chosen in s ways each; z is then uniquely determined. Hence each of the $v_1(v_1-1)/6$ blocks of A leads to s^2 additional blocks of D.

From (i) we get a subsystem isomorphic to C, on the v_3 elements of S_0. From (ii) we get altogether v_1 subsystems, each isomorphic to B, on

12.2 Subsystems

the v_2 elements of $S_0 \cup S_i$ for each i; each of these subsystems contains that of (i) as a subsystem of itself. The blocks constructed in (iii), containing only $n_{1s}, n_{2s}, \ldots, n_{v_1 s}$, form a subsystem isomorphic to A.

Obviously, D has $v_3 + v_1(v_2 - v_3)$ treatments and constant block size 3. We check on balance and verify that any two treatments mutually belong to exactly one triple of D: A triple containing two elements of S_0 in fact contains three elements of S_0, uniquely determined by (i); a triple containing one element of S_0 and one of S_i, $i \neq 0$, contains another element of S_i, uniquely determined by (ii) for each i; a triple containing two elements of S_i, $i \neq 0$, contains either an element of S_0 or a third element of the same S_i, uniquely determined by (ii); a triple containing n_{jx} and n_{ky}, $j \neq k$, contains also $n_{\ell z}$, uniquely determined by (iii). Hence $\lambda = 1$ for D, and D is a $T(1, v)$. ∎

As an example, we construct a $T(1,15)$ using $v_1 = 1$, $v_2 = 7$, and $v_3 = 3$. We have $s = 4$. The constituent subsystems are:

A, on 3 treatments, with block $\{1,2,3\}$;
B, on 7 treatments, with blocks $\{1,2,3\}$, $\{1,4,5\}$, $\{1,6,7\}$, $\{2,4,6\}$, $\{2,5,7\}$, $\{3,4,7\}$, $\{3,5,6\}$;
C, on 3 treatments, with block $\{1,2,3\}$.

Then

$$S_0 = \{m_1, m_2, m_3\},$$
$$S_1 = \{n_{11}, n_{12}, n_{13}, n_{14}\},$$
$$S_2 = \{n_{21}, n_{22}, n_{23}, n_{24}\},$$
$$S_3 = \{n_{31}, n_{32}, n_{33}, n_{34}\}.$$

The block of type (i) is $\{m_1, m_2, m_3\}$. To form the blocks of type (ii) from $S_0 \cup S_i$, we set up the correspondence

$$(1,2,3,4,5,6,7) \longleftrightarrow (m_1, m_2, m_3, n_{11}, n_{12}, n_{13}, n_{14}),$$

so that the blocks of the $T(1,7)$ are $\{m_1, m_2, m_3\}$, $\{m_1, n_{11}, n_{12}\}$, $\{m_1, n_{13}, n_{14}\}$, and so on. The first block is deleted, and the remaining blocks are

$\{m_1, n_{11}, n_{12}\}$, $\{m_1, n_{13}, n_{14}\}$, $\{m_2, n_{11}, n_{13}\}$,
$\{m_2, n_{12}, n_{14}\}$, $\{m_3, n_{11}, n_{14}\}$, $\{m_3, n_{12}, n_{13}\}$.

Replacing S_1 by S_2 and then by S_3 we obtain the blocks

$\{m_1, n_{21}, n_{22}\}$, $\{m_1, n_{23}, n_{24}\}$, $\{m_2, n_{21}, n_{23}\}$,

$\{m_2, n_{22}, n_{24}\}$, $\{m_3, n_{21}, n_{24}\}$, $\{m_3, n_{22}, n_{23}\}$,

$\{m_1, n_{31}, n_{32}\}$, $\{m_1, n_{33}, n_{34}\}$, $\{m_2, n_{31}, n_{33}\}$,

$\{m_2, n_{32}, n_{34}\}$, $\{m_3, n_{31}, n_{34}\}$, $\{m_3, n_{32}, n_{33}\}$.

Finally the system on v_1 treatments has just one block $\{1, 2, 3\}$; so we take all the blocks of the form $\{n_{1x}, n_{2y}, n_{3z}\}$, where $x + y + z \equiv 0 \pmod 4$, as the blocks of type (iii). If we choose x and y, then z is uniquely determined, so these blocks are

$\{n_{11}, n_{21}, n_{32}\}$, $\{n_{11}, n_{22}, n_{31}\}$, $\{n_{11}, n_{23}, n_{34}\}$, $\{n_{11}, n_{24}, n_{33}\}$,

$\{n_{12}, n_{21}, n_{31}\}$, $\{n_{12}, n_{22}, n_{34}\}$, $\{n_{12}, n_{23}, n_{33}\}$, $\{n_{12}, n_{24}, n_{32}\}$,

$\{n_{13}, n_{21}, n_{34}\}$, $\{n_{13}, n_{22}, n_{33}\}$, $\{n_{13}, n_{23}, n_{32}\}$, $\{n_{13}, n_{24}, n_{31}\}$,

$\{n_{14}, n_{21}, n_{33}\}$, $\{n_{14}, n_{22}, n_{32}\}$, $\{n_{14}, n_{23}, n_{31}\}$, $\{n_{14}, n_{24}, n_{34}\}$.

These 35 blocks form the required $T(1, 15)$.

THEOREM 12.4 Given a $T(1, v)$, there is a $T(1, 2v + 1)$ with the original system as a subsystem.

Proof: Suppose that the v treatments of the $T(1, v)$ are $\{1, 2, \ldots, v\}$. Select a one-factorization of the K_{v+1} with the vertices $\{v + 1, v + 2, \ldots, 2v + 1\}$; say that the factors are $M_1, M_2, \ldots, M_v$. (Such a factorization exists, since v is necessarily odd.) The $T(1, 2v + 1)$ has as its triples all the triples in the $T(1, v)$ together with all the triples of the form $\{i, j, k\}$, where $1 \leq i \leq v$ and $\{j, k\}$ is an edge in M_i. It is easy to see that the design is balanced: if i and j are in the range $1 \leq i \leq v$, $1 \leq j \leq v$, then they occur together in one triple from the original system and no other; if $1 \leq i \leq v$ and $j > v$, j occurs in exactly one edge of M_i, so i and j occur together in exactly one block derived from M_i; if both $j > v$ and $k > v$, then $\{j, k\}$ is an edge in exactly one M_i and $\{i, j, k\}$ is the unique triple containing j and k. ∎

The preceding theorem can be applied even in quite small cases. For example, using the $T(1, 3)$ with one block $\{1, 2, 3\}$ and the one-factorization

$M_1 = 45, 67 \quad M_2 = 46, 57 \quad M_3 = 47, 56$,

12.2 Subsystems

we obtain the $T(1,7)$ with blocks

$$\{1,2,3\}, \{1,4,5\}, \{1,6,7\}, \{2,4,6\}, \{2,5,7\}, \{2,4,7\}, \{3,5,6\}.$$

There is a corresponding theorem in which we embed a $T(1,v)$ in a $T(1, 2v+7)$. We first need a preliminary result.

LEMMA 12.5 Given a set of $2n$ treatments where $n \geq 4$, one can find $2n$ 3-sets and $2n - 7$ one-factors which between them contain every pair of treatments precisely once.

Proof: We identify the treatments with the integers $1, 2, \ldots, 2n$, and write P_k for the set of all pairs with difference k:

$$P_k = \left\{ \{i,j\} \ : \ i - j \equiv \pm k \pmod{2n} \right\}.$$

The triples $\{i, i+1, i+3\}$, for $1 \leq i \leq 2n$, contain precisely the pairs in $P_1 \cup P_2 \cup P_3$. So we can prove the lemma by showing that $P_4 \cup P_5 \cup \cdots P_n$ can be partitioned into one-factors, or equivalently that the corresponding graph has a one-factorization.

Suppose that k is odd, $k < n$. If m is the greatest common divisor of k and $2n$, $m = (k, 2n)$, then m is odd and $2n/m$ is even. Considered as a graph, the set P_k consists of m component cycles of length $2n/m$:

$$\{i, i+k\}, \{i+k, i+2k\}, \ldots, \{i + [2n/m - 1]k, i\}$$

is a cycle of length $2n/m$ because $i + xk \equiv i$ cannot occur unless x is a multiple of $2n/m$. One can form two one-factors by taking alternate members of the cycles: They are

$$M_1 = \{1, 1+k\}, \{1+2k, 1+3k\}, \ldots, \{1-2k, 1-k\}, \{2, 2+k\},$$
$$\{2+2k, 2+3k\}, \ldots, \{2-2k, 2-k\}, \ldots, \{k, 2k\}, \{3k, 4k\}, \ldots, \{2n-k, 2n\}$$

and

$$M_2 = \{1+k, 1+2k\}, \{1+3k, 1+4k\}, \ldots, \{1-k, 1\}, \{2+k, 2+2k\},$$
$$\{2+3k, 2+4k\}, \ldots, \{2-k, 2\}, \ldots, \{2k, 3k\}, \{4k, 5k\}, \ldots, \{2n, k\}.$$

In the same way, if k is even, write $m = (k, 2n)$. If $2n/m$ is even, P_k breaks up into two one-factors in exactly the same way. So $P_{2x} \cup P_{2x+1}$ can be partitioned into four one-factors when $2n/(2x, 2n)$ is even (except in the case $2x + 1 = n$, which must be considered separately).

Now consider the case where k is even, say $k = 2x$, and $m = (2x, 2n)$ is such that $2n/m$ is odd. The cycle decomposition now gives odd cycles; we

cannot get one-factors by taking every second pair. We proceed as follows:
First, write down the m cycles in pairs, with the edges of every second
cycle shifted once:

$\{i, 2x + i\}, \{2x + i, 4x + i\}, \ldots, \{-4x + i, -2x + i\}, \{-2x + i, i\};$
$\{2x + 1 + i, 4x + 1 + i\}, \{4x + 1 + i, 6x + 1 + i\}, \ldots, \{-2x + 1 + i, 1 + i\},$
$\{1 + i, 2x + 1 + i\}.$

Now turn these cycles into one large cycle by deleting the edges $\{-2x + i, i\}$ and $\{1 + i, 2x + 1 + i\}$ and inserting instead the edges $\{i, 2x + 1 + i\}$ and $\{-2x + i, 1 + i\}$. The big cycle has an even number of edges, so it can be split into two equal parts by taking alternate edges. If we do this to every pair of cycles, we get two one-factors, M_3 and M_4. However, they do not contain precisely all the pairs in P_{2x}. The pairs in the list

$A = \{1, 2x + 2\}, \{3, 2x + 4\}, \ldots, \{m - 1, m + 2x\}, \{1 - 2x, 2\},$
$\{3 - 2x, 4\}, \ldots, \{m - 1 - 2x, m\},$

which are all members of P_{2x+1}, have been included, and the pairs in

$B = \{1, 1 - 2x\}, \{3, 3 - 2x\}, \ldots, \{m - 1, m - 1 - 2x\}, \{2, 2 + 2x\},$
$\{4, 4 + 2x\}, \ldots, \{m, m - 2x\}$

have been omitted.

Now consider the one-factors M_1 and M_2, obtained in the case $k = 2x + 1$, and M_3 and M_4. These contain all of $P_{2x} \cup P_{2x+1}$ except for B. All of the members of A are repeated in M_1. We form a new one-factor M_1^* by deleting A from M_1 and replacing it by B. Sets A and B contain the same set of symbols, so M_1^* is in fact a matching; and

$$P_{2x} \cup P_{2x+2} = M_1^* \cup M_2 \cup M_3 \cup M_4.$$

We have shown how to replace all pairs $P_{2x} \cup P_{2x+1}$ by one-factors except for $P_{n-1} \cup P_n$ when n is odd. In that case the same switching operation works. When n is even, P_n has not been covered by the cases above; but P_n is itself a one-factor. ∎

To illustrate this lemma, consider the case $2n = 12$. We write 10, 11, and 12 as T, E, D. We have

12.2 Subsystems

$P_1 = 12, 23, 34, 45, 56, 67, 78, 89, 9T, TE, ED, D1$
$P_2 = 13, 24, 35, 46, 57, 68, 79, 8T, 9E, TD, E1, D2$
$P_3 = 14, 25, 36, 47, 58, 69, 7T, 8E, 9D, T1, E2, D3$

and the 12 triangles are

$$\{1,2,4\},\ \{2,3,5\},\ \{3,4,6\},\ \{4,5,7\},\ \{5,6,8\},\ \{6,7,9\},$$
$$\{7,8,T\},\ \{8,9,E\},\ \{9,T,D\},\ \{T,E,1\},\ \{E,D,2\},\ \{D,1,3\}. \quad (12.1)$$

The only "k is odd" case is P_5, which has only one component cycle since 5 and 12 are coprime. We get the one-factors

$M_1 = 16,\ E4,\ 92,\ 7D,\ 5T,\ 38;$
$M_2 = 6E,\ 49,\ 27,\ D5,\ T3,\ 81.$

For P_4, observe that $(12,4) = 4$; we get cycles of length $12/4 = 3$. They pair up as

$$\begin{cases} 15,\ 59,\ 91 \\ 6T,\ T2,\ 26 \end{cases} \text{ and } \begin{cases} 37,\ 7E,\ E3 \\ 8D,\ D4,\ 48 \end{cases}$$

which give

$M_3 = 16,\ 59,\ T2,\ 38,\ 7E,\ D4$
$M_4 = 92,\ 15,\ 6T,\ E4,\ 37,\ 8D.$

The sets A and B are

$A = 16,\ 38,\ 92,\ E4 \quad B = 19,\ 62,\ 3E,\ 48$

so

$M_1^* = 19,\ 62,\ 3E,\ 48,\ 7D,\ 5T.$

Finally, P_6 is the one-factor

$17,\ 28,\ 39,\ 4T,\ 5E,\ 6D.$

So the decomposition consists of the 12 triangles in (12.1) and the factors

19, 62, 3E, 48, 7D, 5T
6E, 49, 27, D5, T3, 81
16, 59, T2, 38, 7E, D4
92, 15, 6T, E4, 37, 8D
17, 28, 39, 4T, 5E, 6D.

THEOREM 12.6 If there is a T(1,v), there is a T(1, 2v + 7) with the original T(1,v) as a subsystem.

Proof: Suppose that S is a T(1,v) on the v symbols $\{1, 2, \ldots, v\}$ and T is a decomposition such as outlined in Lemma 12.5 on the 2n = v + 7 symbols v + 1, v + 2, ..., 2v + 7, into factors $M_1, M_2, \ldots, M_v$ and triples. The triples for the new system are:

the triples of S;
the triples of T;
the triples of $\{x,y,z\} : \{y,z\} \in M_x$, $1 \leq x \leq v$. ∎

EXERCISES

12.2.1 From Theorem 12.3, in the case where $v_3 = 1$, construct two T(1, 19) designs, using the two equations

$$19 = 1 + 9(3 - 1), \quad 19 = 1 + 3(7 - 1).$$

Are the designs isomorphic?

12.2.2 Prove that no T(1, 13) can contain a subsystem T(1, 9).

12.2.3 Prove that if $u \equiv 1$ or 3 (mod 6), then there exists a v, $v \equiv 1$ or 3 (mod 6), such that u = 2v + 1 or u = 2v + 7. Hence show that assuming Theorems 12.4 and 12.6 and the existence of a T(1,1) and a T(1,3), one can prove the existence of Steiner triple systems of all possible orders without using Theorem 12.2.

12.2.4 Find a T(1, 13) with a subsystem T(1, 3), using Theorem 12.6.

12.2.5 (i) Consider the complete graph K_{6n} based on the vertices x^i, where $1 \leq x \leq 2n$ and $0 \leq i \leq 2$. The graph G is formed from K_{6n} by deleting all the edges $x^i x^j$, where $i \neq j$. Select any one-one-factorization $\{F_1^i, F_2^i, \ldots, F_{2n-1}^i\}$ of the K_{2n} with vertices $1^i, 2^i, \ldots, (2n)^i$; define one-factors

$$H_j^i = \{x^{i+1}(x+j)^{i+2} : 1 \leq x \leq 2n\} \cup F_j^i$$

12.3 Simple Triple Systems

for $0 \leq i \leq 2$, $1 \leq j \leq 2n - 1$. (Additions are reduced modulo 2n, when necessary.) Prove that the H_j^i form a one-factorization of G.

(ii) Use the factorization above, in the case $6n = v + 3$, to prove that when $v \equiv 3 \pmod{6}$, any T(1, v) can be embedded as a subsystem of a T(1, 2v + 3).

12.3 SIMPLE TRIPLE SYSTEMS

A balanced incomplete block design is called <u>simple</u> if it has no repeated block. Simple designs are sometimes preferred in geometrical and statistical design theory. For this reason it is interesting to know when simple designs exist. We restrict ourselves to the one case of simple triple systems T(2, v). (The designs with $\lambda = 1$ are necessarily simple.)

We use a special design which we call a <u>skew-transversal square</u>: This is an idempotent Latin square $L = (\ell_{ij})$ with the property that $\ell_{ij} = \ell_{ji}$ is never true unless $i = j$.

LEMMA 12.7 There is a skew-transversal square of side n for every positive n except 2 or 3.

Proof: We first define a Latin square $L = (\ell_{ij})$ of odd-side n by $\ell_{ij} \equiv 2i - j \pmod{n}$. If n is even, $n \neq 2$, we construct L_n from L_{n-1} by replacing (1,2), (2,3), ..., (n - 1, 1) entries by n and appending last row (n - 2, n - 1, 1, ..., n - 3, n) and last column (n - 1, 1, 2, ..., n - 2, n). Then L_n is obviously a transversal square. When n is odd, L_n will be skew unless $2i - j \equiv 2j - i \pmod{n}$, and so will be L_{n+1}. So we have skew-transversal squares of every side n except $n = 3m$ or $3m + 1$, m odd, and $n = 2$.

Side 4 and 9 are covered by examples in Figure 12.1. In the remaining odd cases, $n = 3m$, we can write $n = pq$, where p is a power of 3, $q > 3$, and a skew-transversal square S of side q is already known. We shall need a symmetric idempotent Latin square R of side p. But p is be odd, so such

```
1 4 2 3     1 8 7 6 9 3 5 4 2
3 2 4 1     4 2 9 8 3 7 1 5 6
4 1 3 2     5 6 3 2 7 9 4 1 8
2 3 1 4     3 5 6 4 8 2 9 7 1
            2 9 4 7 5 1 8 6 3
            7 1 9 5 2 6 3 9 4
            9 4 2 1 6 8 7 3 5
            6 3 5 9 1 4 2 8 7
            8 7 1 3 4 5 6 2 9
```

Figure 12.1

a square was constructed in part (ii) of the proof of Theorem 12.2. We write T^k to mean T with every entry increased by k and then reduced modulo q to the usual range $\{1, 2, \ldots, q\}$. The required skew-transversal square is a p × p array of q × q blocks; the ith diagonal block is $S + (i-1)q$, and if $i \neq j$, the (i,j) block is $T^k + (r_{ij} - 1)q$, where $k \in \{0, 1, 2\}$ and $k \equiv i - j$ (mod 3). This is obviously a skew-transversal square.

The remaining even cases $n = 3m + 1$ are easily solved from the preceding construction. The diagonals of the (1,2), (2,3), ..., (p,1) blocks are replaced by the symbol $3m + 1$, and the deleted entries are transported to a new last row and column just as in the construction of L_n for even n. ∎

THEOREM 12.8 There is a simple T(2, v) whenever $v \equiv 1$ or 3 (mod 6).

Proof: We divide the proof into three parts.

(i) Case $v \equiv 1$ (mod 3). We construct an S(2, v) on the 3n symbols x^i, where $1 \leq x \leq n$, $0 \leq i \leq 2$, and a symbol ∞. Suppose that $L = (\ell_{ij})$ is a skew-transversal square of side n. Then if superscripts are reduced modulo 3 when necessary, the triples

$$\{x^0, x^1, x^2\} : 1 \leq x \leq n;$$

$$\{x^i, x^{i+1}, \infty\} : 1 \leq x \leq n; 0 \leq i \leq 2; \tag{12.4}$$

$$\{x^i, y^i, \ell_{xy}^{i+1}\} : 1 \leq x \leq n, 1 \leq y \leq n, x \neq y; 0 \leq i \leq 2$$

form a simple $S_2(2, 3, 3n + 1)$. To handle the case $n = 2$, we exhibit the triples: 123, 145, 167, 246, 257, 347, 356, 124, 137, 156, 235, 267, 346, 357. For $n = 3$ we could also exhibit the triples. However, writing $n = 3$ and

$$L = (\ell_{ij}) = L_3 = \begin{bmatrix} 1 & 3 & 2 \\ 3 & 2 & 1 \\ 2 & 1 & 3 \end{bmatrix}$$

we have the following interesting formulation:

$$\{x^0, x^1, x^2\} : 1 \leq x \leq n;$$

$$\{x^i, x^{i+1}, \infty\} : 1 \leq x \leq n; 0 \leq i \leq 2;$$

$$\{x^i, y^i, \ell_{xy}^{i+1}\} : 1 \leq x < y \leq n; 0 \leq i \leq 2; \tag{12.5}$$

$$\{x^i, y^i, \ell_{xy}^{i+2}\} : 1 \leq y < x \leq n; 0 \leq i \leq 2.$$

12.3 Simple Triple Systems

This could be used whenever there is a <u>symmetric</u> idempotent Latin square of order n; but such designs can obviously exist only when n is odd, so we would still need formula (12.2) for the cases $n \equiv 1$ (mod 6).

(ii) <u>Case $v \equiv 0$ (mod 6)</u>. For $v = 6n$ we use symbols x^i for $1 \leq x \leq n$ and $0 \leq i \leq 5$, and any idempotent square (ℓ_{ij}) of side n. The triples are:

$$\{x^1, x^3, x^5\} : 1 \leq x \leq n;$$

$$\left.\begin{array}{l} \{x^i, x^{i+1}, x^{i+3}\} \\ \{x^i, x^{i+3}, x^{i+4}\} \\ \{x^i, y^{i+4}, x^{i+5}\} \end{array}\right\} \quad 1 \leq x \leq n; \quad i = 0, 2, 4;$$

$$\left.\begin{array}{l} \{x^i, y^i, \ell_{xy}^{i+2}\}, \{x^i, y^i, \ell_{yx}^{i+3}\} \\ \{x^i, y^{i+1}, \ell_{xy}^{i+3}\}, \{x^i, y^{i+1}, \ell_{yx}^{i+2}\} \\ \{x^{i+1}, y^i, \ell_{xy}^{i+3}\}, \{x^{i+1}, y^i, \ell_{yx}^{i+2}\} \\ \{x^{i+1}, y^{i+1}, \ell_{xy}^{i+2}\}, \{x^{i+1}, y^{i+1}, \ell_{yx}^{i+3}\} \end{array}\right\} \quad 1 \leq x < y \leq n; \quad i = 0, 2, 4.$$

(12.6)

The case $n = 2$, $v = 12$, must be treated separately. If $T(a, b, c, d)$ denotes the set of four possible triples with elements in $\{a, b, c, d\}$, then

$$T(j, j+3, j+6, j+9) : j = 0, 1, 2;$$
$$\{3, 3y+1, 3z+2\} : x, y, z \in \{0, 1, 2, 3\}, x+y+z \text{ odd}$$

forms a simple $T(2, 12)$ on $\{0, 1, \ldots, 11\}$.

(iii) <u>Case $v \equiv 3$ (mod 6)</u>. We use the symbols x^i for $1 \leq x \leq n$ and $i = 0, 1, 2$, where n is odd, and a symmetric transversal square (ℓ_{ij}) of order n. The triples are

$$\{x^1, x^2, x^3\}, \{x^1, x^2, (x+1)^3\} : 1 \leq x \leq n;$$

$$\left.\begin{array}{l} \{x^1, y^1, \ell_{xy}^2\}, \{x^1, y^1, (\ell_{xy}+1)^3\} \\ \{x^2, y^2, \ell_{xy}^3\}, \{x^2, y^2, \ell_{xy}^1\} \\ \{x^3, y^3, \ell_{xy}^1\}, \{x^3, y^3, (\ell_{xy}-1)^2\} \end{array}\right\} \quad 1 \leq x < y \leq n$$

where additions and subtractions are carried out modulo n.

12.4 DIFFERENCE TRIPLES AND CYCLIC TRIPLE SYSTEMS

A <u>difference triple</u> modulo v is a set of three integers x, y, and z such that either x + y + z is congruent to zero or else one of x, y, and z is congruent to the sum of the other two (modulo v). A <u>difference partition</u> of a set S is a partition of S into difference triples.

Difference partitions are of importance because of their relationship to a certain type of Steiner triple system. If a design D is based on the first v positive integers, we say it is <u>cyclic</u> if increasing every element of a block by 1 (mod v) always results in another block. For example, any design generated from initial blocks in the integers modulo v (but not including an infinity element) is cyclic, and similarly partial circulation modulo v yields a cyclic design (see Section 4.3).

LEMMA 12.9 If $v \equiv 1$ (mod 6) and if there is a difference partition of $S = \{1, 2, \ldots, (v-1)/2\}$, there is a cyclic Steiner triple system of order v. If $v \equiv 3$ (mod 6) and if there is a difference partition of $S' = \{1, 2, \ldots, (v-1)/2\}\backslash(v/3)$, there is a Steiner triple system of order v.

Proof: Suppose that P is a difference partition of S, in the case $v \equiv 1$ (mod 6). If $\{x, y, z\}$ is a member of P with the property that some member is congruent to the sum of the other two, suppose without loss of generality that $z \equiv x + y$, so that every triple satisfies

$$z \equiv \pm(x + y) \pmod{v}.$$

With the triple $\{x, y, z\}$ we associate the triple $\{0, x, x+y\}$, and write

$$B = \left\{ \{0, x, x+y\} : \{x, y, z\} \in P \right\}.$$

Then it is easy to see that the triples $\{j, x+j, t+j\}$, where j ranges through $\{0, 1, \ldots, v-1\}$ and $\{0, x, t\}$ ranges through B, form a triple system T(1, v). (Either $t = z$ or $t = -z$.) The set B is a set of initial blocks, in the sense of Section 4.3.

In the case $v \equiv 3$ (mod 6), say $v = 3k$, the same construction is used, and the blocks $\{j, k+j, 2k+j\}$ are also included for $0 \leq j < k$.

In both cases the triple system is obviously cyclic. ∎

It is a straightforward matter to prove that no difference partition exists for $v = 9$, and there is no cyclic triple system of that order (see Exercise 12.4.1). It is easy to construct difference partitions of small orders: as an example we consider the case of $v = 15$. The set S' is $\{1, 2, 3, 4, 6, 7\}$. There must be a triple containing 1. It might have the form $\{1, x, x+1\}$ or $\{1, x, 15 - (x+1)\}$ or $\{x, 15 + 1 - x, 1\}$. Since no

12.4 Difference Triples and Cyclic Triple Systems

$v = 19$: $\{1,5,6\}$ $\{2,8,9\}$ $\{3,4,7\}$

$v = 27$: $\{1,12,13\}$ $\{2,5,7\}$ $\{3,8,11\}$ $\{4,6,10\}$

$v = 45$: $\{1,11,12\}$ $\{1,17,19\}$ $\{3,20,22\}$ $\{4,10,14\}$
$\{5,8,13\}$ $\{6,18,21\}$ $\{7,9,16\}$

$v = 63$: $\{1,15,16\}$ $\{2,27,29\}$ $\{3,25,28\}$ $\{4,14,18\}$ $\{5,26,31\}$
$\{6,17,23\}$ $\{7,13,20\}$ $\{8,11,19\}$ $\{9,24,30\}$ $\{10,12,22\}$

Figure 12.2 Difference partitions.

term can be greater than 7, the only possibility is $\{1, x, x+1\}$ with $x + 1 \leq 7$; since 5 is not in S', x must equal 2, 3, or 6. So the candidates are

$\{1,2,3\}$, $\{4,6,7\}$
$\{1,3,4\}$, $\{2,6,7\}$
$\{1,6,7\}$, $\{2,3,4\}$.

By inspection the only case in which both triples are difference triples is the second one:

$\{1,3,4\}$ $\{2,6,7\}$.

The cases $v = 7$ and $v = 13$ are left as exercises. Figure 12.2 shows difference partitions for $v = 19, 27, 45,$ and 63.

THEOREM 12.10 There is a difference partition of S for every case of $v \equiv 1 \pmod 6$ and a difference partition of S' for every case of $v \equiv 3 \pmod 6$ except $v = 9$.

Proof: Figure 12.3 gives constructions for difference partitions of S whenever $v \equiv 1 \pmod 6$, $v \neq 7, 13, 19$, and of S' whenever $v \equiv 3 \pmod 6$, $v \neq 9, 15, 27, 45, 63$. Together with the partitions in Figure 12.2 and the examples for $v = 15$ above and for $v = 7$ and 13 in Exercise 12.4.2, this proves the theorem. ■

COROLLARY 12.10.1 There is a cyclic triple system of every possible order except 9. ■

$v = 18s+1$, $s \geq 2$: $(3r+1,\ 4s-r+1,\ 4s+2r+2)$ $r = 0,\ldots,s-1$
$(3r+2,\ 8s-r,\ 8s+2r+2)$ $r = 0,\ldots,s-1$
$(3r+3,\ 6s-2r-1,\ 6s+r+2)$ $r = 0,\ldots,s-1$
$(3s,\ 3s+1,\ 6s+1)$

$v = 18s+3$, $s \geq 1$: $(3r+1,\ 8s-r+1,\ 8s+2r+2)$ $r = 0,\ldots,s-1$
$(3r+2,\ 4s-r,\ 4s+2r+2)$ $r = 0,\ldots,s-1$
$(3r+3,\ 6s-2r-1,\ 6s+r+2)$ $r = 0,\ldots,s-1$

$v = 18s+7$, $s \geq 1$: $(3r+1,\ 8s-r+1,\ 8s+2r+4)$ $r = 0,\ldots,s-1$
$(3r+2,\ 6s-2r+1,\ 6s+r+3)$ $r = 0,\ldots,s-1$
$(3r+3,\ 4s-r+1,\ 4s+2r+4)$ $r = 0,\ldots,s-1$
$(3s+1,\ 4s+2,\ 7s+3)$

$v = 18s+9$, $s \geq 4$: $(3r+1,\ 4s-r+3,\ 4s+2r+4)$ $r = 0,\ldots,s$
$(3r+2,\ 8s-r+2,\ 8s+2r+4)$ $r = 2,\ldots,s-2$
$(3r+3,\ 6s-2r+1,\ 6s+r+4)$ $r = 1,\ldots,s-2$
$(2, 8s+3, 8s+5)$ $(3, 8s+1, 8s+4)$ $(5, 8s+2, 8s+7)$
$(3s-1, 3s+2, 6s+1)$ $(3s, 7s+3, 8s+6)$

$v = 18s+13$, $s \geq 1$: $(3r+2,\ 6s-2r+3,\ 6s+r+5)$ $r = 0,\ldots,s-1$
$(3r+3,\ 8s-r+5,\ 8s+2r+8)$ $r = 0,\ldots,s-1$
$(3r+1,\ 4s-r+3,\ 4s+2r+4)$ $r = 0,\ldots,s$
$(3s+2,\ 7s+5,\ 8s+6)$

$v = 18s+15$, $s \geq 1$: $(3r+1,\ 4s-r+3,\ 4s+2r+4)$ $r = 0,\ldots,s$
$(3r+2,\ 8s-r+6,\ 8s+2r+8)$ $r = 0,\ldots,s$
$(3r+3,\ 6s-2r+3,\ 6s+r+6)$ $r = 0,\ldots,s-1$

Figure 12.3 Families of difference partitions.

EXERCISES

12.4.1 Prove that there is no difference partition of S' in the case $v = 9$.

12.4.2 Find difference partitions of S when $v = 7$ and $v = 13$.

12.4.3 Verify that the sets shown in Figures 12.2 and 12.3 are difference partitions.

12.5 KIRKMAN TRIPLE SYSTEMS

Recall from Section 6.4 that the first mention of resolvable designs is the question posed by the Reverend Kirkman in 1850: "Fifteen young ladies in a school walk out three abreast for seven days in succession: it is required to arrange them daily, so that no two shall walk twice abreast."

In Section 6.4 a <u>parallel class</u> of blocks in a design was defined to be a set of blocks which between them contain every element precisely once. A design is called <u>resolvable</u> when its blocks admit of a partition into parallel classes. So Kirkman was in fact asking: Is there a resolvable Steiner triple system on 15 treatments?

More generally, Kirkman asked: For which v does there exist a resolvable $T(1,v)$? We know that a $T(1,v)$ must have $v \equiv 1$ or 3 (modulo 6).

12.5 Kirkman Triple Systems

Moreover, for resolvability, the block size 3 must divide v. So a necessary condition is $v \equiv 3$ (modulo 6). We shall show that this condition is in fact sufficient—a fact that was not proven until 120 years after Kirkman originally posed the problem.

Suppose that $v = 6m + 3$. Then the parameters of a KTS(v) are

$$(6m + 3, \ (2m + 1)(3m + 1), \ 3m + 1, \ 3, \ 1)$$

for integer m. There are $2m + 1$ blocks in each parallel class, so there are $3m + 1$ parallel classes.

It is instructive to have some examples of Kirkman triple systems, and some small examples will also be useful later in this section. For this reason we present Kirkman triple systems on $v = 9$, 15, and 21 treatments in Figure 12.4. (The case of 15 treatments is of course a solution to Kirkman's original problem.)

Another interesting small case is $v = 39$. The treatments are represented by the integers modulo 39. Each parallel class consists of the 13 blocks

$$\{x + i, \ i + i, \ z + i\} \ : \ i = 0, \ 3, \ 6, \ 9, \ 12, \ 15, \ 18, \ 21, \ 24, \ 27, \ 30, \ 33, \ 36,$$

where $\{x, y, z\}$ is a starter block for the parallel class; the starter blocks are

$\{0,1,2\}$, $\quad \{3,9,27\}$, $\quad \{4,10,28\}$, $\quad \{5,11,29\}$,
$\{6,15,18\}$, $\quad \{7,16,19\}$, $\quad \{8,17,20\}$, $\quad \{12,31,38\}$,
$\{13,32,36\}$, $\quad \{14,30,37\}$, $\quad \{21,25,35\}$, $\quad \{22,26,33\}$,
$\{23,24,34\}$, $\quad \{12,32,37\}$, $\quad \{13,30,0\}$, $\quad \{14,31,36\}$,
$\{21,26,34\}$, $\quad \{22,24,35\}$, $\quad \{23,25,33\}$.

LEMMA 12.11 Suppose that there exists a PB(v;K;1), and that there exists a KTS(2k + 1) for every k in K. Then there is a KTS(2v + 1).

Proof: Let us write V for the set $(1 \cdots v)$; we shall assume that the PB(v;K;1) has treatment set V. If X is any set, we write X_i for the set resulting from subscripting every element of X with an i. In this notation we exhibit a KTS(2v + 1) with treatment set $\{\infty\} \cup V_1 \cup V_2$.

Suppose that B is any block of the PB(v;K;1); suppose that $|B| = k$. Then there is a KTS(k); take its treatment set as $\{\infty\} \cup B_1 \cup B_2$, and permute the treatments if necessary so as to ensure that $\{\infty, x_1, x_2\}$ is a triple for every x in B. We denote the resulting KTS(2k + 1) by K(B), and write T(B) for the set of triples of K(B). Then we claim that the union of the T(B) is the required KTS(v).

It is easy to verify that the triples form a T(1, 2v + 1): The pair $\{\infty, x_i\}$ appears in the triple $\{\infty, x_1, x_2\}$ and no other, and the same is true of $\{x_1, x_2\}$.

12. Triple Systems

KTS(9):

1 2 3	1 4 7	1 5 9	1 6 8
4 5 6	2 5 8	2 6 7	2 4 9
7 8 9	3 6 9	3 4 8	3 5 7

KTS(15):

7 8 9	1 8 10	2 8 11	4 6 9	5 7 10	2 3 9	1 5 9
2 6 10	3 7 11	1 4 12	3 8 12	4 8 13	1 6 11	3 4 10
4 5 11	5 6 12	6 7 13	2 5 13	3 6 14	5 8 14	2 7 12
1 3 13	2 4 14	3 5 15	1 7 14	1 2 15	4 7 15	6 8 15
12 14 15	9 13 15	9 10 14	10 11 15	9 11 12	10 12 13	11 13 14

KTS(21):

1 2 3	4 5 6	7 8 9	10 11 12	13 14 15
4 7 13	7 10 16	10 13 19	1 13 16	4 16 19
5 8 14	8 11 17	11 14 20	2 14 17	5 17 20
6 9 15	9 12 18	12 15 21	3 15 18	6 18 21
10 17 21	3 13 20	2 6 16	5 9 19	1 8 12
11 18 19	1 14 21	3 4 17	6 7 20	2 9 10
12 16 20	2 15 19	1 5 18	4 8 21	3 7 11

16 17 18	19 20 21	10 18 20	11 16 21	12 17 19
1 7 19	1 4 10	2 13 21	3 14 19	1 15 20
2 8 20	2 5 11	3 5 16	1 6 17	2 4 18
3 9 21	3 6 12	6 8 19	4 9 20	5 7 21
4 11 15	7 14 18	1 9 11	2 7 12	3 8 10
5 12 13	8 15 16	4 12 14	5 10 15	6 11 13
6 10 14	9 13 17	7 15 17	8 13 18	9 14 16

Figure 12.4

(The triple $\{\infty, x_1, x_2\}$ will arise in more than one of the T(B), but since we took the union of the block sets, these multiple occurrences are not counted.) If $x \neq y$, then $\{x_i, y_j\}$ will arise in precisely one member of T(B), where B is the unique block containing $\{x,y\}$ in the PB(v;K;1).

Now suppose that x is any member of V. Say that s is the number of blocks of the pairwise balanced design which contain x; write B_{x1}, B_{x2}, ..., B_{xs} for those blocks of the pairwise balanced design which contain x. Write P_{xi} for the parallel class containing the triple $\{\infty, x_1, x_2\}$ in $K(B_{xi})$, and Q_x

12.5 Kirkman Triple Systems

for the union of the P_{xi} for $1 \leq i \leq s$ (with the triple $\{\infty, x_1, x_2\}$ only counted once, of course). Every member y of V other than x will belong to exactly one of the B_{xi}, so y_1 and y_2 will belong to exactly one block in Q_x. Clearly, ∞, x_1, and x_2 occur exactly once. So Q_x is a parallel class. The sets Q_x and Q_y have no common triple when $x \neq y$, and there are exactly v of these sets—exactly the number of parallel classes in a KTS(2v + 1). So the Q_x partition the blocks of the T(1, 2v + 1) into parallel classes, and we have a Kirkman triple system. ■

In particular, suppose that $K = \{4, 7, 10, 19\}$. Then $\{2k + 1 : k \in K\} = \{9, 15, 21, 39\}$, and we know Kirkman triple systems of those four orders. So any PB(v;K;1) gives rise to a KTS(2v + 1). In particular, if v ranges through all the positive integers congruent to 1 (modulo 3), 2v + 1 will range through all the positive integers congruent to 3 (modulo 6). Therefore, we investigate the existence of PB(v; $\{4, 7, 10, 19\}$;1).

LEMMA 12.12 There is a PB(v; $\{4, 7, 10, 19\}$;1) when v = 1, 4, 7, 10, 13, 16, 19, 22, 25, 28, 31, 34, 37, or 40.

Proof: The case v = 1 is trivial and v = 4, 7, 10, and 19 are realized by one-block designs. From Corollary 10.11.1 we obtain a PB(28; $\{4, 7\}$;1), a PB(37; $\{4, 10\}$;1), and a PB(40; $\{4, 10\}$;1). When v = 13, 16, or 25, there are balanced incomplete block designs with k = 4 and λ = 1: the cases v = 13 and v = 16 are the projective and affine planes of parameters 3 and 4, respectively, while a (25, 50, 8, 4, 1)-design was exhibited in Section 4.3. (We could also have used this approach for v = 28, since suitable block design was constructed in Section 6.4.)

When v = 22, we use a Kirkman triple system on 15 treatments; suppose that its treatment set is (1 $\cdots$ 15). Say that $\mathcal{T}_1$, $\mathcal{T}_2$, ..., $\mathcal{T}_7$ and the seven parallel classes in a resolution of the design. We define a pairwise balanced design with treatment set $\{1, 2, \ldots, 15, t_1, t_2, \ldots, t_7\}$ as follows. Each block in $\mathcal{T}_i$ is augmented to a 4-set by adding treatment t_i to it. Then block $\{t_1, t_2, t_3, t_4 t_5, t_6, t_7\}$ is added on. The result is a PB(22; $\{4, 7\}$;1) with exactly one 7-block. A similar construction works for v = 31 (see Exercise 12.5.1).

When v = 34, we first construct a PB(28; $\{3, 4\}$;1) whose treatments are the ordered pairs (x,y), where x is an integer modulo 3 and y is an integer modulo 9, together with ∞. We write

$$B_1 = \{00, 12, 22, 23\}, \quad B_2 = \{00, 13, 15\},$$
$$B_3 = \{00, 14, 18\}, \quad B_4 = \{00, 03, 06, \infty\}.$$

Then we define

$$B_{ijk} = B_k + ij = \{xy + ij : xy \in B_k\}$$

[where $xy + ij = (x + i, y + j)$, except for the rule $\infty + ij = \infty$], and

$$\mathcal{B}_k = \{B_{ijk} : 0 \leq i \leq 2, 0 \leq j \leq 8\}.$$

The blocks $\mathcal{B}_1 \cup \mathcal{B}_2 \cup \mathcal{B}_3 \cup \mathcal{B}_4$ form the PB(28;$\{3,4,\}$;1). Moreover, if we define

$$\mathcal{S}_h = \{B_{ij2} : j - 1 \equiv h \pmod{3}\}$$
$$\mathcal{T}_h = \{B_{ij3} : j \equiv h \pmod{3}\},$$

then clearly the $\mathcal{S}_h$ and the $\mathcal{T}_h$ partition the blocks of size 3 into six parallel classes. We define six new treatments; we append s_h to all blocks in $\mathcal{S}_h$, and t_h to all blocks in $\mathcal{T}_h$. Then a block $\{\infty, s_1, s_2, s_3, t_1, t_2, t_3\}$ is added. We have a PB(34;$\{4, 10\}$;1). ∎

In Section 10.4 we used simple group-divisible designs or SGDDs. Recall that an SGDD (X, ,) consists of a set of treatments X and sets of subsets of X called groups ( ) and blocks ( ) such that the groups are a partition of X and such that any two treatments belong to precisely one group or precisely one block, but not both. We shall sometimes refer to the SGDD as a GD(v;G;A), where G and A are the sets of cardinalities of members of $\mathcal{G}$ and $\mathcal{A}$, respectively. (If G or A has only one element, we write that number, omitting set brackets.) If $\mathcal{G} \cup \mathcal{A}$ is interpreted as a set of blocks, we have a pairwise balanced design: If there exists a GD(v;G;A), then there exists a PB(v;G $\cup$ A;1).

To investigate the existence of PB(v;$\{4, 7, 10, 19\}$;1), we shall use two specific group-divisible designs repeatedly: a GD(12;3;4) and a GD(15;3;4). From Theorem 10.11, the existence of two orthogonal Latin squares of side 3 implies the existence of a transversal design TD(4, 3), which is a GD(12;3;4). If the technique of puncturing a balanced incomplete block design (as described in Exercise 10.4.4) is applied to a (16, 20, 5, 4, 1)-design, we obtain a GD(15;3;4). The same technique applied to a (28, 63, 9, 4, 1)-design yields a GD(27;3;4).

We also need the following lemma, which is a special case of a more general theorem (see Exercise 12.5.2).

LEMMA 12.13 Let w be a fixed positive integer. Suppose that there exists a GD(v;G;A) and that for every k belonging to A, there exists a GD(wk;w;A_k), for some set A_k. Then there exists a GD(wv;G^*;A^*), where

$$G^* = \{wg : g \in G\},$$

$$A^* = \bigcup_{k \in A} A_k.$$

12.5 Kirkman Triple Systems

Proof: Suppose that the original SGDD is $(X, \mathcal{G}, \mathcal{A})$. For each x in X, define $S_x = \{x_1, x_2, \ldots, x_w\}$. If B is a block of size k, construct a GD(wk;w;A_k) whose treatments are the elements $x_1, x_2, \ldots, x_w$ for x in B and whose groups are $\{S_x : x \in B\}$. Write $\mathcal{A}(B)$ for the set of blocks of this design. Now consider the design $(X^*, \mathcal{G}^*, \mathcal{A}^*)$:

$$X^* = \bigcup_{x \in X} S_x;$$

if H is any member of $\mathcal{G}$, then $H^* = \bigcup_{x \in H} S_x$ and

$$\mathcal{G}^* = \{H^* : H \in \mathcal{G}\};$$
$$\mathcal{A}^* = \bigcup_{B \in \mathcal{A}} \mathcal{A}(B).$$

It is easy to verify that $(X, \mathcal{G}^*, \mathcal{A}^*)$ is a simple group-divisible design with the required parameters. ∎

COROLLARY 12.13.1 If there exists a GD(v;g;4), then there exists a GD(3v;3g;4) and a PB(3v + 1; $\{3g + 1, 4\}$; 1).

Proof: We take w = 3, G = $\{g\}$, and A = $\{4\}$. The condition "for every k belonging to A there exists a GD(wk;w;A_k)" now reduces to "there exists a GD(12;3;A_4)." We have constructed a GD(12;3;4). So the lemma applies with $A_4 = \{4\}$, and we have a GD(3v;3g;4). If an additional treatment ∞ is added to each group and these augmented groups are treated as blocks, we have a PB(3v + 1; $\{3g + 1, 4\}$; 1). ∎

COROLLARY 12.13.2 There is a PB(v; $\{4, 7, 10, 19\}$;1) when v = 43, 46, 79, or 82.

Proof: We apply Corollary 12.13.1 to designs of the following kind: GD(14;2;4), GD(15;3;4), GD(26;2;4), and GD(27;3;4).

To construct a GD(14;2;4), take as treatments the set of integers modulo 14, as groups the sets $\{x, x + 7\}$ and as blocks

$$\{i, 2 + i, 5 + i, 6 + i\} : 0 \leq i \leq 13.$$

The construction of a GD(26;2;4) is analogous (see Exercise 12.5.3). The other two designs were constructed earlier in this section. ∎

THEOREM 12.14 There exists a PB(v; $\{4, 7, 10, 19\}$;1) whenever v ≡ 1 (mod 3).

Proof: The theorem is true when v ≤ 46 or v = 79 or 82, from Lemma 12.12 and Corollary 12.13.2. So suppose that v is not one of those values. Then we can write

$v = 12m + 3t + 1,$

where m and t are integers such that $0 \leq t \leq m$ and such that there exist three pairwise orthogonal Latin squares of side m: from Theorem 10.20 it is sufficient to require that $m \geq 4$, $m \neq 6$, and $m \neq 10$. This is easy whenever $v \geq 133$, since we can select $m \geq 11$. For $49 \leq v \leq 58$, we use $m = 4$, $0 \leq t \leq 3$. For $61 \leq v \leq 76$, we use $m = 5$, $0 \leq t \leq 5$. For $85 \leq v \leq 106$, we use $m = 7$, $0 \leq t \leq 7$. The cases $109 \leq v \leq 130$ are handled by $m = 9$, $0 \leq t \leq 7$.

We now proceed by induction. Suppose that $v \equiv 1 \pmod 3$ and that there is a PB(v'; $\{4,7,10,19\}$;1) whenever $1 \leq v' < v$ and $v' \equiv 1 \pmod 3$. If $v \leq 49$ or $v = 79$ or 82, there is nothing to prove. So suppose that

$v = 12m + 3t + 1,$

where $0 \leq t \leq m$ and three pairwise orthogonal Latin squares of side m exist. From Theorem 10.11 we can construct a GD(5m, m, 5). From one of the groups of this design, delete m - t elements. The result is a GD(4m + t, $\{m, t\}$, $\{4, 5\}$) with just one block of size t.

Now apply Lemma 12.13 in the case w = 3. There exist group-divisible designs GD(12, 3, 4) and GD(15, 3, 4). So we can construct a GD(12m + 3t, $\{3m, 3t\}$, 4). Adding one new point to each group we obtain a PB(12m + 3t + 1; $\{3m + 1, 3t + 1, 4\}$; 1).

Finally, we know that there are simple pairwise balanced designs on 3m + 1 and 3t + 1 treatments with block sizes $\{4, 7, 10, 19\}$, by the induction hypothesis. We replace each block of those sizes by the appropriate pairwise balanced design (the construction of Theorem 2.1). The result is a PB(v; $\{4, 7, 10, 19\}$;1). ■

COROLLARY 12.14.1 There is a Kirkman triple system on v treatments for all $v \equiv 3 \pmod 6$. ■

EXERCISES

12.5.1   (i) Suppose that there exists a Kirkman triple system with 6n + 3 treatments, and consequently 3n + 1 parallel classes. Prove that there exists a PB(9n + 4; $\{4, 3n + 1\}$; 1) with exactly one block of size 3n + 1. Hence prove that there is a PB(31; (4, 10);1).
  (ii) Given that there exists a Kirkman triple system with v = 27, prove that there exists a balanced incomplete block design with parameters (40, 130, 13, 4, 1).
  (iii) Assuming further the existence of a Kirkman triple system with v = 81, prove the existence of a balanced incomplete block design with parameters (121, 1210, 40, 4, 1).

12.5.2 Suppose that $(X, \mathcal{G}, \mathcal{A})$ is an SGDD. With each treatment x associate a positive integer weight w_x, and select a collection of pairwise disjoint sets $\{S_x : x \in X\}$ where S_x has size w_x. Given a block $B \in \mathcal{A}$, define

$$S(B) = \bigcup_{x \in B} S_x,$$
$$\mathcal{G}(B) = \{S_x : x \in B\},$$

and for each B select a collection $\mathcal{A}(B)$ of subsets of $S(B)$ such that $(S(B), \mathcal{G}(B), \mathcal{A}(B))$ is an SGDD. If G is any member of $\mathcal{G}$, then define

$$G^* = \bigcup_{x \in G} S_x.$$

If $X^* = \bigcup_{x \in X} S_x$, $\mathcal{G}^* = \{G^* : G \in \mathcal{G}\}$, $\mathcal{A}^* = \bigcup_{B \in \mathcal{A}} \mathcal{A}(B)$, prove that $(X^*, \mathcal{G}^*, \mathcal{A}^*)$ is an SGDD.

12.5.3 A design is constructed using as treatments the set of integers modulo 26, groups $\{x, x+13\}$ and blocks

$$\{i, 6+i, 8+i, 9+i\} : 0 \leq i \leq 25$$
$$\{i, 4+i, 11+i, 16+i\} : 0 \leq i \leq 25.$$

Prove that the design is a GD(26;2;4).

12.5.4 (i) Prove that there exists a $PB(v; \{4,3\}; 1)$ if and only if $v \equiv 0$ or 1 (mod 3), $v \neq 6$.
(ii) For what values of v does there exist a $PB(v; \{4,3\}; 1)$ which contains at least one block of size 4 and at least one block of size 3?

BIBLIOGRAPHIC REMARKS

It appears that triple systems were first discussed by Woolhouse [16]; the main early papers were those of Kirkman [4] and Steiner [12]. The easy construction of T(1,n) follows constructions given by Bose [1] and Skolem [10]—but the inherent simplicity of their methods was really pointed out by Lindner, in his lectures and in papers such as [6]. Subsequently, the easy constructions for $\lambda > 1$ were given by Stinson and Wallis [13]. The subsystem approach, using one-factorizations, is due to Stanton and Goulden [11]. Section 12.3 is based on [14].

Difference partitions were discussed by Heffter [3] and the problems of constructing them in the two cases $v \equiv 1$ (mod 6) and $v \equiv 3$ (mod 6) are often called Heffter's difference problems. The solutions were given—essentially as we have done here—by Peltesohn [8].

Kirkman posed his "schoolgirl problem" in [5], and it was first solved by Ray-Chaudhuri and Wilson [9]. Lemma 12.11 appears in [7], Lemma 12.12 is a result of Wilson [15]. The $PG(34; \{4,7\}; 1)$ is due to Brouwer [2]. The general approach in Section 12.5 is due to S. A. Vanstone.

1. R. C. Bose, "On the construction of balanced incomplete block designs," Annals of Eugenics 9(1939), 353-399.
2. A. E. Brouwer, "Optimal packings of K_4's into a K_n," Journal of Combinatorial Theory 26A(1979), 278-279.
3. L. Heffter, "Uber Tripelsysteme," Mathematische Annalen 49(1897), 101-112.
4. T. P. Kirkman, "On a problem in combinations," Cambridge and Dublin Mathematical Journal 2(1847), 191-204.
5. T. P. Kirkman, "Query VI," Lady's and Gentleman's Diary (1850), 48.
6. C. C. Lindner, "A survey of embedding theorems for Steiner systems," Annals of Discrete Mathematics 7(1980), 175-202.
7. R. C. Mullin and S. A. Vanstone, "Steiner systems and Room squares," Annals of Discrete Mathematics 7(1980), 95-104.
8. R. Peltesohn, "Eine Losung der beiden Heffterschen Differenzenprobleme," Compositio Mathematica 6(1939), 251-267.
9. D. K. Ray-Chaudhuri and R. M. Wilson, "Solution of Kirkman's schoolgirl problem," Proceedings of Symposia in Pure Mathematics 19, Combinatorics (American Mathematical Society, Providence, R.I., 1971), 187-204.
10. T. Skolem, "Some remarks on the triple systems of Steiner," Mathematica Scandinavica 6(1958), 273-280.
11. R. G. Stanton and I. P. Goulden, "Graph factorisation, general triple systems, and cyclic triple systems," Aequationes Mathematicae 22 (1981), 1-28.
12. J. Steiner, "Combinatorische Aufgabe," Journal fur der Reine und Angewante Mathematik 45(1853), 181-182.
13. D. R. Stinson and W. D. Wallis, "Snappy constructions for triple systems," Gazette of the Australian Mathematical Society 10(1983), 84-88.
14. D. R. Stinson and W. D. Wallis, "Two-fold triple systems without repeated blocks," Discrete Mathematics 47(1983), 125-128.
15. R. M. Wilson, "Construction and uses of pairwise balanced designs," Mathematical Centre Tract 55(1974), 18-41.
16. W. S. B. Woolhouse, "Prize question 1733," Lady's and Gentleman's Diary (1844).

13
Room Squares

13.1 BASIC IDEAS

A <u>Room square</u> of side $2n - 1$ is a square array having $2n - 1$ cells in each row and each column and such that each cell is either empty or contains an unordered pair of symbols chosen from a set of $2n$ elements. Without loss of generality, we can take these elements as the numbers $1, 2, \ldots, 2n - 1$, ∞. Each row and each column of the design contains $n - 1$ empty cells and n cells each of which contains a pair of symbols. Each row and each column contains each of the $2n$ symbols exactly once and each of the $n(2n - 1)$ possible distinct pairs of symbols is required to occur exactly once in a cell of the square. An example, of side 7, is shown in Figure 13.1.

The first mention of a Room square occurred in relation to Kirkman's schoolgirl problem. Consider the square of Figure 13.1, with its columns labeled a, b, c, d, e, f, g as shown. From it we construct a triple system as follows. From the column headed "a" we construct the four triples a47, a∞5, a12, and a35, which are obtained from the pairs in the column by

a	b	c	d	e	f	g
		35	17	∞2	46	
	26	∞4			15	37
	13	57	∞6	24		
47		16		∞3		25
∞5		23	14		67	
12	∞7			56	34	
36	45		27			∞1

Figure 13.1

appending a to them; proceeding similarly with the other columns, we form 28 triples. To each row there corresponds one more triple, made up of the labels on the three columns that have no entries in the row. It will be observed that this is in fact a Kirkman triple system: Each parallel class consists of the triple corresponding to a row and the four triples which came from pairs in that row: the classes are

$\{abc\}$, $\{d35\}$, $\{e17\}$, $\{f\infty 2\}$, $\{g46\}$;
$\{ade\}$, $\{b26\}$, $\{c\infty 4\}$, $\{f15\}$, $\{g37\}$;
$\{afg\}$, $\{b13\}$, $\{c57\}$, $\{d\infty 6\}$, $\{e24\}$;
$\{bdf\}$, $\{a47\}$, $\{c16\}$, $\{e\infty 3\}$, $\{g25\}$;
$\{bge\}$, $\{a\infty 5\}$, $\{c23\}$, $\{d14\}$, $\{f67\}$;
$\{cdg\}$, $\{a12\}$, $\{b\infty 7\}$, $\{e56\}$, $\{f34\}$;
$\{cef\}$, $\{a36\}$, $\{b45\}$, $\{d27\}$, $\{g\infty 1\}$.

It is clear that the rows of a Room square are all one-factors, as are the columns. To construct one we must first find a one-factorization and allocate one of its factors to each row; then we must distribute the entries in each row into columns so that the columns are all factors as well. So a Room square can be interpreted as two related one-factorizations, a <u>row</u> factorization and a <u>column</u> factorization. The relationship is as follows: If R_i is any factor from the row factorization and C_j is any factor from the column factorization, then either $R_i \cap C_j$ is empty or it contains precisely one pair. Two such factorizations are called <u>orthogonal.</u> Conversely, given two orthogonal one-factorizations, one can construct a Room square— the (i,j) cell contains $R_i \cap C_j$.

The side of the Room square is necessarily odd. It is natural to ask which odd numbers occur as sides. The 1×1 array

$\infty 1$

is a Room square of side 1. There can be no Room square of side 3, for if such a square existed, it would have $\infty 1$ as one of its entries. Assume (without loss of generality) that it lies in the $(1,1)$ cell. The square has the form

$\infty 1$	?	?
?	?	?
?	?	?

.

13.2 Starter Constructions

Now each of ∞, 1, 2, and 3 must appear once in the first row. So 23 must occur in one of the cells of that row. Similarly, 23 must occur in the first column. But this is impossible, since the pair cannot be repeated and the common cell of row 1 and column 1 is already occupied.

A similar argument can be used to show that no Room square of side 5 exists. However, this fact also follows directly from the discussion of one-factorizations of K_6 in Section 11.3: since any two factorizations always have a common factor, they cannot be orthogonal.

Sides 3 and 5 are exceptional. We shall prove in Section 13.4 that there is a Room square of every odd side greater than 5.

The properties of Room squares are not dependent on the set of symbols, and the ordering of rows and columns is a matter of convenience. So we define two Room squares to be _isomorphic_ if one can be obtained from the other by permuting rows and columns and relabeling the elements. In particular, it is possible in this way to produce a square with diagonal entries

$(\infty 1, \infty 2, \ldots, \infty r)$

from any square of side r. We say that such a Room square is _standardized_.

EXERCISES

13.1.1 Suppose that there is a Room square of side 5. Without loss of generality, one may assume that its first row is

| $\infty 1$ | 23 | 45 | — | — |

Prove by exhaustion that this first row cannot be completed to a Room square.

13.1.2 Find a standardized Room square of side 7.

13.2 STARTER CONSTRUCTIONS

Consider the Room square of Figure 13.2. It has been constructed as follows: The $(i+1, j+1)$ entry was obtained from the (i, j) entry by adding 1 to each member, subject to the rules of addition modulo 7 and "$\infty + 1 = \infty$". That is, both one-factorizations come from starters in GF(7). Initially, the starter $\{16, 25, 34\}$ has been placed in the columns of the last row in such a way that the resulting last column is also a starter.

Can this be done more generally? Any starter in any Abelian group of order $2n - 1$ will satisfy the conditions that every pair appears exactly once

$\infty 1$	45	27	--	36	--	--
--	$\infty 2$	56	31	--	47	--
--	--	$\infty 3$	67	42	--	51
62	--	--	$\infty 4$	71	53	--
--	73	--	--	$\infty 5$	12	64
75	--	14	--	--	$\infty 6$	23
34	16	--	25	--	--	$\infty 7$

Figure 13.2

and that every symbol appears exactly once per row. To satisfy the column condition, one must make some more restrictions on the starter. So we make the following definition.

DEFINITION. Suppose that $S = \{x_1 y_1, x_2 y_2, \ldots, x_{n-1} y_{n-1}\}$ is a starter in an Abelian group G of order $2n - 1$. An <u>adder</u> A_S for S is an ordered set of $n - 1$ nonzero elements $a_1, a_2, \ldots, a_{n-1}$ of G, all different, with the property that the $2n - 2$ sums $x_1 + a_1, y_1 + a_1, x_2 + a_2, \ldots, y_{n-1} + a_{n-1}$ include each nonzero element of G once.

THEOREM 13.1 If there is a starter S in a group of order $2n - 1$ with an adder A_S, then there is a Room square of side $2n - 1$.

Proof: Label the elements of G as $g_1, g_2, \ldots, g_{2n-1}$ in some order, where g_1 is the identity element. We construct an array R. In the first row, column 1 contains $\{\infty, g_1\}$; column j contains an empty cell if $-g_j$ is <u>not</u> a member of A_S, but if $-g_j = a_k$, then column j contains $\{x_k, y_k\}$. The other rows are constructed from row 1: The cell in the $(1, j)$ position gives rise to the cell in the (i, h) position, where $g_h = g_j + g_i$; if the former is empty, then so is the latter, while if cell $(1, j)$ contains $\{x_k, y_k\}$, then cell (i, h) contains $\{x_k + g_i, y_k + g_i\}$. The diagonal element in row g_i is $\{\infty, g_i\}$. Then R is easily seen to be a Room square. ∎

If we replace g_i by i throughout the foregoing construction, the resulting Room square is standardized.

Given the starter $\{x_1, y_1\}, \{x_2, y_2\}, \ldots, \{x_{n-1}, y_{n-1}\}$, consider the numbers $a_i = -x_i - y_i$. We have

$$x_i + a_i = -y_i, \quad y_i + a_i = -x_i$$

for every i; the $2n - 1$ sums $x_i + a_i$ and $y_i + a_i$ will be distinct and nonzero because the elements of pairs in a starter are distinct and nonzero. So,

13.2 Starter Constructions

provided that the elements a_i (or equivalently, the elements $-a_i$) are themselves distinct and nonzero, we have an adder. We say a starter is **strong** if the sums $x_i + y_i$ are distinct and nonzero; what we have just said means that every strong starter has an adder.

THEOREM 13.2 If p is an odd prime such that $p^n = 2^k t + 1$ where t is an odd integer greater than 1, then there is a strong starter in $GF(p^n)$.

Proof: Suppose that x is a primitive element in $GF(p^n)$ (see Section 3.2). For convenience write δ for 2^{k-1}; then $x^{2\delta t} = 1$, but $x^\alpha \neq 1$ when $1 \leq \alpha < 2\delta t$. We shall verify that X,

$$X = \{X_{ij} : 0 \leq i \leq \delta - 1, \; 0 \leq j \leq t - 1\},$$

is a strong starter, where

$$X_{ij} = \{x^{i+2j\delta}, x^{i+(2j+1)\delta}\}.$$

It is easy to see that the totality of the entries in the X_{ij} constitute all nonzero elements of $GF(p^n)$—in fact, the left-hand members of $X_{00}, X_{10}, \ldots, X_{\delta-1,0}$ are $x^0, x^1, \ldots, x^{\delta-1}$, and the right-hand members are $x^\delta, x^{\delta+1}, \ldots, x^{2\delta-1}$; similarly, $\{X_{i1}\}$ contains the members $x^{2\delta}, x^{2\delta+1}, \ldots, x^{4\delta-1}$, and so on. The differences between the elements of X_{ij} are $\pm x^{i+2j\delta}(x^\delta - 1)$. Since $x^\delta - 1$ is nonzero and $GF(p^n)$ is a field, the entries $x^{i+2j\delta}(x^\delta - 1)$ and $x^{k+2\ell\delta}(x^\delta - 1)$ will be equal if and only if $x^{i+2j\delta} = x^{k+2\ell\delta}$, and this can occur only when $i = k$ and $j = \ell$; so the δt differences with a + sign are all different. Similarly, the differences with a − sign are all different. Moreover, if

$$x^{i+2j\delta}(x^\delta - 1) = -x^{k+2\ell\delta}(x^\delta - 1)$$

then

$$x^{i+2j\delta - k - 2\ell\delta} = -1$$

and squaring yields

$$x^{2(i-k)+4\delta(j-\ell)} = 1.$$

This must mean that

$$2(i - k) + 4\delta(j - \ell) \equiv 0 \pmod{2\delta t}$$

since $2\delta t$ is the order of x. In particular,

$$2(i - k) \equiv 0 \pmod{2\delta}$$

and since $0 \leq i < \delta$ and $0 \leq k < \delta$, we must have $i - k = 0$. Therefore,

$$4\delta(j - \ell) \equiv 0 \pmod{2t}$$

from which

$$2(j - \ell) \equiv 0 \pmod{t};$$

but $-t < j - \ell < t$, so $j = \ell$ or $2(j - \ell) = \pm t$. The second case contradicts the fact that t is odd, and if $j = \ell$, then we have

$$2x^{i+2j\delta}(x^\delta - 1) = 0$$

which is impossible since p^n is odd and consequently 2, x, and $x^\delta - 1$ are all nonzero in $GF(p^n)$. So no two of the $2\delta t$ differences can be equal. Therefore, the set of differences constitutes all nonzero elements of $GF(p^n)$.

This means that the set X satisfies the definition of a starter. To show that the starter is strong, we have to show that the δt elements formed by adding the two members of X_{ij} are different. The sum resulting from X_{ij} is

$$x^{i+2j\delta}(x^\delta + 1),$$

and these objects are all different because $x^\delta + 1$ is nonzero. ∎

THEOREM 13.3 If k is an integer greater than 1, then there is a strong starter in the additive group of integers modulo $2^{2k} + 1$.

Proof: Since $m \geq 2$, we can write $2^{2k} = 16t^2$, where $t = 2^{k-2}$ is a positive integer. The required starter consists of the following four pairs, for each i and j satisfying $i \leq i \leq 2t$, $0 \leq j \leq t - 1$:

$(i + 4jt, 4it - j)$

$(j - 4it, -8t^2 - 4jt - 2t - i)$

13.3 Subsquare Constructions

$$(8t^2 + 4jt + 2t + i,\ 8t^2 + 4it - 2t - j)$$
$$(-8t^2 - 4it + 2t + j,\ -i - 4jt).$$

The required verification is straightforward. ∎

Observe that Theorems 13.2 and 13.3 between them provide starters for Room squares of all prime orders except 3 and 5: Theorem 13.2 covers all primes that are not of the form $2^n + 1$; Theorem 13.3 provides all primes greater than 5 of the form $2^n + 1$ for n even; and if n is odd, $2^n \equiv 2 \pmod{3}$, so 3 divides $2^n + 1$ and there is no prime greater than 3 of this type.

EXERCISES

13.2.1 Verify that the Room squares of side 7 given in Sections 13.1 and 13.2 are nonisomorphic.

13.2.2 Find adders for the three starters in the cyclic group of order 7.

13.2.3 Prove that there is no adder for the patterned starter in either group of order 9.

13.2.4 Prove that the patterned starter is never strong.

13.2.5 Verify that the pairs given in Theorem 13.3 do in fact form a strong starter.

13.2.6 A <u>skew Room square</u> R is a standardized Room square with the property that when $i \neq j$, either the (i,j) or the (j,i) cell of R is occupied, but not both.
   (i) Suppose that R is a Room square constructed from a starter and adder. Prove that in the notation used in Theorem 12.1, the $(i,1)$ cell of R is empty if and only if g_i is not a member of A_S. Consequently, prove that the Room square constructed from a starter and adder is skew if and only if the adder never contains the negative of one of its elements.
   (ii) Prove that the Room squares developed from strong starters on Theorems 13.2 and 13.3 are skew.
   (iii) If R is a skew Room square based on $\{\infty, 1, 2, \ldots, v\}$, consider the set of all blocks $\{x, y, i, j\}$, where $1 \leq x < y \leq v$ and $\{x, y\}$ lies in the (i, j) cell of R. Prove that these blocks form a $(2n+1, 2n^2+n, 4n, 4, 6)$-balanced incomplete block design.

13.3 SUBSQUARE CONSTRUCTIONS

In a Room square of side r we say that a set of s rows and s columns forms a <u>subsquare of side s</u> if the s × s array formed by their intersection is itself precisely a Room square based on some s + 1 of the symbols.

THEOREM 13.4 Suppose that there is a Room square of side r and a Room square of side s with a subsquare of side t. Suppose further that there exist orthogonal Latin squares of side s − t. Then there is a Room square of side v,

$$v = r(s - t) + t,$$

which has subsquares of sides r, s, and t.

Proof: We construct a Room square based on the v + 1 symbols in the set

$$X = \{\infty, 1_1, 1_2, \ldots, 1_r, 2_1, 2_2, \ldots, \ldots n_r, n + 1, \ldots, s\}$$

where n = s − t. We assume the existence of standardized Room squares R of side r based on $\{\infty, 1, 2, \ldots, r\}$ and S of side s based on $\{\infty, 1, 2, \ldots, s\}$. We further assume S to have the form

$$S = \begin{array}{|c|c|} \hline A & B \\ \hline C & T \\ \hline \end{array}$$

where the subarray T is a Room square of side t based on $\{\infty, n + 1, n + 2, \ldots, s\}$. L and M are orthogonal Latin squares of side n. We write L_i to mean L with every entry given a subscript i: If L has (1,1) entry x, then L_i has (1,1) entry x_i. M_j is defined similarly. N_{ij} denotes the n × n array of unordered pairs whose (k, ℓ) entry consists of the (k, ℓ) entries of L_i and μ_j. We also subscript the arrays A, B, and C; for example, in A_i, the pair $\{x, y\}$ from A would be replaced by $\{x_i, y_i\}$, where x = x when x = ∞ or x > n, and the empty cells of A would correspond to empty cells in A_i.

We first convert R into an $nv_1 \times nv_1$ array U be replacing each of its cells by an n × n array. An empty cell is replaced by an empty n × n array; the entry $\{\infty, i\}$ is replaced by A_i; and the entry $\{i, j\}$, $i \neq 0$, $j \neq 0$, is replaced by N_{ij}.

Now write

$$V = \begin{array}{|c|c|} \hline & B_1 \\ & B_2 \\ U & \vdots \\ & B_r \\ \hline C_1\ C_2\ \cdots\ C_r & T \\ \hline \end{array}$$

13.3 Subsquare Constructions

Row i of R contains $\{\infty, i\}$ and a set of entries $\{j, k\}$ whose elements collectively exhaust $\{1, \ldots, i-1, i+1, \ldots, r\}$. So the ith row of blocks of V, $i \leq i \leq r$, contains A_i, B_i, and a set of N_{jk} such that the collection of all the j and k is $\{0, 1, \ldots, i-1, i+1, \ldots, r\}$. Each row of N_{jk} contains all the x_j and y_k, $1 \leq x, y \leq n$, once each. Each row of

$$\boxed{A_i \mid B_i}$$

contains all the objects

$$\{0_i, 1_i, \ldots, s_i\},$$

that is,

$$\{0, 1_i, \ldots, n_i, n+1, \ldots, s\},$$

once each. So every row in the ith row of blocks in V contains every element of X once, for $1 \leq i \leq r$. In the last row of blocks, since each row of C_i contains $\{1_i 2_i, \ldots, n_i\}$ once each, $\{C_1, C_2, \ldots, C_r\}$ will contain every element of X except for $0, n+1, n+2, \ldots,$ and s precisely once, and the missing elements are contained once in each row of T. Consequently, every row of V contains every member of X once. A similar proof applies to columns.

If we take the whole of V, we have every entry of N_{ij} once, for $1 \leq i \leq r$ and $i \leq j \leq r$. So every possible unordered pair of the form $\{x_i, y_j\}$, $1 \leq x$, $y \leq n$, $1 \leq i, j \leq r$, appears once. We also have entries of A_i, B_i, and C_i, $i \leq i \leq r$; for a fixed i, these three arrays will contain every pair of the form $\{x_i, y_i\}$, $1 \leq x, y \leq n$, and every entry $\{x_i, z\}$, $1 \leq x \leq n$, $z = 0$ or $n \leq z \leq s$. T will contain all the pairs $\{z, t\}$, where both z and t belong to $\{0, n+1, n+2, \ldots, s\}$. So V contains every unordered pair of members of X.

Finally, A_i and B_i together contain $(s+1)/2$ pairs per row, and each row of each N_{ij} contains n pairs, C_i contains $n/2$ pairs in each row, and T_3 has $(t+1)/2$. So the number of pairs in V is

$$rn\left[\frac{1}{2}(r-1)n + \frac{1}{2}(s+1)\right] + t\left[\frac{1}{2}rn + \frac{1}{2}(t+1)\right]$$

$$= \frac{1}{2}\left[rn(rn - n + n + t + 1) + t(rn + t + 1)\right]$$

$$= \frac{1}{2}(rn + t)(rn + t + 1)$$

$$= \frac{1}{2}v(v + 1).$$

This is the number of unordered pairs that can be chosen from X, so each pair must appear precisely once. Therefore, V is a Room square.

If we take the intersection of the first n and last t rows with the corresponding columns and delete everything else, we have

A_1	B_1
C_1	T

,

which is essentially S. T is exhibited as a subsquare in the last t rows and columns. To discover R, take the intersection of rows 1, n + 1, ..., n n(r - 1) + 1 and the corresponding columns. The array formed has entry $\{0, 1_i\}$ where R had $\{0, 1\}$ and $\{1_i, 1_j\}$ where R had $\{i, j\}$. ■

Although there cannot be a subsquare of side 0, the proof above works when t = 0. If t = 1, the subsquare size is 1 and any occupied cell is a subsquare of side 1. The only problem will arise when n = 2 or 6—when t = 0, this can never occur, and when t = 1, only n = 6 (or s = 7) can occur. So we have:

COROLLARY 13.4.1 If there are Room squares of sides r and s, then there is a Room square of side rs with subsquares of sides r and s, and there will also be a Room square of side r(s - 1) + 1 with subsquares of sides r and s except possibly when s = 7. ■

EXERCISE

13.3.1 Suppose that there is a pairwise balanced design D with parameters (v;K;1), and that there is a Room square R_k of side k for every k ∈ K. Without loss of generality, one may assume that each R_k is standardized. If B = $\{b_1, b_2, \ldots, b_k\}$ is a block of D of size k, then define a set $f(b_i, b_j)$ as follows:
(a) If the (i, j) cell of R_k is empty, $f(b_i, b_j)$ equals the empty set.
(b) If the (i, j) cell of R_k contains $\{p, q\}$, then $f(b_i, b_j)$ equals the set $\{b_p, b_q\}$.

13.4 The Existence Theorem

(i) Prove that if x and y are any distinct integers between 1 and v, then f(x,y) is defined exactly once by the rules above.
(ii) A v × v array S is defined by:
 (a) if $x \neq y$, then $s_{xy} = f(x,y)$;
 (b) $s_{xx} = \{\infty, x\}$.
Prove that S is a Room square of side v with subsquares of side k for every k in K.

13.4 THE EXISTENCE THEOREM

From Section 13.3, we know that the existence of Room squares of sides r and s implies the existence of a square of side rs. Now, in Section 13.2 we constructed squares of all prime sides except 3 and 5. So repeated multiplication will provide a Room square of side m, provided that $(m, 15) = 1$.

Theorem 13.2 provides squares of sides $25 = 5^2$ and $125 = 5^3$, so we can construct a square of side 5^k for every $k > 1$. There is a square of side 27, from Theorem 13.2 again, and there are Room squares of side 9: for example, one can be constructed by cyclically developing the first row

∞1 69 48 — — — — 57 23.

Similarly, a square of side 15 can be developed from

∞1 38 — — — 1114 — — 713 56 — 415 1012 — 29.

So we have

LEMMA 13.5 There is a Room square of every odd integer side r except possibly when $r = 3v$ or $r = 5v$, where $(v, 15) = 1$. ∎

If $(v, 15) = 1$, we can construct a Room square of side v, so the cases missing in Lemma 13.5 are all three or five times the side of a Room square. Therefore, it will be sufficient for our purposes to provide a theorem that allows us to produce squares of sides 3r and 5r from a given square of side r.

Suppose that n is an odd integer. Write A_n for the n × n array of ordered pairs chosen from the n-set $N = \{1, 2, \ldots, n\}$ of the first n natural numbers, whose (i, j) entry is the ordered pair a_{ij},

$$a_{ij} \equiv (j - i + 1, i + j - 1) \pmod{n}.$$

Any ordered pair of members of N appears precisely once in A_n: The pair (i, j) appears in position (i, j), where $1 \leq i, j \leq n$ and

$$i \equiv \frac{1}{2}(n + 1)(y - x) + 1 \pmod{n}$$

$$j \equiv \frac{1}{2}(n + 1)(x + y) \pmod{n}.$$

Consider row i. The left-hand members of the entries of A_n (mod n) are

$$\{2 - i, 3 - i, \ldots, -i, 1 - i\}$$

and the right-hand members are

$$\{i, i + 1, \ldots, i - 2, i - 1\}.$$

In column j, the left-hand members are

$$\{j, j - 1, \ldots, j + 2, j + 1\},$$

while the right-hand positions contain

$$\{j, j + 1, \ldots, j - 2, j - 1\}.$$

In every case the set is N.

Suppose that the pair (x,y) occurs in column j of A_n. That is,

$$j = \frac{1}{2}(x + y).$$

Then the pair (y, x) will also lie in column j, since

$$\frac{1}{2}(y + x) = \frac{1}{2}(x + y).$$

These properties are summarized in the following lemma.

LEMMA 13.6 The entries of A_n consist of the ordered pairs of members of N taken once each. The entries in a given row or column of A_n contain between them every member of N once as a left member and once as a right member. If the pair (x,y) occurs in a given column, then (y, x) also occurs in that column. ∎

Now suppose that R is a given Room square of side v, based on the symbols $\{\infty, 1, \ldots, v\}$, and that n is an odd integer less than v. We shall construct a Room square of side nv based on the symbol set

$$S = \{\infty, 1_1, 2_1, \ldots, v_1, 1_2, 2_2, \ldots, v_n\}.$$

13.4 The Existence Theorem

For convenience assume that R is standardized. We define an array R_{ij}, for $1 \leq i \leq n$ and $1 \leq j \leq n$, as follows:

(i) Delete all diagonal entries of R;
(ii) if $x < y$, replace the entry $\{x,y\}$ of R by $\{x_i, y_j\}$.

The arrays R_{ij} will contain between them all unordered pairs of the symbols other than ∞ except for the pairs $\{x_i, x_j\}$. If R_{ij} and R_{jk} are placed next to one another, then row x of the resulting array will contain each of $1_j, 2_j, \ldots, v_j$ except for x_j precisely once; a similar remark applies to columns when R_{ij} is placed above R_{jk}.

LEMMA 13.7 Given a Room square of R of side v, where $v = 2s + 1$, there are s permutations $\varphi_1, \varphi_2, \ldots, \varphi_s$ of $\{1, 2, \ldots, v\}$ with the properties that $k\varphi_i = k\varphi_j$ never occurs unless $i = j$, and that cell $(k, k\varphi_i)$ is empty for $1 \leq k \leq v$, $1 \leq i \leq s$.

Proof: Consider the $r \times r$ matrix M whose (k, ℓ) position contains 1 if R has its (k, ℓ) cell empty, but is 0 otherwise. M is a matrix of zeros and ones with every row and column sum equal to s. So by Theorem 1.5 M is a sum of s permutation matrices, say

$$M = P_1 + P_2 + \cdots + P_s.$$

We define φ_i to be the permutation corresponding to P_i: If P_i has its (k, ℓ) entry equal to 1, then $k\varphi_i = \ell$. If $k\varphi_i = k\varphi_j$ occurred when $i \neq j$, then P_i and P_j would both have 1 in position $(k, k\varphi_i)$, so M would have an entry equal to 2 or more. That the $(k, k\varphi_i)$ cell of R is empty follows from the definition of M. ∎

THEOREM 13.8 If v and n are odd integers such that $v \geq n$, and if there is a Room square R of side v, then there is a Room square of side nv.

Proof: For convenience write $v = 2s + 1$ and $n = 2t + 1$. We construct the Room square by replacing every entry of A_n by a $v \times v$ block.

For a given j, select permutations $\varphi_{j1}, \varphi_{j2}, \ldots, \varphi_{jn}$ satisfying the following conditions:

(i) $\varphi_{jk} = \varphi_{j\ell}$ if and only if (k, ℓ) and (ℓ, k) appear in column j of A_n;
(ii) if the entry in row j and column j of A_n is (x,y), then φ_{jx} and φ_{jy} equal the identity permutation;
(iii) all the φ_{jk} except for the identity permutation are selected from the set of s permutations associated with R according to Lemma 13.7.

(This will be possible since $v \geq n$ and consequently $s \geq t$.) Now replace the entry (k, ℓ) in column j of A_n by the array $R_k \ell \varphi_{jk}$ which is obtained by performing the permutation φ_{jk} on the columns of R_k.

From Lemma 13.6 and the properties of the arrays $R_k \ell$ it follows that the resulting array will contain every number from 1 to rn in each row and each column, except that every x_k is missing from every row x_j and that when j is such that (k, ℓ) is an entry in column j of A_n, then x_k and x_ℓ are missing from column $(x\varphi_{jk})_j$. Moreover, the array contains every unordered pair of members of $S \setminus \{\infty\}$ except for the pairs of the form $\{x_k, x_\ell\}$, and contains each precisely once.

We now insert some more entries into the jth diagonal block of the new array. For each k, if (k, ℓ) was an entry of column j of A_n, we place $\{x_k, x_\ell\}$ in the $(x, x\varphi_k)$ position of this block [i.e., in the $(x_j, (x\varphi_k)_j)$ position of the new square] except that $\{\infty, x_j\}$ is used instead of $\{x_j, x_j\}$ in the relevant position. Lemma 13.7 tells us that we shall not finish with two entries in the same cell, so this step is possible. The finished array now contains every entry from S once per row and once per column, and contains every unordered pair from that set precisely once. So it is the required Room square. ∎

Consequently, there exist Room squares of all odd sides except 3 and 5.

EXERCISE

13.4.1 Suppose that x and y are integers modulo n, where n is odd. Prove that the unique integers i and j modulo n which satisfy

$$j - i + 1 \equiv x, \quad i + j - 1 \equiv y$$

are given by

$$i \equiv \frac{1}{2}(n + 1)(y - x) + 1,$$

$$j \equiv \frac{1}{2}(n + 1)(x + y).$$

13.5 SUBSQUARES OF ROOM SQUARES

Subsquares were defined in Section 13.3. There has been some interest in finding the values (r, s) for which there is a Room square of side s with a subsquare of side r.

13.5 Subsquares of Room Squares

THEOREM 13.9 If there is a Room square of side s with a subsquare of side r, where r < s, then

$$s \geq 3r + 2.$$

Proof: Suppose that there is a Room square S of side s based on $\{\infty, 1, 2, \ldots, s\}$, and that it has a subsquare R of side r based on $\{\infty 1, 2, \ldots, r\}$, where r < s. Moreover, suppose that R is located in the top left corner of S: say

$$S = \begin{array}{|c|c|} \hline A & B \\ \hline C & T \\ \hline \end{array}$$

Select any integer x with $r < x \leq s$. There must be a pair involving x in each of rows 1, 2, ..., r; suppose that $\{x, y_i\}$ is in row i. Since x > r, this pair must lie in the subarray A, so $y_i > r$. Similarly, there is a pair $\{x, z_i\}$ in column i for i = 1, 2, ..., r, and $z_i > r$. All of these numbers must be different. So we know at least $3r + 2$ members of $\{\infty, 1, 2, \ldots, s\}$:

$$\infty, 1, 2, \ldots, r, x, y_1, y_2, \ldots, y_r, z_1, z_2, \ldots, z_r.$$

This proves that $s \geq 3r + 1$. Now both s and r must be odd, so $s \geq 3r + 2$. ∎

This lower bound can in fact be obtained in every case where r and $3r + 2$ are possible sides of Room squares. The construction involves a new definition. A Room square of side $r = 2n - 1$ will be called <u>special</u> if it has n - 1 cells, occupied by n - 1 disjoint pairs of objects, which all lie in different rows and all lie in different columns. By row and column permutation it is easy to move these n - 1 cells into the first n - 1 diagonal positions; then, by relabeling the objects we can impose the condition that the diagonal begins with the pairs

$$\{\infty, 2n - 1\}, \{1, n\}, \{2, n + 1\}, \ldots, \{n - 2, 2n - 3\},$$

and that the other two symbols are n - 1 and 2n - 1; such a special Room square is said to be "in standard form." It may be shown that a special Room square exists of every odd side other than 3 and 5; the proof is beyond our scope here. However, our next theorem shows that there are special Room squares of all prime power orders other than 3, 5, and 9.

THEOREM 13.10 A Room square constructed from a strong starter is special.

Proof: Suppose that R is a Room square of side 2n - 1 developed from the strong starter

$$\{x_1, y_1\}, \{x_2, y_2\}, \ldots, \{x_{n-1}, y_{n-1}\}$$

in the group $G = \{g_1, g_2, \ldots, g_{2n-1}\}$, where g_1 is the zero. We prove that the pairs $\{z, -z\}$, where z ranges through the nonzero elements of G, all lie in different rows and columns. From the proof of Theorem 13.1, the (i, j) cell in R is occupied if and only if $g_i - g_j$ is an element of the adder; if $g_i - g_j = a_k$, the adder element for $\{x_k, y_k\}$, then the (i, j) cell contains $\{x_k + g_i, y_k + g_i\}$. So $\{z, -z\}$ lies in cell (i, j) where

$$z = x_k + g_i$$

$$-z = y_k + g_i,$$

so

$$2g_i = -x_k - y_k = a_k$$

and

$$g_j = g_i - a_k = g_i - 2g_i = -g_i.$$

Therefore, $\{z, -z\}$ lies in cell (i, j) where $g_j = -g_i$, for some i. The n - 1 cells must cover n - 1 different rows and n - 1 different columns. ∎

THEOREM 13.11 Suppose that R is a special Room square of side r, $r \neq 1$, and $r \neq 11$. Then there is a Room square of side 3r + 2 with R as a subsquare.

Proof: We assume that R is based on $\{0, 1, \ldots, r\}$, and that R is in standard form. (It will be convenient to use 0 rather than ∞.) Write r = 2n - 1; and let N_i denote the set of n consecutive integers starting at ni for i = 0, 1, 2, 3, 4, 5.

Let S be formed from R by adding 2n to each element. Then form an array T as follows: Delete the (i, i) entry $\{i + 1, n + i\}$ from R for $1 \leq i \leq r - 1$; append to the left of R a new column which has ith entry $\{i + 1, n + i\}$ for $1 \leq i \leq n - 1$; and add 4n to each element.

Since r is the side of a Room square and $r \neq 1$ or 11, r = 7 or r = 9 or $r \geq 13$, so n = 4 or n = 5 or $n \geq 7$. So there exist orthogonal idempotent Latin squares of side n. Let L and M be such squares, with diagonal (0, 1, ..., n - 1). Let $\tilde{L}, \tilde{M}$ stand for L, M with entry labels permuted so

13.5 Subsquares of Room Squares

that each first column becomes $(0, 1, \ldots, n-1)$; write L^1 for L with 1 added to each entry (mod n). For any of these arrays, a subscript i means that every entry is increased by ni [e.g., M_2 is based on $\{2n, 2n+1, \ldots, 3n-1\}$]. The join of two arrays X and Y is the array with (i, j) entry $\{x_{ij}, y_{ij}\}$; we define three special arrays:

A_{ij} is the join of $\tilde{L}_i$ and $\tilde{M}_j$, with the first column deleted;

B_{03} is the join of L and M_3, with the diagonal replaced by empty cells;

C_{12} is the join of L_1^1 and M_2, with the diagonal replaced by empty cells, and then columns $n-1$ and n removed and placed as columns 1 and 2, respectively.

We write D_{ij} for the column of length n with elements $(\{ni, nj\}, \{ni+1, nj+1\}, \ldots, \{ni+n-1, nj+n-1\})$. E_{12} is the column $(\{n+1, 2n\}, \{n+2, 2n+1\}, \ldots, \{2n, 3n-1\})$.

To complete the description we need an $n \times n$ array F, defined as follows: The (1,1), (1,2), and (1,3) entries are $\{\infty, 3n\}$, $\{n, 4n\}$, and $\{2n, 5n\}$, respectively; all other cells in the first row are empty; if cell (i, j) is empty, then cell (i+1, j+1) is empty; if cell (i, j) contains $\{x, y\}$, then cell (i+1, j+1) contains $\{x+1, y+1\}$ (where column numbers exceeding n are reduced modulo n). Then the array Z shown in Figure 13.3 is the required Room square.

To verify this, we first check that all pairs are present. Since R, S, and T contain all pairs with both elements in $N_0 \cup N_1$, in $N_2 \cup N_3$, or in $N_4 \cup N_5$, we need only check the pairs with one element in each of two of these sets. If A_{ij} and D_{ij} are both present, then all pairs with one element in each of N_i and N_j are covered; so only the pairs that would be in A_{03}, A_{12}, D_{03}, D_{12}, D_{14}, and D_{25} need be found. But B_{03} has the same entries as A_{03} (in different order), C_{12} and E_{12} cover A_{12} and D_{12}, and F covers the remainder.

Z =

R				S				T	
A_{24}			D_{34}	A_{05}	D_{15}			D_{02}	A_{13}
A_{35}	D_{35}		E_{12}		D_{04}	A_{04}	C_{12}		
		A_{34}				A_{15}	F		A_{02}
	D_{24}	A_{25}	D_{05}	A_{14}			B_{03}	D_{13}	

Figure 13.3

∞C	3R	--	--	--	2Q	1M	--	--	4I	7K	--	--	9Z	AX	--	--	8S	6H	--	BT	--	--	5G	FJ	--	--	EL	--	UW	--	VY	DO	NP	--
3M	∞D	--	2C	--	1R	--	--	8W	--	--	BY	5L	--	--	6J	4E	--	--	AV	--	7F	9U	--	--	TX	SZ	--	GI	--	HK	--	PQ	--	NO
NQ	OP	∞M	--	1C	--	2R	3D	6L	--	9X	--	--	7I	--	8Y	--	BU	--	5F	4H	--	--	AS	--	VW	GK	--	EJ	--	TZ	--	--	--	--
2D	--	1P	∞N	--	--	3O	--	--	AZ	--	5J	BW	--	4K	--	7G	--	9T	--	--	8V	6E	--	HI	--	--	UY	--	SX	--	FL	--	CM	QR
--	2M	--	--	∞O	3N	--	1Q	7S	--	4L	--	AY	--	--	9W	5H	--	--	BS	6G	--	8T	--	--	EK	FI	--	VZ	--	--	UX	--	DR	CP
--	1N	2O	--	3Q	∞P	--	--	--	BX	--	8Z	--	6K	5I	--	--	9V	7E	--	--	AU	--	4F	SY	--	--	TW	--	HL	GJ	--	CR	--	DM
1O	--	--	3P	MR	CD	∞Q	2N	--	5K	AW	--	4J	--	BZ	--	6F	--	8U	--	9S	--	7H	--	--	GL	VX	--	--	TY	--	EI	--	--	--
--	--	3C	1D	2P	--	--	∞R	9Y	--	--	6I	--	8X	--	7L	--	AT	--	4G	--	5E	--	BV	UZ	--	--	HJ	FK	--	SW	--	MN	OQ	--
4W	--	6Z	--	7X	--	5Y	--	∞E	3V	--	--	--	2U	1G	--	9P	--	BO	--	8Q	--	AR	--	--	IM	--	DK	LN	--	--	CJ	FS	HT	--
--	8I	--	AL	--	BJ	--	9K	3G	∞F	--	2E	--	1V	--	--	--	5N	--	7M	--	4C	--	6D	OW	--	RY	--	--	PZ	QX	--	TU	--	HS
--	7Y	5X	--	4Z	--	6W	--	HU	ST	∞G	--	1E	--	2V	3F	8R	--	AQ	--	--	9O	BP	--	CL	--	JN	--	--	DI	--	KM	--	--	--
BK	--	--	9J	--	8L	--	AI	2F	--	1T	∞H	--	--	3S	--	--	4D	--	6C	5M	--	--	7N	--	QZ	--	PX	RW	--	OY	--	--	EG	UV
5Z	--	7W	--	6Y	--	4X	--	--	2G	--	--	∞S	3H	--	1U	AO	--	--	8P	--	BR	--	9Q	--	DJ	LM	--	CK	--	IN	--	--	FV	ET
--	9L	--	BI	--	AK	--	8J	--	1H	2S	--	3U	∞T	--	--	--	6M	4N	--	7D	--	5C	--	RX	--	--	OZ	--	QY	--	PW	EV	--	FG
--	6X	4Y	--	5W	--	7Z	--	1S	--	--	3T	GV	EF	∞U	2H	BQ	--	--	9R	AP	--	--	8O	KN	--	--	CI	--	JM	DL	--	--	--	--
AJ	--	--	8K	--	9I	--	BL	--	--	3E	1F	2T	--	--	∞V	--	7C	5D	--	--	6N	4M	--	--	PY	QW	--	OX	--	--	RZ	GH	SU	--
BE	--	--	7U	9H	--	--	6T	5P	--	--	4R	--	AM	--	BC	∞I	3Z	--	--	--	2Y	1K	--	--	OS	--	QV	DF	--	--	GN	JW	LX	--
--	4S	BG	--	--	5V	AF	--	--	9N	8D	--	6O	--	7Q	--	3K	∞J	--	2I	--	1Z	--	--	EM	--	CH	--	--	RT	PU	--	XY	--	LW
--	AH	9F	--	--	7S	--	4U	BM	--	6Q	--	8N	--	--	5R	LY	WX	∞K	--	1I	--	2Z	3J	DG	--	PT	--	--	OV	CE	--	--	--	--
6V	--	--	5T	BE	--	8G	--	--	7O	--	AC	--	4P	9D	--	2J	--	1X	∞L	--	--	3W	--	--	RU	--	FN	HM	--	--	QS	--	IK	YZ
--	BF	8H	--	--	6U	--	5S	4Q	--	--	9M	7R	--	--	AN	--	2K	--	--	∞W	3L	--	1Y	PV	--	DE	--	CG	--	--	OT	--	JZ	IX
7T	--	--	4V	AG	--	9E	--	--	8C	5O	--	--	BD	6P	--	--	1L	2W	--	3Y	∞X	--	--	--	HN	--	RS	--	QU	FM	--	IZ	--	JK
9G	--	AE	--	--	4T	--	7V	--	6R	BN	--	--	5Q	8M	--	1W	--	--	3X	KZ	IJ	∞Y	2L	--	CF	OU	--	--	PS	--	DH	--	--	--
--	5U	--	6S	8F	--	BH	--	AD	--	--	7P	9C	--	--	4O	--	--	3I	1J	2X	--	--	∞Z	QT	--	--	GM	EN	--	RV	--	KL	WY	--
--	JT	IV	--	KS	--	LU	--	--	MW	CZ	--	DX	--	MY	--	--	EO	GR	--	--	HP	FQ	--	∞1	--	69	--	--	--	35	27	AB	--	48
FX	--	--	HW	--	EY	--	GZ	IO	--	--	LQ	--	JR	--	KP	MS	--	--	DU	NV	--	--	CT	59	∞2	--	7A	--	--	--	46	38	1B	--
LS	--	KU	--	IT	--	JV	--	--	DY	--	NX	MZ	--	CW	--	--	HQ	FP	--	ER	--	GO	--	--	6A	∞3	--	8B	--	--	--	57	49	12
--	EZ	--	GY	--	FW	--	HX	KR	--	JP	--	--	LO	--	IQ	CV	--	--	NT	--	DS	--	MU	23	--	7B	∞4	--	19	--	--	--	68	5A
IU	--	JS	--	LV	--	KT	--	NZ	--	--	DW	--	CY	--	MX	--	FR	--	HO	--	GQ	--	EP	6B	34	--	18	∞5	--	2A	--	--	--	79
--	GW	--	EX	--	HZ	--	FY	--	LP	IR	--	KQ	--	JO	--	DT	--	MV	--	CU	--	NS	--	8A	17	45	--	29	∞6	--	3B	--	--	--
HY	--	--	FZ	--	GX	--	EW	--	JQ	--	KO	IP	--	LR	--	NU	--	CS	--	--	MT	DV	--	--	9B	28	56	--	3A	∞7	--	14	--	--
--	KV	LT	--	JU	--	IS	--	CX	--	MY	--	--	MW	--	DZ	--	GP	--	EQ	FO	--	--	HR	--	--	1A	39	67	--	4B	∞8	--	25	--
--	--	NR	MQ	--	--	DP	CO	--	--	HV	GU	--	--	FT	ES	--	--	LZ	KY	--	--	JX	IW	--	--	--	2B	4A	78	--	15	∞9	--	36
--	--	DQ	OR	--	--	CN	MP	--	--	FU	SV	--	--	EH	GT	--	--	JY	WZ	--	--	IL	KX	47	--	--	--	13	5B	89	--	26	∞A	--
PR	CQ	--	--	DN	MO	--	--	TV	EU	--	--	FH	GS	--	--	XZ	IY	--	--	JL	KW	--	--	--	58	--	--	--	24	16	9A	--	37	∞B

Figure 13.4

13.5 Subsquares of Room Squares

Next, it is easy to see that every element occurs exactly once in each row, and also exactly once in each column (E_{12} is carefully chosen to fit with C_{12}). The only ones worth checking are columns $4n + 1$ to $5n$. In the first half of T every element of $N_4 \cup N_5$ occurs once, with the following exceptions:

$5n - 1$ and $6n - 2$ are missing from column 1;
$4n$ and $6n - 1$ are missing from column 2;

and for $3 \leq j \leq n$,

$4n + j - 2$ and $5n + j - 3$ are missing from column j.

C_{12} covers N_1 and N_2, except that

$2n - 1$ and $3n - 2$ are missing from column 1;
n and $3n - 1$ are missing from column 2;

and for $3 \leq j \leq n$,

$n + j - 2$ and $2n + j - 3$ are missing from column j.

B_{03} covers N_0 and N_3, except that

$j - 1$ and $3n + j - 1$ are missing from column j, $1 \leq j \leq n$.

Now the entries in F precisely fill in the gaps.

So the array is a Room square of the required size, and it clearly has R as a subsquare in the top left $r \times r$ square. ∎

There is in fact a square of side 35 with a subsquare of side 11; one is shown in Figure 13.4. So the bound of $3r + 2$ can be attained for a subsquare of side r, except when $r = 1$.

Various other subsquare constructions exist—see, for example, Theorem 13.4. However, the question of determining all pairs (r, s) such that there is a square of side s with a subsquare of side r is far from resolved.

EXERCISES

13.5.1 Prove that if there is a Room square of side s with a subsquare of side r, and there is a Room square of side t with a subsquare of side s, then there is a Room square of side t with a subsquare of side r.

13.5.2 Exhibit a special Room square of side 9.

BIBLIOGRAPHIC REMARKS

Room squares were so named because they were introduced in a short note by T. G. Room [7]. However, the equivalent structures had been in use in planning whist and bridge tournaments for many years. The first appearance of a Room square seems to be in 1850, in [5], where the relationship with Kirkman triple systems was first observed; this was rediscovered in [1]. Starters were introduced in [8], although their full generality was not obvious in that paper. Theorem 13.2 is due to Mullin and Nemeth [6], and Theorem 13.3 to Chong and Chan [2]. The general approach of Theorem 13.4 was outlined by Horton [4], who proved that Theorem and several others. Theorem 13.8 is due to Wallis [10].

Many authors have studied the subsquare problem. We believe the lower bound of Theorem 13.9 was first discovered by Collens and Mullin [3]. Theorem 13.11 was proven by Wallis [11], [12], and it has been significantly extended by Stinson and Wallis [9].

1. A. Cayley, "On a tactical theorem relating to the triads of fifteen things," Edinburgh and Dublin Philosophical Magazine and Journal of Science (4)25(1863), 59-61.
2. B. C. Chong and K. M. Chan, "On the existence of normalized Room squares," Nanta Mathematica 7(1974), 8-17.
3. R. J. Collens and R. C. Mullin, "Some properties of Room squares— A computer search," Congressus Numerantium 1(1970), 87-111.
4. J. D. Horton, "Variations on a theme by Moore," Congressus Numerantium 1(1970), 146-166.
5. T. P. Kirkman, "Note on an unanswered prize question," Cambridge and Dublin Mathematical Journal 5(1850), 255-262.
6. R. C. Mullin and E. Nemeth, "An existence theorem for Room squares," Canadian Mathematical Bulletin 12(1969), 493-497.
7. T. G. Room, "A new type of magic square," Mathematical Gazette 39 (1955), 307.
8. R. G. Stanton and R. C. Mullin, "Techniques for Room squares," Congressus Numerantium 1(1970), 445-464.
9. D. R. Stinson and W. D. Wallis, "On an even-side analogue of Room squares," Aequationes Mathematicae 27(1984), 201-213.
10. W. D. Wallis, "On the existence of Room squares," Aequationes Mathematicae 9(1973), 260-266.
11. W. D. Wallis, "A new Room square," Congressus Numerantium 34 (1982), 457-460.
12. W. D. Wallis, "All Room squares have minimal supersquares," Congressus Numerantium 36(1982), 3-14.

14
Asymptotic Results on Balanced Incomplete Block Designs

14.1 INTRODUCTION

We have seen various ways of constructing balanced incomplete block designs. On the other hand, apart from the obvious necessary conditions (1.1) and (1.2) and Fisher's inequality, the only results which show that certain designs are impossible are Theorem 6.4 (the Bruck-Ryser-Chowla theorem) and Corollary 6.4.2. There are two possible viewpoints: One is that all possible design parameters can be realized, with the exception of those ruled out by the theorems mentioned above and possibly a few other such theorems; the other is the belief that relatively few of the admissible parameters can be realized.

To be more specific, let us write $B(v, k, \lambda)$ to mean a balanced incomplete block design with parameters v, k, λ, and with

$$r = \frac{\lambda(v-1)}{k-1},$$

$$b = \frac{\lambda v(v-1)}{k(k-1)};$$

write $B(k, \lambda)$ for the set of all v such that a $B(v, k, \lambda)$ exists. In order to eliminate the redundant parameters r and b, we can replace (1.1) and (1.2) by the conditions

$$\lambda(v-1) \equiv 0 \pmod{k-1}, \tag{14.1}$$

$$\lambda v(v-1) \equiv 0 \pmod{k(k-1)}. \tag{14.2}$$

Then the former viewpoint is expressed by the following conjecture.

THE EXISTENCE CONJECTURE. Given positive integers k and λ, there exists a constant $v_0 = v_0(k,\lambda)$ such that $v \in B(k,\lambda)$ for every integer $v \geq v_0$ that satisfies (14.1) and (14.2).

The existence conjecture was proven in the 1970s. While references are given in the Bibliographic Remarks to this chapter, we shall say here that the main worker was Richard M. Wilson. His proof is quite long; it makes use of various constructions of transversal designs, group-divisible designs, and block designs due to Hanani and others, and various deep mathematical results, mainly from number theory, such as the theorem of Dirichlet that every arithmetical progression d, a + d, a + 2d, ... in which a and d are coprime will contain an infinitude of prime numbers.

One of Wilson's most interesting ideas was the idea of closure. If S is any set of positive integers, we say that S is PBD-closed, or simply closed, if it has the following property: Whenever K is a set of elements of S, and v is any integer such that a PB(v;K;1) exists, then v is in S; or, in symbols,

$$\left. \begin{array}{l} K \subseteq S, \ v \in Z, \\ PB(v;K;1) \text{ exists} \end{array} \right\} \implies v \in S.$$

Not all sets of positive integers are closed. For example, suppose that the set 2Z of all even integers were closed. We put $K = \{4,2\}$; as K is in 2Z, and as there is a PB(5;$\{4,2\}$;1), closure implies that $5 \in 2Z$, which is false. So 2Z is not closed.

However, there are many closed sets. One easy but important example follows from Theorem 2.1. Suppose that $P(K,\lambda)$ is the set of all integers v such that a PB(v;K;λ) exists. Then $P(K,\lambda)$ is PBD-closed. In particular, $B(k,\lambda)$ is a closed set.

Suppose that S is any set of positive integers. Define an integer $\beta(S)$ as follows: $\beta(S)$ is the greatest common divisor of all the numbers s (s - 1), where s is in S:

$$\beta(S) = GCD\{s(s-1) : s \in S\}.$$

It is usually quite easy to calculate the function β. Wilson proved that if S is any closed set, then there exists a positive integer N such that whenever $n > N$ and n is congruent (modulo $\beta(S)$) to some member of S, then n belongs to S.

As an example, suppose that both k and k - 1 are prime powers: for example, k = 3, or 4, or 8, or 9. Then we know of a B($k^2 - k + 1$, k, 1) (a projective plane of parameter k - 1) and a B(k^2, k, 1) (an affine plane of parameter k). So both $k^2 - k + 1$ and k^2 belong to B(k,1). So $\beta(B(k,1))$ is the greatest common divisor of a set containing both $(k^2 - k + 1)(k^2 - k)$ and $k^2(k^2 - 1)$. It is easy to see that $k^2 - k$ is the greatest common divisor of these two numbers. So (B, (k, 1)) is a divisor of k(k - 1). But (14.2) tells

14.2 Designs with Large λ

us that k(k - 1) divides v(v - 1) for every v in B(k, 1). So we have determined $\beta(B(k,1))$. If we construct a B(v, k, 1) for some v in each possible congruence class modulo k(k - 1) which is possible subject to (14.1) and (14.2) (and this will be done later in this chapter), we have proven the existence conjecture for designs B(v, k, 1).

Unfortunately, the theory becomes more complicated than this. It can be shown that $\beta(B(k,1))$ = k(k - 1) for any k, and congruence classes modulo k(k - 1) are very important. We shall define a k-<u>fiber</u> to be a congruence class modulo k(k - 1) such that there is an integer v in the class which satisfies (14.1) and (14.2) in the case λ = 1. The fiber that contains 1 is called the <u>principal fiber</u>. To prove the existence conjecture for λ = 1, one starts by constructing an example of a design in each fiber. Then B(k, 1) is shown to contain all sufficiently large members of the fiber, by recursive methods involving the existence of group divisible designs. The process for higher values of λ is similar.

These general proofs are too long to include here. However, we give the flavor of these asymptotic results about designs by proving that every k-fiber contains at least one value v for which a B(v, k, 1) exists. (in fact, we provide infinitely many designs.) We first prove that there is a B(v, k, λ) whenever (14.1) and (14.2) are satisfied and λ is sufficiently large (we actually prove the corresponding result for t-designs in general).

14.2 DESIGNS WITH LARGE λ

Recall the necessary conditions for a t-(v, k, λ) to exist, which follow from Theorem 2.15:

$$\binom{k-x}{t-x} \text{ divides } \lambda \binom{v-x}{t-x}, \quad 0 \leq x \leq t. \tag{14.3}$$

In this section we prove a result that can be stated informally as follows: Provided that λ is sufficiently large, the necessary condition (14.3) is sufficient. The proof uses the ideas of linear algebra, so we first restate the definition of a t-design in vector space terminology.

If P is any v-set, we shall write V(P) to mean the vector space of dimension v over the rational numbers which is indexed by the members of P. If x is any member of P, we write ϵ_x for the vector with entry 1 in position x and every other entry 0: the standard basis for V(P) is $\{\epsilon_x : x \in P\}$. The incidence map I_P is the map from the subsets of P to V(P), defined by

$$I_P(X) = \sum_{x \in X} \epsilon_x;$$

$I_P(X)$ has entry 1 in the places corresponding to a member of X and entry 0 otherwise.

We shall assume that a v-set P of treatments has been chosen. Since we are interested in subsets of P of fixed size, we write P_k for the set of all k-subsets of P; V_k denotes $V(P_k)$, and I_k is the map I_{P_k}. We indicate that basis elements refer to the set P_k by writing ϵ_B^k for the basis element of V_k corresponding to k-set B. The subscript or superscript 1 (referring to P itself) is suppressed.

We now define a t-(v,k,λ) design to be a vector $\delta = \{d_B : B \in P_k\}$ in V_k whose entries are nonnegative integers such that if T is any t-set on P, then

$$\sum_{B:T\subseteq B} d_B = \lambda. \qquad (14.4)$$

(The summation symbol means "sum over all B such that $T \subseteq B$.") To avoid triviality we demand that $v > k$ and $\lambda > 0$. To obtain the usual definition, one interprets the number d_B as the number of copies of the block B. Equation (14.4) can be rewritten in terms of the incidence map as

$$\sum_{B\subseteq P_k} d_B I_t(B) = \lambda I_t(P) \qquad (14.5)$$

An interesting consequence of this way of viewing designs is that we need not restrict the entries of δ to be nonnegative. We shall define a generalized t-(v,k,λ) design, or $G_t(v,k,\lambda)$, in the same way as a t-(v,k,λ) design, but the entries of δ are only required to be integers. In effect, negative multiplicities are allowed. Not all generalized designs are designs; a simple example is the $G_2(7,3,1)$, whose blocks are 123, 145, 167, 146, 157, 247, 347, 256, and 356 each with multiplicity 1, and 147 and 156 each with multiplicity -1. More generally, a generalized 2-design need not satisfy Fisher's inequality—see Exercise 14.2.3 for an example.

Suppose that S is any s-subset of P. We define the replication number of S to be

$$\sum_{B:S\subseteq B} d_B.$$

It is not hard to verify that in a generalized design there is a constant r_s for each $s \leq t$ such that every s-set has replication number r_s, and

$$r_s = \lambda \frac{\binom{v-s}{t-s}}{\binom{k-s}{t-s}}$$

14.2 Designs with Large λ

(For t-designs, this is just Theorem 2.15.) So the necessary condition (14.3) applies to generalized t-designs also.

LEMMA 14.1 Suppose that $\alpha = \{a_y : y \in P_s\}$ is any member of V_s, and suppose that H is any subset of P. Then

$$\sum_{Y:Y \cap H=\emptyset} a_Y = \sum_{X:X \subseteq H} \left[(-1)^x \sum_{Y:X \not\subseteq Y} a_Y \right] \qquad (14.6)$$

(where the first and third sums are over all appropriate Y in P_s and the second is over all subsets X of H; x is the order of X).

Proof: Suppose that Y is an s-subset of P such that $Y \cap H$ has n elements. Then the number of x-subsets X of H which are contained in Y is $\binom{n}{x}$. So a_Y arises $\binom{n}{h}$ times in computing the right-hand side of (14.6), each time with coefficient $(-1)^x$. So the coefficient of a_Y in the total is

$$\sum_{x=0}^{n} (-1)^x \binom{n}{x} = (1-1)^n,$$

which equals 0 if $n \geq 1$ but equals 1 if $n = 0$. So we have the left-hand side. ∎

It will be convenient to define linear maps f_X from P_t to Q as follows:

$$f_X(\epsilon_T^t) = \begin{cases} 1 & \text{if } X \subseteq P, \\ 0 & \text{if } X \not\subseteq P. \end{cases}$$

By linearity,

$$f_X(I_t(B)) = f_X\left(\sum_{T \subseteq B} \epsilon_T^t \right)$$

$$= \sum_{T \subseteq B} f_X(\epsilon_T^t).$$

If $X \subseteq B$, then $f_X(I_t(B)) = 0$. If X is a subset of B of size x (where $0 \leq x \leq t$), then

$$f_X(I_t(B)) = \binom{k-x}{t-x}.$$

In effect, f_X sums the entries corresponding to t-sets which contain the subset X.

LEMMA 14.2 If $v \leq k + t$, then the vectors $\{I_t(B) : B \in P_k\}$ are linearly independent. If $v = k + t$, then they form a basis for P_t.

Proof: Suppose that we are given a zero linear combination of the $I_t(B)$: say

$$\sum_{B \in P_k} a_B I_t(B) = 0.$$

We must prove that each a_B is zero.

Suppose that X is any x-subset of P, where $0 \leq x \leq t$. Since f is linear, $f(0) = 0$. So

$$0 = f_X\left(\sum a_B I_t(B)\right) = \binom{k - x}{t - x} \sum_{X \subseteq B} a_B. \tag{14.7}$$

Thus $\sum_{X \subseteq B} a_B$ is zero for all subsets X of P of size t or less.

Now select a particular k-set B_0 and write $H = P \setminus B_0$. Since any other k-subset of P must intersect H in at least one point,

$$\sum_{B: B \cap H = \emptyset} a_B = a_{B_0}.$$

Applying Lemma 14.1, we have

$$a_{B_0} = \sum_{X: X \subseteq H} \left[(-1)^x \sum_{B: X \subseteq B} a_B\right].$$

Since $v \leq k + t$, H has at most t elements, so any subset X of H has at most t elements, and (14.7) applies; so $a_{B_0} = 0$. Since B_0 was an arbitrary k-set, each coefficient a_B is zero. So the $I_t(B)$ are independent.

There are $\binom{v}{k}$ vectors $I_t(B)$, and P_t has dimension $\binom{v}{t}$. When $v = k + t$ these numbers are equal, so the $I_t(B)$ form a basis in that case. ∎

THEOREM 14.3 If $\alpha = \{a_T : T \in P_t\}$ is any vector in V_t, and $v \geq k + t$, then a necessary and sufficient condition for α to be an integer linear combination of the vectors $\{I_t(B) : B \in P_k\}$ is that

14.2 Designs with Large λ

$$\sum_{T:X\subseteq T} a_T \equiv 0 \left[\mathrm{mod}\binom{k-x}{t-x}\right] \tag{14.8}$$

for any x-subset X of P with $0 \leq i \leq t$.

Proof: We first prove necessity. Then we prove sufficiency in three parts: We deal with the special cases $v = k + t$ and $t = 0$ separately and then proceed with an induction in the general case.

(i) <u>Necessity</u>. Assume that α is a linear combination of the $I_t(B)$: say

$$\alpha = \sum_{B \in P_k} d_B I_t(B).$$

Select an x-subset X of P. Then

$$\sum_{T:X\subseteq T} a_T = \sum_{T\subseteq P_T} a_T f_X(\epsilon_T^t)$$

$$= f_X\left(\sum_T a_T \epsilon_T^t\right)$$

$$= f_X(\alpha)$$

$$= f_X\left[\sum_{B\subseteq P_k} d_B I_t(B)\right]$$

$$= \sum_B d_B f_X(I_t(B))$$

$$= \binom{k-x}{t-x} \sum_{B:X\subseteq B} d_B.$$

Now assume that each d_B is an integer. Then

$$\sum_{T:X\subseteq T} a_T \equiv 0 \left[\mathrm{mod}\binom{k-x}{t-x}\right],$$

as required.

(ii) <u>Case $v = k + t$.</u> Assume that $v = k + t$ and that (14.8) is always satisfied. By Lemma 14.2 the $I_t(B)$ form a basis for P_t, so we can again assume that there exist rational numbers d_B such that

290   14. Asymptotic Results on BIBD

$$\alpha = \sum_{B \in P_k} d_B I_t(B),$$

and we can deduce that

$$\sum_{T: X \subseteq T} a_T = \binom{k-x}{t-x} \sum_{B: X \subseteq B} d_B, \qquad (14.9)$$

just as before. Of course, we cannot now assume that the d_B are integers, but we know from (14.8) that the left hand of (14.9) is an integer multiple of $\binom{k-x}{t-x}$, so $\sum_{B: X \subseteq B} d_B$ is an integer.

Select a particular k-set B_0 and put $H = P \backslash B_0$. Then B_0 is the only k-set whose intersection with H is empty, so using Lemma 14.1 we have

$$d_{B_0} = \sum_{B: B \cap H = \emptyset} d_B = \sum_{X: X \subseteq H} \left[(-1)^x \sum_{B: X \subseteq B} d_B \right].$$

The right hand is always an integer, so d_{B_0} is integral for every k-set B_0.

(iii) <u>Case t = 0</u>. In this case α consists of one member $\alpha_\emptyset$. The condition (14.8) amounts to saying that $\alpha_\emptyset$ is an integer. Since $I_0(B)$ is just the integer 1 for every B, one can take

$$\alpha = \alpha_\emptyset I_0(B)$$

for any k-set B, and the conditions are trivially satisfied.

(iv) <u>Sufficiency in general</u>. Suppose that the theorem is true for every v, k, and t with $v < \bar{v}$ and for every α in V_t. Suppose that we are given a vector $\bar{\alpha} = \{\bar{a}_T\}$ for the case $v = \bar{v}$, $k = \bar{k}$, $t = \bar{t}$, where $v > k + t$ and $t > 0$, and (14.8) is satisfied. Select an element w of P. We define two new vectors:

$\alpha^1 = \left\{ a^1_{T^1} \right\}$, defined on the (t - 1)-subsets of $P \backslash \{w\}$ by $a^1_{T^1} = \bar{a}_{T^1 \cup \{w\}}$;

$\alpha^2 = \left\{ a^2_{T^2} \right\}$, defined on the t-subsets of $P \backslash \{w\}$ by $a^2_{T^2} = \bar{a}_{T^2}$

In other words, if T is a t-subset of P that contains w, define T^1 to be the (t - 1)-subset $T \backslash \{2\}$ and $a^1_{T^1} = \bar{a}_T$; if T is a t-subset of P that does not

14.2 Designs with Large λ

contain w, define T^2 to be the same as T and $a^2_{T^2} = \overline{a_T}$. (This is rather like the construction of the leave of a t-design.)

We first show that the vector α^1 satisfies (14.8) in the case $v = \overline{v} - 1$, $k = \overline{k} - 1$, $t = \overline{t} - 1$. If X is any subset of size x in $P\setminus\{w\}$, then $X \cup \{w\}$ is a subset of size $x + 1$ in P, so

$$\sum_{X \subseteq T} a^1_T = \sum_{X \cup \{w\} \subseteq T \cup \{w\}} \overline{a}_{T \cup \{w\}}$$

$$\equiv 0 \left[\mod \binom{k - (x+1)}{t - (x+1)} \right]$$

$$\equiv 0 \left[\mod \binom{(k-1) - x}{(t-1) - x} \right]$$

So, by part (iii) if $\overline{t} - 1 = 0$ or else by induction, we can find integer coefficients such that

$$\alpha^1 = \sum d^1_{B^1} I_{t-1}(B^1),$$

where the sum is taken over the $(k-1)$-subsets B^1 of $P\setminus\{w\}$.

Also, α^2 satisfies (14.8) in the case $v = \overline{v} - 1$, $k = \overline{k}$, $t = \overline{t}$. [To check this, observe that we only need to verify (14.8) for all sets X that do not contain w—a subset of the cases.] So by induction or by part (ii) we can find integers such that

$$\alpha^2 = \sum d^2_{B^2} I_t(B^2),$$

where B^2 ranges through the k-subsets of $P\setminus\{w\}$.

We now construct a set of integer coefficients $\{d_B : B \in P_k\}$ as follows. If $w \in B$, then write $B^1 = B\setminus\{w\}$ and $d_B = d^1_{B^1}$. If $w \notin B$, write $B^2 = B$ and $d_B = d^2_{B^2}$. Clearly,

$$\overline{\alpha} = \sum_{B \in P_k} d_B I_t(B). \blacksquare$$

COROLLARY 14.3.1 A necessary and sufficient conditions for the existence of a $G_t(v, k, \lambda)$-design with $v \geq k + t$ is that

$$\binom{k-x}{t-x} \text{ divides } \lambda\binom{v-x}{t-x}, \quad 0 \leq x \leq t \tag{14.3}$$

Proof: A $G_t(v,k,\lambda)$ is equivalent to expressing the vector $(\lambda, \lambda, \ldots, \lambda)$ as an integer linear combination of the $I_t(B)$. Now each a_T equals λ. If X is any x-set, there are $\binom{v-x}{t-x}$ t-sets T that contain it, so

$$\sum_{T: X \subseteq T} a_T = \lambda \binom{v-x}{t-x}$$

and the result follows from (14.8). ■

It is interesting to observe that it is much easier to prove Theorem 14.3 in its general form that it is to prove a version with each a_T equal to λ.
We now proceed to the main result.

THEOREM 14.4 Suppose that v, k, and t are given and $v \geq k + t$. There exists a positive integer $\lambda_0 = \lambda_0(v,k,t)$ such that if $\lambda \geq \lambda_0$ and if equation (14.3) is satisfied, then there is a t-(v,k,λ)-design.

Proof: Suppose that v, k, and t are given. We shall work modulo the integer $m = \binom{v-t}{k-t}$. One can easily calculate all the possible residue classes modulo m in which that it is ever possible to find a solution λ to the conditions (14.3); suppose that $\lambda_1, \lambda_2, \ldots, \lambda_\alpha$ are small positive members of all those classes. In each case, find a $G_t(v,k,\lambda_i)$-design δ, as guaranteed by Corollary 14.3.1: say $\delta = \{d_B : B \in P_k\}$. Suppose that the smallest of all those d_B is d^i. (Usually d^i will be negative.) Then consider the design $\{c_B\}$, defined by

$$c_B = d_B - d^i.$$

(Essentially, we are saying "add $-d^i$ copies of the design constructed in Exercise 2.2.2 to the design δ.") Then $\{c_B\}$ is a t-design with parameters

(v, k, $\lambda_i - d^i m$),

and $\{c_B + n\}$ is a design with parameters

(v, k, $\lambda_i - d^i m + nm$)

14.3 Designs in the Principal Fiber for $\lambda = 1$

for every positive integer n.
If we put $\lambda_0 = \max_i(\lambda_i - d^i m + m)$, we have the result. ∎

We have not covered designs with $v < k + t$ in our discussion. However, this is no problem, as it is easy to show that such a design (or one with $v = k + t$) must consist of all possible k-sets taken equally often. For details, see Exercise 14.2.4.

EXERCISES

14.2.1 Verify that the $G_2(7, 3, 1)$-design presented in the text does satisfy equation (14.4).

14.2.2 Prove that any generalized balanced incomplete block design with $k = 2$ is in fact a balanced incomplete block design.

14.2.3 Prove that the parameters $v = 16$, $k = 6$, $\lambda = 1$, $t = 2$ satisfy (14.8), so that a $G_2(16, 6, 1)$-design exists, but no 2-design with these parameters exists.

14.2.4 Suppose that a $G_t(v, k, \lambda)$-design δ exists with $v \leq k + t$. Prove that

$$\sum_{B \in P_k} c_B I_t(B) = \lambda I_t(P)$$

and that

$$\sum_{B \in P_k} I_t(B) = \binom{v - t}{k - t} I_t(P).$$

Use Lemma 14.2 to prove that

$$c_B = \frac{\lambda}{\binom{v - t}{k - t}},$$

so that $\binom{v - t}{k - t}$ divides λ and the design consists of equally many copies of all possible k-sets. Thus prove the equivalent of Theorem 14.4 for the case where $v \leq k + t$.

14.3 DESIGNS IN THE PRINCIPAL FIBER FOR $\lambda = 1$

In this section we prove the existence of a $B(v, k, 1)$ whenever v is a prime power congruent to 1 modulo $k(k - 1)$ and v is sufficiently large. We shall assume that k is a positive integer and v is a prime power such that $v = tk(k-1)/2 + 1$ for some even positive integer t. It is convenient to write $m = k(k-1)/2$. We denote the multiplicative group of nonzero elements of GF(v) by G; x is a primitive element of GF(v), so G is the cyclic group generated by r (see Section 3.2).

In general, if v is a prime power and $v = mt + 1$, the <u>cyclotomic classes</u> of v of <u>index</u> m are defined to be the sets $C_0, C_1, \ldots, C_{m-1}$, where

$$C_j = \{x^{im+j} : 0 \leq i < t\}.$$

So C_0 is the subgroup of index m generated by x^m, and C_j is $x^j C_0$.

In the particular case where $m = k(k-1)/2$, we shall define a <u>base block</u> B of order k in GF(v) to be a set

$$B = \{a_1, a_2, \ldots, a_k\}$$

with the property that the differences $a_j - a_i$, where $j > i$, form a system of distinct representatives for the cyclotomic classes $C_0, C_1, \ldots, C_{m-1}$. This will be possible only when m is a triangular number. Base blocks are of interest because of the following lemma.

LEMMA 14.5 Suppose that v is a prime power of the form $tk(k-1)/2 + 1$, where t is even, and that B is a base block of order k in GF(v). Then there is a B(v,k,1) on the elements of GF(v).

Proof: Since t is even, the element $-1 = x^{(1/2)mt}$ is in C_0. So the elements of C_0 can be partitioned into $t/2$ pairs of the form $\{y, -y\}$. Arbitrarily select a set E of order $t/2$ which contains one element from each of these pairs. Now consider the blocks $eB + f$, where e ranges through the elements of E and f ranges through GF(v). We have $tv/2$ blocks of size k. We shall verify that they form the required design.

From (1.1) and (1.2), the design would have $r = tk/2$ and $b = tv/2$. So the number of blocks is correct. If we show that every pair of treatments occurs together in at least one block, then a simple counting argument shows that each pair must occur together in precisely one block and we have $\lambda = 1$; from Theorem 2.7 this proves the design is a balanced incomplete block design.

Consider any two distinct elements x^c and x^d of GF(v). There will be some cyclotomic class, say C_j, which contains $x^d - x^c$. Say $x^d - x^c = x^{im+j}$. Suppose that the elements of B whose difference belongs to C_j are a_p and a_q, and $q > p$: $a_q - a_p = x^{hm+j}$ for some h. So

$$x^d - x^c = x^{(i-h)m}(a_q - a_p).$$

14.3 Designs in the Principal Fiber for $\lambda = 1$

Since $x^{(i-h)m}$ belongs to C_0, there is an element e of E such that $\pm e = x^{(i-h)m}$. So either

$$x^d - x^c = e(a_q - a_p)$$

or

$$x^c - x^d = e(a_q - a_p).$$

Without loss of generality, assume the former case. If we write

$$f = x^c - ea_p,$$

then we have

$$x^c = ea_p + f,$$
$$x^d = ea_q + f,$$

and both x^c and x^d belong to the block $eB + f$. ∎

In order to investigate the existence of base blocks, we shall need some elementary facts about means and variances of sets of numbers. If $c_1, c_2, \ldots, c_n$ are n real numbers, their mean μ and variance V are defined by

$$\mu = \frac{c_1 + c_2 + \cdots + c_n}{n},$$

$$V = \frac{(c_1 - \mu)^2 + (c_2 - \mu)^2 + \cdots + (c_n - \mu)^2}{n}.$$

It is convenient to observe that

$$nV = \sum_{i=1}^{n} (c_i - \mu)^2$$

$$= \sum_{i=1}^{n} c_i^2 - 2\mu \sum_{i=1}^{n} c_i + \sum_{i=1}^{n} \mu^2$$

$$= \sum_{i=1}^{n} c_i^2 - 2\mu(n\mu) + n\mu^2$$

so

$$\sum_{i=1}^{n} c_i^2 = nV + n\mu^2. \qquad (14.10)$$

Moreover, since V is nonnegative, we can deduce that

$$\sum_{i=1}^{n} c_i^2 \geq n \left(\frac{\sum_{i=1}^{n} c_i}{n} \right)^2$$

so

$$n \sum_{i=1}^{n} c_i^2 \geq \left(\sum_{i=1}^{n} c_i \right)^2. \qquad (14.11)$$

Slightly less obvious is the following result.

LEMMA 14.6 If $c_1, c_2, \ldots, c_n$ have mean μ and variance V, and if $m \leq n$, then

$$(c_1 + c_2 + \cdots + c_m - m\mu)^2 \leq m(n-m)V. \qquad (14.12)$$

Proof: Let us write $d_i = c_i - \mu$. Then the left-hand side above is $(d_1 + d_2 + \cdots + d_m)^2$. From (14.10),

$$nV = \sum_{i=1}^{n} d_i^2$$

$$= \sum_{i \leq m} d_i^2 + \sum_{i > m} d_i^2$$

$$\geq \frac{1}{m} \left(\sum_{i \leq m} d_i \right)^2 + \frac{1}{n-m} \left(\sum_{i > m} d_i \right)^2$$

using (14.11); so

$$\left(\sum_{i > m} d_i \right)^2 \leq (n-m)nV.$$

14.3 Designs in the Principal Fiber for $\lambda = 1$

But the sum of the d_i is zero, so

$$(d_1 + d_2 + \cdots + d_m)^2 = (-d_{m+1} - d_{m+2} - \cdots - d_n)^2$$

$$= (d_{m+1} + d_{m+2} + \cdots + d_n)^2$$

$$\leq (n - m)nV. \qquad \blacksquare$$

We now wish to prove that provided v is sufficiently large, GF(v) admits of a base block. We prove a more general result.

LEMMA 14.7 Suppose that k and m are given, and v is a prime power congruent to 1 (modulo m). Let $C_{12}, C_{13}, \ldots, C_{k-1,k}$ be any ordered selection of $k(k-1)/2$ cyclotomic classes C_{ij} of index m, where $i < j$. Then there exists a k-tuple $(a_1, a_2, \ldots, a_k)$ of elements of GF(v) such that $a_j - a_i \in C_{ij}$ for all $i < j$, provided that v is sufficiently large.

Proof: We shall prove (for $h < k$) that (provided v is large enough) for any sequence of $h(h-1)/2$ cyclotomic classes C_{ij} there is a sequence $(a_1, a_2, \ldots, a_h)$ of elements such that $a_j - a_i = C_{ij}$, with the following additional property: Given a further sequence $C_{1,h+1}, C_{2,h+1}, \ldots, C_{k,h+1}$ there will be an element a_{h+1} such that $a_{h+1} - a_i = C_{i,h+1}$ for each i. Then the solution for $h = k - 1$, with a_{h+1} appended, will be a k-tuple of the form required by the lemma.

We define numbers $N_D(A)$: If A is any sequence $(a_1, a_2, \ldots, a_h)$ of elements of GF(v) and $D = (d_1, d_2, \ldots, d_h)$ is any sequence of cyclotomic classes of index m in GF(v), then $N_D(A)$ is the number of elements y of GF(v) such that

$$y - a_1 \in C_{d_1}, \; y - a_2 \in C_{d_2}, \; \ldots, \; y - a_h \in C_{d_h}.$$

We need to show that $N_D(A) > 0$ for any given sequence of subscripts D.

We shall need to calculate some statistics on the numbers $N_D(A)$. The number of sequences D is m^h, and the number of sequences A equals the falling factorial $v^{(h)} = (v!)/(v-h)!$. So there are $n = m^h v^{(h)}$ numbers $N_D(A)$. The sum of them equals $v^{(h+1)}$, since this equals the number of sequences of $h + 1$ distinct elements on GF(v). [Each sequence $(a_1, a_2, \ldots, a_k, y)$ gives one solution if the elements are distinct, and none if there are repetitions.] So their mean is

$$\mu = \frac{v^{(h+1)}}{v^{(h)} m^h} = \frac{v - h}{m^h}.$$

14. Asymptotic Results on BIBD

We do not need the exact value of the variance V, but we shall show that $V < \mu$. We first sum $N_D(A)(N_D(A) - 1)$ over all A and D. For a given A, summing $N_D(A)(N_D(A) - 1)$ over all D is the same as counting all the ordered pairs (p,q) of elements of GF(q) such that $p \neq q$ and such that $p - a_j$ and $q - a_j$ belong to the same cyclotomic class for all a_j. When we sum this over all A, it is the same as counting, for all distinct p and q, those sequences of h elements a_j such that $p - a_j$ and $q - a_j$ are in the same cyclotomic class for all j. Given p and q, the number of elements c such that $p - c$ and $q - c$ are in the same class is $(v - 1)/m - 1$, because they are in the same class if and only if

$$\frac{x - c}{y - c} \in C_0 \setminus \{1\}.$$

So the number of possible sequences is

$$\left(\frac{v - 1}{m} - 1\right)^{(h)}$$

for a given p and q. There are $v(v - 1)$ choices for the ordered pair (x,y). Therefore,

$$\sum N_D(A)(N_D(A) - 1) = \left(\frac{v - 1}{m} - 1\right)^{(h)} v(v - 1).$$

Since $N_D(A)$ is a nonnegative integer, this sum is nonnegative, and it will be positive for sufficiently large v. We now divide $\sum N_D(A)(N_D(A) - 1)$ into $n\mu^2$: the answer is

$$\frac{m^h v^{(h)}(v - h)^2}{m^{2h}\left(\frac{v - 1}{m} - 1\right)^{(h)} v(v - 1)}$$

$$= \frac{v(v - 1) \cdots (v - h + 1)(v - h)(v - h)}{m^h \frac{(v - 1 - m)}{m} \frac{(v - 1 - 2m)}{m} \cdots \frac{(v - 1 - hm)}{m} v(v - 1)}$$

$$= \frac{(v - 2)(v - 3) \cdots (v - h + 1)(v - h)(v - h)}{(v - 1 - m)(v - 1 - 2m) \cdots (v - 1 - hm)}.$$

Clearly, each of the upper terms is greater than the corresponding lower term. So

$$\sum N_D(A)(N_D(A) - 1) < n\mu^2. \tag{14.13}$$

14.3 Designs in the Principal Fiber for $\lambda = 1$

Now from (14.10)

$$V = \frac{1}{n} \sum N_D(A)^2 - \mu^2$$

$$= \frac{1}{n} \sum N_D(A)[N_D(A) - 1] + \mu - \mu^2$$

$$\leq \mu,$$

using (14.13).

Now suppose that we have in mind a particular ordered set of cyclotomic classes C_{ij}. We write $\mathcal{B}_h$ for the set of order b_h of all sequences of length h which have appropriate differences: $a_j - a_i \in C_{ij}$. Then each member of $\mathcal{B}_{h+1}$ corresponds to a sequence A in $\mathcal{B}_h$ to which has been added a further element a_{h+1} such that $a_{h+1} - a_i = c_{i,h+1}$ for each i. The number of such sequences for a given A is $N_D(A)$, where $D = (d_1, d_2, \ldots, d_h)$ is defined by

$$C_{d_i} = C_{i, h+1},$$

Therefore,

$$b_{h+1} = \sum_{A \in \mathcal{B}_h} N_D(A)$$

where D is defined as above. We apply Lemma 14.4 to these b_h numbers $N_D(A)$. Equation (14.12) is

$$(b_{h+1} - b_h \mu)^2 \leq b_h(n - b_h)V.$$

Substituting for μ and n and using the fact that $V < \mu$, we have

$$\left(b_{h+1} - b_h \frac{v-h}{m^h}\right)^2 < b_h(m^h v^{(h)} - b_h) \frac{v-h}{m^h}.$$

Clearly, $m^h v^{(h)} - b_h < m^h v^h$ and $v - h < v$, so

$$\left(b_{h+1} - b_h \frac{v-h}{m^h}\right)^2 < b_h m^h v^h \frac{v}{m^h} = b_h v^{h+1}. \qquad (14.14)$$

We now prove by induction that as v approaches infinity, b_h is asymptotically equal to $v^{(h)}m^{-h(h-1)/2}$. This is clearly true for sequences of length 1, for $b_1 = v$. Assume that it is true for length h. Then for sufficiently large v, $b_h < v^h$, so (14.14) yields

$$\left(b_{h+1} - b_h \frac{v-h}{m^h}\right)^2 < v^{2h+1}.$$

So b_{h+1} differs from $b_h(v-h)/m^h$ by less than $v^{h+1/2}$. Now by hypothesis

$$b_h \frac{v-h}{m^h} \sim \frac{v^{(h)}(v-h)}{m^{h(h-1)/2}m^h} = \frac{v^{(h+1)}}{m^{h(h+1)/2}}.$$

Since this is order v^{h+1}, a term of order $v^{h+1/2}$ can be ignored asymptotically, so

$$b_{h+1} \sim \frac{v^{(h+1)}}{m^{h(h+1)/2}},$$

and the induction proof is complete.

It follows that for sufficiently large v, b_h is greater than zero for all h, and $b_k > 0$. ∎

From Lemmas 14.5 and 14.7 we deduce the following result.

THEOREM 14.8 Suppose that v is a prime power congruent to 1 modulo $k(k-1)$. If v is sufficiently large, then there is a $B(v, k, 1)$. ∎

14.4 DESIGNS IN OTHER FIBERS FOR $\lambda = 1$

We now show how to construct balanced incomplete block designs with $\lambda = 1$, where v is not necessarily congruent to 1 modulo $k(k-1)$.

THEOREM 14.9 Suppose that q is a prime power, u is a positive integer such that $q \geq u + 2$, and $d = u(u-1)/2$. Suppose that there exist balanced incomplete block designs $B(u, k, q)$ and $B(q^d, k, 1)$. Then there is a $B(uq^d, k, 1)$.

Proof: We assume that the $B(q^d, k, 1)$ has treatment set $F = GF(q^d)$ and the $B(u, k, q)$ has treatment set $S = (1 \cdots u)$ and block set $\mathcal{B}$. We construct a balanced incomplete block design with treatment set $S \times F$. It is convenient to note the number of blocks in each of the designs under consideration [using (1.1) and (1.2)]:

14.4 Designs in Other Fibers for $\lambda = 1$

$B(q^d, k, 1)$ has $q^d(q^d - 1)/k(k - 1)$ blocks;
$B(u, k, q)$ has $qu(u - 1)/k(k - 1)$ blocks;
$B(uq^d, k, 1)$ will have $uq^d(uq^d - 1)/k(k - 1)$ blocks.

We start by constructing u copies of the $B(q^d, k, 1)$. For each i in S we define blocks

$$\{(i, Z) : Z \text{ is a block of } B(q^d, k, 1)\}.$$

The block $Z = \{z_1, z_2, \ldots, z_k\}$ of $B(q^d, k, 1)$ gives rise to u blocks $\{\{(i, z_1), (i, z_2), \ldots, (i, z_k)\} : 1 \leq i \leq u\}$. This gives

$$\frac{uq^d(q^d - 1)}{k(k - 1)}$$

blocks. There remain to be constructed

$$b' = \frac{uq^d(uq^d - 1)}{k(k - 1)} - \frac{uq^d(q^d - 1)}{k(k - 1)} = \frac{qu(u - 1)}{k(k - 1)} q^d q^{d-1}$$

further blocks. If we construct exactly this many blocks, and they contain between them every pair of the form $\{(i, x), (j, y)\}$ where i and j belong to S, x and y belong to F and $i \neq j$, then a counting argument shows that we have the required design; we do not need to prove that each pair occurs at most once.

Every element of F can be written as a vector of length d over $GF(q)$. Since $d = u(u - 1)/2$, we can label the coordinates of these vectors with the unordered pairs of members of S: The field element h has a representation

$$h = (h_{12}, h_{13}, \ldots, h_{u-1, u}).$$

Let us define a subset H of F by

$$H = (h : h_{12} + h_{13} + \cdots + h_{u-1, u} = 0\}$$

(where the addition takes place over $GF(q)$). Then H has q^{d-1} elements. We observe that

$$b' = |\mathcal{B}| \cdot |F| \cdot |H|.$$

We shall set up b' blocks by constructing one block for each triple (B, z, h), where B belongs to $\mathcal{B}$, z to F, and h to H. We do this by defining maps

$T_i: F \longrightarrow F$, defined for each $i \in S$
$m_B: B \longrightarrow F$, defined for each $B \in \mathcal{B}$.

Then for each B in $\mathcal{B}$, z in F, and h in H, we define a block $A(B, z, h)$ by

$$A(B, z, h) = \{(i,\ z + T_i(h) + m_B(i)) : i \in B\}.$$

It is necessary to show that these maps can be chosen in such a way that the blocks A(B,z,h) between them contain all pairs $\{(i,x),(j,y)\}$, where $x \in F$, $y \in F$, and $i \neq j$. In other words, for every i and j, $i \neq j$, we need to show that x - y covers all members of F: As h ranges through H and B through all blocks containing $\{i, j\}$,

$$\alpha(B, h) = T_i(h) - T_j(h) + f_B(i) - f_B(j) \tag{14.15}$$

must cover all of F.

Since the blocks $\mathcal{B}$ form a design based on S with $\lambda = q$, each pair $\{i, j\}$ of elements of S is contained in exactly q of the blocks $\mathcal{B}$. So there is a one-to-one correspondence between these blocks and the members of GF(q). For each $\{i, j\}$, choose a bijection

$$\ell_{ij} : \{B : \{i, j\} \subseteq B \in \mathcal{B}\} \longrightarrow GF(q).$$

Then, given B, define $m_B: B \longrightarrow F$ as follows: $m_B(i) = (m_{12}, m_{13}, \ldots, m_{u-1,u})$ has all components zero unless the smaller subscript is i, and if $i < j$, then

$$m_{ij} = \begin{cases} \ell_{ij}(B) & \text{if } \{i, j\} \subseteq B, \\ 0 & \text{otherwise}. \end{cases}$$

Next we specify the maps T_i. Let x be a primitive element in GF(q). Since $q \leq u + 2$, the powers $x, x^2, \ldots, x^u$ are all distinct and different from 1. Now if $z = (z_{12}, z_{13}, \ldots, z_{u-1,u})$ is any element of F, define $T_i(z)$ by

$$(T_i(z))_{rs} = \begin{cases} z_{is} & \text{if } i = r \\ z_{ri} & \text{if } i = s \\ z_{rs} x^i & \text{otherwise} \end{cases}$$

Then when i and j are different,

14.4 Designs in Other Fibers for λ = 1

$$(T_i(z) - T_j(z))_{rs} = \begin{cases} 0 & \text{if } \{r,s\} = \{i,j\} \\ z_{is}(1 - x^j) & \text{if } r = i,\ s \neq j; \\ z_{ri}(1 - x^j) & \text{if } s = i,\ r \neq j; \\ z_{js}(x^i - 1) & \text{if } r = j,\ s \neq i; \\ z_{rj}(x^i - 1) & \text{if } s = j,\ r \neq i; \\ z_{rs}(x^i - x^j) & \text{otherwise.} \end{cases}$$

Now suppose that an element α of F is given. We wish to prove that $\alpha = \alpha(B, h)$ for some B and h. We assume that i and j are given. If $i < j$ and $\{i, j\} \subseteq B$, then $m_B(i) - m_B(j)$ has $\{i, j\}$ entry

$$(m_B(i) - m_B(j))_{ij} = \ell_{ij}(B).$$

Since ℓ_{ij} is a bijection, it takes all values in GF(q) as B ranges through all blocks containing $\{i, j\}$. So there is a block B such that $\ell_{ij}(B) = \alpha_{ij}$. We shall use that value of B. Define β by

$$\beta = \alpha - f_B(i) + f_B(j).$$

Then β is a member of F with $\{i, j\}$ entry 0. Finally, define h by

$$h_{is} = \beta_{is}(1 - x^j)^{-1} \quad \text{if } s \neq j,$$
$$h_{ri} = \beta_{ri}(1 - x^j)^{-1} \quad \text{if } r \neq j,$$
$$h_{js} = \beta_{js}(x^i - 1)^{-1} \quad \text{if } s \neq i,$$
$$h_{rj} = \beta_{rj}(x^i - 1)^{-1} \quad \text{if } r \neq i,$$

and whatever values are taken for the h_{rs} with r and s both different from i and j, select h_{ij} such that

$$\sum_{1 \leq i < j \leq u} h_{ij} = 0.$$

Then h belongs to H and $\alpha = \alpha(B, h)$, as required. ∎

Now suppose that u is any positive integer greater than k - 1 satisfying the congruences

$u \equiv 1 \pmod{(k-1)}$

$u(u-1) \equiv 0 \pmod{k(k-1)}$.

From Theorem 14.4 it follows that there will be a $B(u,k,q)$ design for sufficiently large q, say $q \geq q_1$. From Theorem 14.8 there is a number q_2 such that a $B(q^d, k, 1)$ will exist provided that $q > q_2$ and q is a prime power such that $q^d \equiv 1 \pmod{k(k-1)}$. For the given value of u, put $d = u(u-1)/2$, and select any prime power q such that $q \geq \max(q_1, q_2, u+2)$ and $q^d \equiv 1 \pmod{k(k-1)}$. {The theorem of Dirichlet guarantees that there are infinitely many primes q such that $q \equiv 1 \pmod{k(k-1)}$; all of these satisfy the conditions, but they are not necessarily the only solutions.} Then the required conditions are met, and Theorem 14.9 guarantees a $B(uq^d, k, 1)$. This design has $v \equiv u$. So the u-fiber contains a design (in fact, it contains infinitely many designs). We have:

COROLLARY 14.9.1 For any positive integer k there is a member of $B(k, 1)$ in every fiber of k.

BIBLIOGRAPHIC REMARKS

The existence conjecture is sometimes called Hall's conjecture, because it was first published in the book [4], but it was discussed before that time. Wilson's first published work on the conjecture appears in [6] and [7]; in the latter paper he proves the conjecture for cases in which $k/(k,\lambda)$ is 1 or a prime power, and for λ sufficiently large [the lower bound is slightly less than $k^2/4$]. The theory of closed sets is developed in those papers. He later proved the full conjecture in [11]. Extensive use is made of various other results, especially those in [2], [5], [8], and [9]. A very terse summary of the proof of the existence theorem may be found in [1].

The results of Section 14.2 are taken from [9] (see also [3]). Section 14.3 is based on [10]; the original proof, in [8], was more difficult. The proof of Theorem 14.9 is based on [1]; the original version appeared in [10].

1. A. E. Brouwer, "Wilson's theory," <u>Packing and Covering in Combinatorics</u> (Mathematical Centre Tracts 106) (Mathematische Centrum, Amsterdam, 1979), 75-88.
2. S. Chowla, P. Erdos, and E. S. Straus, "On the maximal number of pairwise orthogonal Latin squares of a given order," Canadian Journal of Mathematics <u>12</u>(1960), 204-208.
3. J. E. Graver and W. B. Jurkat, "The module structure of integral designs," Journal of Combinatorial Theory <u>15A</u>(1973), 75-90.
4. M. Hall, <u>Combinatorial Theory</u>, 1st Ed. (Blaisdell, Waltham, Mass., 1967).

Bibliographic Remarks

5. H. Hanani, "The existence and construction of balanced incomplete block designs," Annals of Mathematical Statistics 32(1961), 361-386.
6. R. M. Wilson, "An existence theory for pairwise balanced designs I. Composition theorems and morphisms," Journal of Combinatorial Theory 13A(1972), 220-245.
7. R. M. Wilson, "An existence theory for pairwise balanced designs II. The structure of PBD-closed sets and the existence conjecture, Journal of Combinatorial Theory 13A(1972), 246-273.
8. R. M. Wilson, "Cyclotomy and difference families in elementary Abelian groups," Journal of Number Theory 4(1972), 17-47.
9. R. M. Wilson, "The necessary conditions for t-designs are sufficient for something," Utilitas Mathematica 4(1973), 207-215.
10. R. M. Wilson, "Constructions and uses of pairwise balanced designs," Combinatorics, Part I (Mathematical Centre Tracts 55) (Mathematische Centrum, Amsterdam, 1974), 18-41.
11. R. M. Wilson, "An existence theory for pairwise balanced designs III. Proof of the existence conjectures," Journal of Combinatorial Theory 18A(1975), 71-79.

Answers, Solutions, and Hints

CHAPTER 1

Section 1.1, page 3

1.1.1 Without loss of generality, one can assume blocks 012, 034, 056, 078, 135. As 146 is impossible (since it would imply 178, i.e., 78 occurs twice), take 147 and 168. This is all unique up to isomorphism. The only possible extra blocks containing 3 are 238, 367. Then 246, 257, 248 are forced.

1.1.2 If 012, 345 were 3-sets, there could be no other. Possibilities are 012, 034, and one or both of 135, 245 (up to isomorphism). Then the 2-sets are completely determined.

Section 1.2, page 5

1.2.4 8; possible edge-lists are ∅; 12; 13; 23; 12 13; 12 23; 13 23; 12 13 23.

Section 1.3, page 9

1.3.1 Subdivide the problem by asking,"How many solutions are there in which some block occurs n times but none more than n?"

n = 3: Say 12 12 12. Must have 34 34 34.
n = 2: Say 12 12. Without loss of generality add 13 24. Must have 34 34.
n = 1: No repeats. Only possibility is 12 13 14 23 24 34.

1.3.3 Parameters are as follows: Asterisk means "noninteger, cannot complete." (15, 35, 7, 3, 1), (9, *, 6, 4, 2), (14, 7, 4, 8, *), (66, 143, 13, 6, 1), (21, 28, 4, 3, *), (17, *, 8, 5, 2), (21, 30, 10, 7, 3), (17, , *, 7, 1)

1.3.4 (i) (16, 16, 6, 6, 2).
(ii) Consider two elements in same row of t × t array. Necessarily $\lambda = t - 2$. But for two elements not in same row or column, $\lambda = 2$. So $t - 2 = 2$, $t = 4$, is necessary for balance.

Answers, Solutions, and Hints

Section 1.4, page 12

1.4.1 123456, 124365, 143265, 143526, 143652, 213546, 214365, 241365, 243561, 314265, 314526, 314562, 321546, 324561, 341265, 341526, 341562.

CHAPTER 2

Section 2.1, page 21

2.1.1 Suppose that a $PB(8;\{4,3\};1)$ exists, with f_4 4-blocks and f_3 3-blocks. Then $6f_4 + 3f_3 = 28$. But 3 does not divide 28.

2.1.3 No. For suppose that 0123 is a block. The only way to place 0 in further blocks size 3 and 4 is to have a block 0456. No completion is possible.

2.1.4 $g^{(4)}(6) = g^4(6) = 8$. [Along the way, one finds that $g^{(2)}(6) = 15$ and $g^{(3)}(6) = 9$.]

2.1.5 Yes.

Section 2.2, page 30

2.2.2 $\left(v\binom{v}{k}, \binom{v-1}{k-1}, k, \binom{v-2}{k-2}\right)$.

2.2.3 No.

2.2.4 $\left(v, \binom{v}{k} - b, \binom{v}{k} - r, k, \frac{v}{k}\binom{v-2}{k-2}\right)$.

2.2.8 $(v, nb, nr, k, n\lambda)$.

Section 2.3, page 34

2.3.4 (i) No (try $s = 2$); (ii) yes; (iii) yes; (iv) no.

2.3.5 Condition (2.18) becomes "$\binom{k-s}{k-1-s}$ divides $\binom{v-s}{k-1-s}$."

CHAPTER 3

Section 3.1, page 41

3.1.1 (i) $1 + 0 = 0$, so $a = a1 = a(1 + 0) = a1 + a0 = a + a0$. So $0 = (-a) + a = (-a) + (a + a0) = ((-a) + a) + a0 = 0 + a0 = a0$.
  (ii) $a + (-a) = 0$ and $a + (-1)a = 1a + (-1)a = (1 + (-1))a = 0a = 0$. So $-a$ and $(-1)a$ are both inverses of a; by uniqueness, $(-1)a = -a$.
  (iii) If $a \neq 0$, then $a \in F^*$, so a^{-1} exists, so $ab = 0 \implies b = a^{-1}ab = a^{-1}0 = 0$.

3.1.4 The multiplicative identity is 6.

3.1.7 The polynomials are $x^3 + 2x + 1$, $x^3 + 2x + 2$, $x^3 + x^3 + 2$, $x^3 + x^2 + x + 2$, $x^3 + x^2 + 2x + 1$, $x^3 + 2x^2 + 1$, $x^3 + 2x^2 + x + 1$, $x^3 + 2x^2 + 2x + 2$.

3.1.9 (i) $(0,0)$, $(1,1)$; (ii) any $(x,0)$ and $(0,y)$, $x \neq 0$, $y \neq 0$.

Section 3.2, page 46

3.2.1 Essentially use $a^q a^r = a^{q+r}$

3.2.3 Express each quadratic element, and -1, as a power of a primitive root.

CHAPTER 4

Section 4.1, page 60

4.1.2 A $(13,4,1)$ difference set must contain two members, s and $s+1$ say, with differences ± 1. By subtracting s (mod 13) from all elements, one obtains an example of the form $\{0,1,x,y\}$. So assume that form. Then $\{x, y, x-1, y-1, y-x\} = \{\pm 2, \pm 3, \pm 4, \pm 5, \pm 6\}$. $x = 2$ is impossible (since $x - 1$ would be outside the permissible range). Try $x = 3$: by trial and error, the only possibility is $y = 9$. Other solutions are $\{x,y\} = \{5,11\}$, $\{8,10\}$.

4.1.3 Similar to Exercise 4.1.2: for example, $\{0,1,2,4\}$.

4.1.4 $\{0,1,2,4,5,8,10\}$; $\{0,1,2,3,4,6,7,8,9,12,13,14,16,18,19,21,23,24, 25,26,27,28,32,36,38,39,41,42,46,48,49,50,52,53,54,56,57,63,64,65, 69,72,73,75,76,78,81,82,83,84,85,91,92,96,98,100,103,104,106,108,109, 112,113,114,117,123,126,128,130,133,138\}$.

4.1.5 Theorem 4.3 gives $t = 1$, 2, 3, 5, 6, 7, 8, 11, 12, 15, 17, 18, 20, 21, 26, 27, 30, 32, 33, 35, 38, 40, 41, 42, 45, 48, 50. Theorem 4.4 gives $t = 4$, 9, 16, 25, 36. Eighteen values remain.

Section 4.2, page 62

4.2.1 From Corollary 4.2.1, $\lambda(v-1) = k(k-1)$. Put $v = 2k$. Then $\lambda \equiv 0$ (mod k). But $k(k-1) < k(2k-1)$, obviously, so $\lambda < k$. So $\lambda = 0$, which is clearly impossible.

Section 4.3, page 67

4.3.2 One can assume that the initial blocks are $\{0,1,x\}$ and $\{0,y,z\}$. Difference ± 2 must arise; if it is from the first block, then $x = \pm 2$ or $x - 1 = \pm 2$, so $x = 2$ or 3 or 11 or 12. If $x = 2$ or 12 difference ± 1 comes up again; by trial and error $x = 3$ and 11 are also impossible. So try $\{0,1,x\}$ and $\{0,2,z\}$. An example that works is $\{0,1,4\}$, $\{0,2,8\}$.

4.3.3 $\{00, 01, 12, 22\}$ is a second block.

4.3.5 $\{\infty, 0, 1\}$, $\{0,1,3\}$, for example.

Answers, Solutions, and Hints

CHAPTER 5

Section 5.1, page 72

5.1.1 Hints and remarks: If there are no points, there can be no lines; if there is one point, either there is a line or not (two cases). For two points, there must be a line joining them; if there is another line, there must be exactly two more. For more than two points, there must be a line that contains all or all-but-one points.

Section 5.2, page 77

5.2.1 The line joining $(0,0)$ to $(1,0)$ consists precisely of the points $x(0,0) + y(1,0)$ [i.e., points like $(y,0)$]. Neither $(0,1)$ nor $(1,1)$ has this form. So neither of the triples $\{(0,0),(1,0),(0,1)\}$, $\{(0,0),(1,0),(1,1)\}$ is collinear. Similarly for the other two triples.

Section 5.3, page 82

5.3.3 (i) 013, 103, 111, 124, 132, 140; (ii) $x_1 + 4x_3 = 0$; (iii) 103.

5.3.6 An example is $\{001, 010, 100, 111\}$.

Section 5.4, page 87

5.4.2 If a cubic polynomial is reducible, it must have a linear factor. If $y - a$ is a factor of $f(y)$, then $f(a) = 0$. If $f(y) = 1 - y + y^3$, then $f(0) = f(1) = f(2) = 1$.

5.4.3 Using $g(y) = 1 + y + y^3$, the line x_2 yields set $\{0,1,3\}$ for PG(2,2). Same polynomial and $x_3 = 0$ yields $\{0,1,3,8,12,18\}$ for PG(2,5).

5.4.4 (ii) $\{0,1,2,3,4,8,9,10,11,13,14,16,17,18,21,24,25,27,30,32,36,41,42,45,46,47,49,51,53,54,61\}$.

CHAPTER 6

Section 6.1, page 98

6.1.1 (i) $(10,15,6,4,3)$, $(6,15,5,2,1)$;
(ii) $(15,21,7,5,2)$, $(7,21,6,2,1)$;
(iii) $(33,44,12,9,3)$, $(12,44,11,3,2)$;
(iv) $(4,12,9,3,6)$, $(9,12,8,6,5)$.

6.1.2 (i) Use $\lambda(v - 1) = k(k - 1)$.

6.1.3 (i) $(n^2 + n + 1, n^2, n^2 - n)$; (ii) $(n + 1, n^2 + n, n^2, n, n^2 - n)$; (iii) it is a multiple of a known design.

Section 6.2, page 105

6.2.2 (i) $28 + 369t + 243t^2$; (ii) generalize the argument which proves that $(141, 21, 3)$ cannot be realized.

Section 6.4, page 115

6.4.3 (i) No.

6.4.4 Every object belongs to $n(3n - 1)(3n - 2)/2$ triples, so this must equal the number of rows and the number of columns in the array. No example exists when $n = 2$; one may be found for $n = 3$.

CHAPTER 7

Section 7.1, page 120

7.1.1 (ii) $(t - 1) - (v - 1, k - 1, \lambda)$

7.1.2 (i) $t - \left(v, k, \binom{v-t}{k-t}\right)$

7.1.3 Proceed as in "$\lambda = 1$" case to eliminate all possibilities except $k = 3, 5, 11,$ and 23. The case $k = 23$ is ruled out by the Bruck-Ryser-Chowla theorem.

7.1.6 Any such design is an extension of $PG(2,2)$; but that extension is unique.

Section 7.2, page 128

7.2.1 Say that (x,y) lies on both circles. Then one can show that $2(a - d)x + 2(b - e)y = h$, for constant h, so all common points lie on this line. It is then easy to show that no straight line and circle have more than two common points.

CHAPTER 8

Section 8.1, page 139

8.1.1 There are four places, and each can hold two values, so the number of matrices is $2^4 = 16$. By inspection half are Hadamard.

8.1.2 One can construct $2^{16} = 65,536$ matrices. To count the Hadamard ones, first count the standardized ones: there are 6. So the number with first row 1111 is 48 (for each of the 2^3 possible first columns, there are 6 matrices, one for each standardized form). Each of the 48 gives rise to 8 matrices with $(1,1)$ entry 1. So there are $8 \cdot 48 = 384$ Hadamard matrices with $(1,1)$ element 1, and 384 more (their negatives) with $(1,1)$ element -1. So there are 768.

8.1.3 Examples: 0 +,   0 + + + + +
              + 0   + 0 + - - +
                       + + 0 + - -
                       + - + 0 + -
                       + - - + 0 +
                       + + - - + 0

Answers, Solutions, and Hints

Section 8.2, page 143

8.2.1 No (you need $BC^T = 0$).

Section 8.3, page 150

8.3.3 To prove the matrix is Hadamard, simply multiply by its transpose. This is essentially the "h = 2" case of Theorem 8.8 with Kronecker products reversed ("$A \otimes B$ changed to $B \otimes A$").

8.3.4 (iii) Get side 8 from part (ii) and use this in part (i).

8.3.5 (iv) Use Exercise 8.3.4(iii).

Section 8.4, page 156

8.4.1 We show first rows. Side 3: one must be +++ or --- (two choices). Other must be chosen from +-- and -++ (four choices: 0, 1, 2, or 3 is +--). Total $2 \cdot 4 = 8$. Side 5: one must be 1---- or -1111; one be 1-11- or -1--1; other two are chosen from $\{11--1, --11-\}$. So $2 \cdot 2 \cdot 3 = 12$ cases.

8.4.2 One example:

$$\begin{vmatrix} A & B & C & D & E & F & G & H \\ -B & A & D & -C & F & -E & -H & G \\ -C & -D & A & B & G & H & -E & -F \\ -D & C & -B & A & H & -G & F & -E \\ -E & -F & -G & -H & A & B & C & D \\ -F & E & -H & G & -B & A & -D & C \\ -G & H & E & -F & -C & D & A & -B \\ -H & -G & F & E & -D & -C & B & A \end{vmatrix}$$

Section 8.5, page 159

8.5.4
$$\begin{vmatrix} J & I & I & I & I \\ I & J & I & L & M \\ I & I & J & M & L \\ I & L & M & J & I \\ I & M & L & I & J \end{vmatrix}$$

8.5.5 (i) Minimum polynomial must divide $x^2 - n$.
(ii) Trace = sum of eigenvalues (since H is a real symmetric matrix).
(iii) Trace = $n = (a - b)\sqrt{n}$, so $a - b = \sqrt{n}$. So $\sqrt{n}$ is integral.

8.5.6 (i) Follows from $\lambda = 1$.
(ii) B_i has u elements and each belongs to 2u further blocks, which must all be different from part (i). So there are $2u^2$ values j such that $a_{ij} = 1$. One can also show that $a_{ij} = a_{ik} = 1$ for u^2 values i when $j \ne k$. So we have a $(4u^2 - 1, 2u^2, u^2)$-SBIBD.

(iii) A is symmetric and has diagonal all zeros. So the corresponding Hadamard matrix is graphical.

8.5.7 (i) The matrix is necessary regular.

(ii) For side 4, use first row +++- (for example). For side 16, observe that the matrix is equivalent to a $(16, 6, 2)$-difference set in Z_{16}; a relatively short exhaustive search shows that none exists.

CHAPTER 9

Section 9.1, page 165

9.1.1 This follows from the uniqueness of a $(7, 3, 1)$-SBIBD.

9.1.3 (i) Take H of side n. Normalize it, and then negate the result. One ends with all the first row and first column negative, and $n/2$ positive elements in each row and column except the first. Now negate rows 2 to n and negate columns 2 to n. The final matrix has $n - 1$ positive entries in row 1 and $(n + 1)/2$ positive entries in each other row. The weight is

$$n - 1 + (n - 1)\left(\frac{1}{2}n + 1\right) = (n - 1)\left(\frac{1}{2}n + 2\right) = \frac{1}{2}(n + 4)(n - 1).$$

(ii) Write $n = 4t$; t must be odd, say $t = 2s + 1$. Again normalize H; reorder the columns so that the first t columns start $+++\ldots$, next t start $++-\ldots$, next t start $+-+\ldots$, last t start $+--\ldots$. Say row i (where $i > 3$) contains q_i entries among the first t columns. It is easy to see that the second, third, and fourth sets each contain q_i 1's also. Either $q_i \leq s$ or $q_i \geq s + 1$. For each i such that $q_i \leq s$, negate row i. The result is a matrix equivalent to H in which three rows have weight $4s + 2$ and the rest have weight at least $4s + 4$. So the weight is at least $3(4s + 2) + (8s + 1)(4s + 4) = 12s + 6 + 32s^2 + 36s + 4 = 32s^2 + 48s + 10 = 2(4s + 5)(4s + 1) = (n + 6)(n - 2)/2$.

Section 9.2, page 172

9.2.1 (i) First, show that the matrix is Z-equivalent to

$$\begin{pmatrix} D & D \\ D & -D \end{pmatrix}$$

(ii) $1, 2^{(4m)}, 4^{(4m-1)}, 4m^{(4m-1)}, 8m^{(4m)}, 16m$.

9.2.4 Say that the invariants are $1, 2^{(a)}, 4^{(b)}, 8^{(c)}, 16^{(b)}, 32^{(a)}, 64$. Then $6 \leq a \leq 31$. For each a there are at most $32 - a$ possible values for b.

Answers, Solutions, and Hints

CHAPTER 10

Section 10.1, page 182

10.1.1 Up to equivalence, one may assume that the first row is 1 2 3 4 and so is the first column. There are only four ways to complete a Latin square:

$$
\begin{array}{cccc}
A = 1234 & B = 1234 & C = 1234 & D = 1234 \\
2143 & 2143 & 2341 & 2413 \\
3412 & 3421 & 3412 & 3142 \\
4321 & 4312 & 4123 & 4321
\end{array}
$$

Exchanging the symbols 3 and 4 in C gives D, and the permutation (243) changes B to C (after suitable row and column permutations in both cases).

10.1.2 Use induction on k.

Section 10.2, page 186

10.2.5 (iii) $\begin{bmatrix} 1 & 3 & 4 & 2 \\ 4 & 2 & 1 & 3 \\ 2 & 4 & 3 & 1 \\ 3 & 1 & 2 & 4 \end{bmatrix}$ $\begin{bmatrix} 1 & 3 & 2 & 5 & 4 \\ 4 & 2 & 5 & 1 & 3 \\ 5 & 4 & 3 & 2 & 1 \\ 3 & 5 & 1 & 4 & 2 \\ 2 & 1 & 4 & 3 & 5 \end{bmatrix}$

Section 10.3, page 191

10.3.1 One example is:

$$\begin{bmatrix} 1 & 5 & 6 & 3 & 2 & 4 \\ 3 & 2 & 1 & 6 & 4 & 5 \\ 5 & 4 & 3 & 2 & 6 & 1 \\ 6 & 1 & 5 & 4 & 3 & 2 \\ 4 & 6 & 2 & 1 & 5 & 3 \\ 2 & 3 & 4 & 5 & 1 & 6 \end{bmatrix}$$

10.3.2 $\begin{bmatrix} 1 & 7 & 4 & 8 & 9 & 0 & 2 & 3 & 5 & 6 \\ 7 & 2 & 0 & 9 & 6 & 8 & 1 & 5 & 4 & 3 \\ 8 & 0 & 3 & 1 & 7 & 9 & 5 & 4 & 6 & 2 \\ 3 & 9 & 8 & 4 & 0 & 7 & 6 & 1 & 2 & 5 \\ 9 & 8 & 7 & 0 & 5 & 2 & 3 & 6 & 1 & 4 \\ 0 & 5 & 9 & 7 & 8 & 6 & 4 & 2 & 3 & 1 \\ 2 & 1 & 5 & 6 & 3 & 4 & 7 & 9 & 0 & 8 \\ 4 & 6 & 1 & 3 & 2 & 5 & 0 & 8 & 7 & 9 \\ 5 & 4 & 6 & 2 & 1 & 3 & 8 & 0 & 9 & 7 \\ 6 & 3 & 2 & 5 & 4 & 1 & 9 & 7 & 8 & 0 \end{bmatrix}$

314   Answers, Solutions, and Hints

10.3.3 Use the squares

$$\begin{bmatrix} 1 & 3 & 4 & 2 \\ 4 & 2 & 1 & 3 \\ 2 & 4 & 3 & 1 \\ 3 & 1 & 2 & 4 \end{bmatrix} \quad \begin{bmatrix} 1 & 5 & 4 & 3 & 2 \\ 3 & 2 & 1 & 5 & 4 \\ 5 & 4 & 3 & 2 & 1 \\ 2 & 1 & 5 & 4 & 3 \\ 4 & 3 & 2 & 1 & 5 \end{bmatrix}$$

and the design

0	1	2	3	4		1	6	11	16		2	7	9	16		3	8	9	14
0	5	6	7	8		1	7	12	14		2	8	11	13		4	5	11	14
0	9	10	11	12		1	8	10	15		3	5	10	16		4	6	9	15
0	13	14	15	16		2	5	12	15		3	6	12	13		4	7	10	13
	1	5	9	13		2	6	10	14		3	7	11	15		4	8	12	16

to obtain

$$\begin{bmatrix}
0 & 4 & 3 & 2 & 1 & 8 & 7 & 6 & 5 & 12 & 11 & 10 & 9 & 16 & 15 & 14 & 13 \\
2 & 1 & 0 & 4 & 3 & 9 & 11 & 12 & 10 & 13 & 15 & 16 & 14 & 5 & 7 & 8 & 6 \\
4 & 3 & 2 & 1 & 0 & 12 & 10 & 9 & 11 & 16 & 14 & 13 & 15 & 8 & 6 & 5 & 7 \\
1 & 0 & 4 & 3 & 2 & 10 & 12 & 11 & 9 & 14 & 16 & 15 & 13 & 6 & 8 & 7 & 5 \\
3 & 2 & 1 & 0 & 4 & 11 & 9 & 10 & 12 & 15 & 13 & 14 & 16 & 7 & 5 & 6 & 8 \\
6 & 13 & 15 & 16 & 14 & 5 & 0 & 8 & 7 & 1 & 3 & 4 & 2 & 9 & 11 & 12 & 10 \\
8 & 16 & 14 & 13 & 15 & 7 & 6 & 5 & 0 & 4 & 2 & 1 & 3 & 12 & 10 & 9 & 11 \\
5 & 14 & 16 & 15 & 13 & 0 & 8 & 7 & 6 & 2 & 4 & 3 & 1 & 10 & 12 & 11 & 9 \\
7 & 15 & 13 & 14 & 16 & 6 & 5 & 0 & 8 & 3 & 1 & 2 & 4 & 11 & 9 & 10 & 12 \\
10 & 5 & 7 & 8 & 6 & 13 & 15 & 16 & 14 & 9 & 0 & 12 & 11 & 1 & 3 & 4 & 2 \\
12 & 8 & 6 & 5 & 7 & 16 & 14 & 13 & 15 & 11 & 10 & 9 & 0 & 4 & 2 & 1 & 3 \\
9 & 6 & 8 & 7 & 5 & 14 & 16 & 15 & 13 & 0 & 12 & 11 & 10 & 2 & 4 & 3 & 1 \\
11 & 7 & 5 & 6 & 8 & 15 & 13 & 14 & 16 & 10 & 9 & 0 & 12 & 3 & 1 & 2 & 4 \\
14 & 9 & 11 & 12 & 10 & 1 & 3 & 4 & 2 & 5 & 7 & 8 & 6 & 13 & 0 & 16 & 15 \\
16 & 12 & 10 & 9 & 11 & 4 & 2 & 1 & 3 & 8 & 6 & 5 & 7 & 15 & 14 & 13 & 0 \\
13 & 10 & 12 & 11 & 9 & 2 & 4 & 3 & 1 & 6 & 8 & 7 & 5 & 0 & 16 & 15 & 14 \\
15 & 11 & 9 & 10 & 12 & 3 & 1 & 2 & 4 & 7 & 5 & 6 & 8 & 14 & 13 & 0 & 16
\end{bmatrix}$$

Section 10.4, page 196

10.4.3 Yes.

10.4.4 (i) Block size = k, group size = k - 1.
(ii) There are r groups. So r = k is necessary (and it is sufficient: the transversal designs are essentially affine planes, with one parallel class interpreted as groups).

Answers, Solutions, and Hints   315

Section 10.5, page 203

10.5.1 (i) $\begin{matrix} [1,2|3,4] & [2,1|4,3] \\ [1,3|2,4] & [3,1|4.2] \\ [1,4|2,3] & [4,1|3,2] \end{matrix}$

(ii) No. First show that up to isomorphism, the first three rounds are two symmetric SR(4)'s glued together. Then completion is impossible.

CHAPTER 11

Section 11.1, page 211

11.1.4 Say x occurs a_x times above the diagonal of A, b_x times below and d_x times on the diagonal. By symmetry, $a_x = b_x$, so $d_x = 2n - 1 - a_x - b_x$ is odd. Since $d_1 + d_2 + \cdots + d_{2n-1} = 2n - 1$, each d_x equals 1.

11.1.5 3-(2n, 4, 3).

Section 11.2, page 214

11.2.1 12; 14 15; 16 25 34; 15 23 46; 13 26 45.

11.2.3 There are nine solutions: 01,02 10,20 11,22 12,21; 01,02 11,21 12,20 10,22; 01,02 12,22 10,21 11,20; 10,11 02,22 01,20 12,21; 10,12 01,21 11,22 02,20; 11,12 10,20 02,21 01,22; 20,21 02,12 11,22 01,10; 20,22 01,11 02,10 12,21; 21,22 10,20 01,12 02,11.

Section 11.3, page 223

11.3.2 Since $(n - 2) 2^{n-2} = (2n - 4) \cdot (2n - 6) \cdots 2$, the number given is actually $(2n - 3) \cdot (2n - 5) \cdots 3 \cdot 1$. This is the number of one-factors that contain a given pair. Using the notation of Theorem 11.2: Say that G is any one-factor containing $\{\infty, 0\}$. One can find a permutation ϕ of $\{1, 2, \ldots, 2n - 2\}$ such that ϕ maps F_0 to G. Then ϕ maps the one-factorization of the theorem to one containing G. So the number of one-factorizations is at least equal to the number of factors containing $\{\infty, 0\}$.

11.3.6 Use Theorem 11.8; they are not isomorphic.

CHAPTER 12

Section 12.1, page 241

12.1.3 (i) $(n - k)/2$; (ii) the answer to part (i) is an integer; (iii) k must equal 1.

12.1.5 Prove that the construction of Theorem 12.2(v) does not give two copies of a design.

12.1.6 Say the $T(v - 3, v - 1)$ has treatment set S and block set B. Let P be the set of all unordered pairs of elements of S; write $Q = \{p \cup \{\infty\} : p \in P\}$. Then $Q \cup B$ is the block set of the required design.

Section 12.2, page 248

12.2.1 Putting $m_1 = 0$, $n_{i1} = i$, $n_{i2} = 9 + i$ we get blocks

0 1 10, 0 2 11, 0 3 12, ..., 0 9 18,

together with 48 more blocks, 4 from each block of a $T(1,9)$: for example, the block 1 2 3 gives rise to

1 2 12, 1 11 3, 10 2 3, 10 11 12.

The other design is derived similarly from the theorem. It is easy enough to show that the design above contains no sub-$T(1, 7)$, while the other will do so; so they are not isomorphic.

12.2.4 1 2 3; 4 5 7, 5 6 8, ..., 10 11 13, 11 12 4, 12 13 5, 13 4 6, 4 8 1, 12 6 1, 10 5 1, 7 11 1, 9 13 1, 8 12 2, 6 10 2, 11 5 2, 13 7 2, 4 9 2, 13 8 3, 6 11 3, 7 12 3, 4 10 3, 5 9 3.

Section 12.4, page 254

12.4.1 $S = \{1, 2, 4\}$, which does not form a difference partition.

12.4.2 $v = 7$: $3 = 1 + 2$; $v = 13$: for example, $1 + 3 = 4$, $2 + 5 + 6 = 13$.

Section 12.5, page 260

12.5.4 (i) Suppose that a $PB(v; \{4, 3\}; 1)$ exists. In the notation of Section 2.1, $12f_4 + 6f_3 = v(v - 1)$; since f_4, f_3, and v are integers, 3 divides $v(v - 1)$, so $v \equiv 0$ or 1 (mod 3). Case $v = 6$ is easily eliminated. If $v \equiv 1$ or 3 (mod 6), use the Steiner triple system on v points. If $v \equiv 4$ (mod 6), take a Kirkman triple system of $v - 1$ points; add a new element A to every block in one parallel class. If $v \equiv 0$ (mod 6), take a $KTS(v - 3)$ and use three new elements A, B, C; for each, select a parallel class, and add the new element to every block. Then add a block $\{A, B, C\}$. [In the latter case one needs at least three parallel classes; since a $KTS(v - 3)$ contains $(v - 4)/2$ parallel classes, the necessary condition is $v \geq 10$.]

(ii) Clearly, $v \equiv 0$ or 1 (mod 3) is still necessary. The constructions above work for $v \equiv 0, 4$ (mod 6), except trivially for $v = 4$. If $v \equiv 1$ (mod 6), add four new elements to a $KTS(v - 4)$, together with the block consisting of all four of them. This is possible provided the $KTS(v - 4)$ contains more than four parallel classes (exactly four classes won't work; you get no 3-blocks), so we need $(v - 5)/2 > 4$, $v > 13$. So $v = 7, 13$ are not covered. If $v \equiv 3$ (mod 6), one can use a $KTS(v - 12)$ with 12 parallel classes; the 12 extra elements are formed into the blocks of a $PB(12; \{4, 3\}; 1)$ and added on.

Answers, Solutions, and Hints   317

This works when $(v - 13)/2 \geq 12$, or $v \geq 37$. (Case $v = 12$ <u>does not</u> have to be excluded this time: The PB(12;$\{4,3\}$;1) includes both a 4-block and a 3-block.) So cases 3, 9, 15, 21, 27, and 35 remain. Of the residue, 3, 4, 6, 7, and 9 are impossible. (For 7, see Exercise 2.1.3; for 9, say 0 belongs to a 4-block. Without loss of generality, blocks containing 0 are 0123, 0456, 078. Every other block will contain exactly one element from each of these three sets; there can be at most six (three containing each of 7 and 8). So $f_4 = 2$, $f_3 \leq 7$, $12f_4 = 6f_3 = 72$—impossible. For $v = 13$ or 21, delete three elements from one block of a $(v + 3)$-treatment BIBD with $k = 4$, $\lambda = 1$. For $v = 15$ or 27, take one treatment from the $(v + 1)$-treatment BIBD. For $v = 34$, use the PB(34;$\{7,4\}$;1) which was constructed in this section; replace the 7-block by a $(7,3,1)$-SBIBD, and delete one element <u>not</u> in the 7-block. So the values are $v \equiv 0$ or 1 (mod 3), $v \neq 3, 4, 6, 7,$ or 9.

CHAPTER 13

Section 13.2, page 269

13.2.2 $\{16,25,34\}$, adder 2, 4, 1; $\{23,46,15\}$, adder 1, 2, 4; $\{45,13,26\}$, adder 5, 3, 6.

Section 13.5, page 281

13.5.2

$$\begin{bmatrix} 01 & 69 & \circled{48} & -- & -- & -- & -- & 57 & 23 \\ 34 & 02 & 17 & \circled{59} & -- & -- & -- & -- & 68 \\ 79 & 45 & 03 & 28 & \circled{16} & -- & -- & -- & -- \\ -- & 18 & 56 & 04 & 39 & \circled{27} & -- & -- & -- \\ -- & -- & 29 & 67 & 05 & 14 & 38 & -- & -- \\ -- & -- & -- & 13 & 78 & 06 & 25 & 49 & -- \\ -- & -- & -- & -- & 24 & 89 & 07 & 36 & 15 \\ 26 & -- & -- & -- & -- & 35 & 19 & 08 & 47 \\ 58 & 37 & -- & -- & -- & -- & 46 & 12 & 09 \end{bmatrix}$$

Author Index

A

Albert, A. A., 94
Anderson, B. A., 235

Bartee, T. C., 53
Baumert, L. D., 69
Belevitch, V., 161
Beth, T., 13
Beutelspacher, A., 235
Bhattacharya, K. N., 117
Birkhoff, G., 53
Bondy, J. A., 13
Bose, R. C., 69, 208, 262
Brouwer, A. E., 262, 304
Bruck, R. H., 117
Bussey, W. H., 94

C

Cameron, P. J., 133
Cayley, A., 282
Chan, K. M., 282
Chong, B. C., 282
Chowla, S., 117, 304
Collens, R. J., 282
Connor, W. S., 117
Cooper, J., 178

D

Dean, R. A., 54
de Bruijn, N. G., 34
Dernbowski, P., 94, 133
Dickson, L. E., 235
Dorwart, H. L., 94
Dulmage, A. L., 208

E

Erdos, P., 34, 304
Euler, L., 208

F

Fisher, R. A., 34

G

Gardner, M., 208
Goethals, J. M., 161
Goulden, I. P., 262
Graver, J. E., 304

H

Hadamard, J., 161
Hall, M., 13, 117, 161, 178, 304
Hall, P., 13
Hanani, H., 117, 305
Harary, F., 13
Hardy, G. H., 54
Hedayat, A., 161
Heffter, L., 262
Hirschfeld, J., 94, 235
Horton, J. D., 282
Hughes, D. R., 94, 133

I

Ito, N., 178

J

Johnson, D. M., 208
Jungnickel, D., 13
Jurkat, W. B., 304

K

Kimura, H., 178
Kirkman, T. P., 262, 282
Konig, D., 235
Kotzig, A., 235

L

Lenz, H., 13, 117
Leon, J. S., 178
Levi, F. W., 208
Lindner, C. C., 235, 262
Longyear, J. L., 178

M

MacNeish, N. L., 208
Mann, H. B., 69
Mendelsohn, E., 235
Mendelsohn, N. S., 208
Milas, J., 178
Mullin, R. C., 262, 282
Murty, U. S. R., 13

N

Nemeth, E., 282

P

Paley, R. E. A. C., 69, 161
Parker, E. T., 208
Peltesohn, 262
Piper, F. C., 94, 133

R

Raghavarao, D., 13
Ray-Chaudhuri, D. K., 117, 262
Room, T. G., 282
Rosa, A., 235
Ryser, H. J., 13, 34, 117

S

Safford, F. H., 235
Sandler, R. F., 94
Scarpis, U., 161
Schellenberg, P. J., 208
Schutzenberger, M. P., 117
Seberry, J. (= J. Wallis), 162
Seidel, J. J., 161

Shrikhande, S. S., 117, 208
Singer, J., 94
Singhi, N. M., 117
Skolem, T., 262
Sprott, D. A., 69
Stanton, R. G., 13, 34, 35, 69, 262, 282
Steiner, J., 262
Stinson, D. R., 35, 262, 282
Storer, T., 69
Straus, E. S., 304
Street, A. P., 13, 35, 162
Sylvester, J. J., 162

T

Tarry, G., 208
Todorov, V., 208

V

Vajda, S., 94
van Lint, J. H., 117

van Rees, G. H. J., 208
Vanstone, S. A., 208, 262
Veblen, O., 94
Vinogradov, I. M., 54

W

Wallis, J. (= J. Seberry), 162, 178
Wallis, W. D., 13, 35, 117, 161, 162, 178, 208, 235, 262, 282
Wang, S. P., 208
Williamson, J., 162
Wilson, R. M., 117, 208, 262, 305
Woolhouse, W. S. B., 262
Wright, E. M., 54

Y

Yates, F., 13

Subject Index

A

Adder, 266
Adjacency, 4
Admissible quadruple, 33-34
Affine design, 110-115
Affine plane, 70 (see also Finite affine plane)
Affine t-design, 132-133
Affine resolvable design, 110-115
AG (see Finite affine geometry, Finite affine plane)
AG(d,n), 121
Antidifference sets, 60
Assignment problem, 11
Automorphism, 216
  of one-factorization, 216

B

Balance, 6
Balanced incomplete block design, 6-9, 21-32, 95-117, 184-185, 254, 283-304
  derived, 95-98
  difference constructions, 56, 62-63, 66-67
  dual, 26, 94
  empty, 6

[Balanced incomplete block design]
  existence theory, 98-108, 283-304
  Hadamard, 139
  index, 6
  and Latin squares, 184-185
  quasi-symmetric, 143
  residual, 95-98, 105
  resolvable, 109-117, 254-262
  symmetric, 26-29, 95-108, 113-115, 128-131
  trivial, 6
Base block, 294-300
Baumert-Hall array, 156
BIBD (see Balanced incomplete block design)
Bipartite graph, 210-211
Block, 5, 14
  of pairwise balanced design, 14
Block design, 5
  balanced, 6
  balanced incomplete, 6
  complete, 6
  covalency, 6
  empty, 6
  frequency, 6
  incidence, 7
  incomplete, 6
  isomorphic, 8
  regular, 5

Subject Index

[Block design]
  replication number, 6
  trivial, 6
Bruck-Ryser-Chowla theorem, 98-108

C

Characteristic, 37-38
Circle, 124
Closed set, 284-285
c-Matrix, 139
Collinear, 71, 81, 125
  diagonals of quadrangle, 81
Combinatorial design, 1
  isomorphic, 2
  linked, 4
Complement
  of design, 29-30
  of difference set, 61
Complementary design, 29-30
Complete bipartite graph, 210
Complete design, 6
Concurrent, 71
Conference matrix, 145-150, 161
  normalized, 145
Congruent matrices, 105
Contraction, 118
Core, 151, 152
Covalency, 6
Cyclic Hadamard matrix, 160
Cyclotomic class, 294-300

D

Derivative, 33, 118
Derived design, 95-98, 131
Desargues' configuration, 90-94
Design
  complementary, 29-30
  dual, 26, 31-32, 94
  generated from set, 55
  Hadamard, 139

[Design]
  multiple, 32
  quasi-residual, 96
  symmetric, 6
Determinant, 23
Diagonal point, 81
Diagonals of quadrangle, 81
Difference partition, 252-254
Difference set, 55-62
  complement, 61
  generalized, 62-69
  Hadamard, 60
  multiple, 62
  multiplier, 62
  shift, 62
  Singer, 60, 83-88
  supplementary, 62-69
Difference triple, 252-254
Direct product, 143
Division, 216-217, 223, 224-228, 232
  maximal, 216
Dual design, 26, 31-32, 94

E

Edge, 4, 210
EG (see Euclidean geometry)
Ellipse, 128
Empty design, 6
Endpoint, 4
Equivalence, 136, 163-178, 180, 182, 214
  Hadamard, 136, 163-166, 175
  integral, 165-173
  of Latin squares, 180, 182
  of starters, 214
Euclidean geometry, 75, 124, 125 (see also Finite affine geometry, Finite affine plane)
Existence conjecture, 284, 304
Extendable, 119
Extension, 118-130
Exterior point, 234

External line, 88, 234

F

Factor, 210
Fiber, 285, 293-304
  principal, 285, 293-300
Field, 36-46
  characteristic, 37-38
  finite, 37-46
  Galois, 38
  identity, 36
  multiplicative group, 42
  prime, 38
  subfield, 38, 41
  zero, 36
Field plane, 81, 91-92
  always Desarguesian, 91-92
Finite affine geometry, 72, 75-77
  prime, 76
  trivial, 72
Finite affine plane, 70-75, 257
  construction, 73-75
  Euclidean, 75
  parameter, 72
Finite field, 37-46
  characteristic, 37-38
  construction, 39-41
  Galois field, 38
  multiplicative group, 42-44
  prime field, 38
  primitive element, 44
  primitive root, 44
  quadratic character, 44-46
  quadratic element, 44
Finite incidence geometry, 73
Finite projective geometry, 80-81
  dimension, 80
Finite projective plane, 77-94, 104-105, 257
  Desarguesian, 90-94
  field plane, 81, 91-92
  non-Desarguesian example, 92-94

[Finite projective plane]
  oval, 88-90
  uniqueness results, 83
Fisher's inequality, 24-26, 32
4-Profile, 175
Four square theorem, 53, 99
Frequency, 6, 17

G

Galois field, 38 (see also Finite field)
GD (see Group-divisible design)
Generalized t-design, 286-293
Geometry, 70
GF (see Galois field)
Graph, 4, 165, 210-216
  bipartite, 210-211
  complete bipartite, 210, 211
  component, 216
  cycle, 216
  edge, 210
  factor, 210
  of Hadamard matrix, 165
  subgraph, 210
  vertex, 210
Graphical Hadamard matrix, 160
Group divisible design, 192, 198, 258-261
$G_t(v, k, \lambda)$-design (see Generalized t-design)

H

Hadamard design, 123, 129, 133, 139
Hadamard difference set, 60
Hadamard equivalence, 136
Hadamard matrix, 134-187
  and block designs, 139-143, 156-157, 159
  cyclic, 160
  graph, 165

[Hadamard matrix]
  graphical, 160
  regular, 156-160
  skew, 145, 151-152
  standardized, 136
  sub-regular, 159
  transpose equivalence, 163
  weight, 165
  Williamson's construction, 152-156
Hadamard's determinant theorem, 160-161
Hall's conjecture, 284, 304
Heffter's difference problems, 252-254, 262

I

Idempotent Latin square, 187-192
Incidence, 1, 7
  matrix, 7, 22
  structure, 70
Incidence matrix, 7, 22, 142
Incidence structure, 70
Incomplete, 6
Index, 6
  of cyclotomic class, 294
Infinity element, 65
Integer, proper representation, 49
Integral equivalence, 165
Intercalate, 182
Inversive planes, 126
Isomorphism, 2, 8, 215-216, 223-233, 265, 269
  of one-factorizations, 215-216, 223, 233
  of Room squares, 265-269

K

Kirkman's schoolgirl problem, 110, 116-117, 254-262, 263
k-Profile, 178

Kronecker product, 143

L

Latin rectangle, 180-181
Latin square, 4, 179-208, 211-212, 237-241, 249-251
  and balanced incomplete block designs, 184-185
  idempotent, 187-192, 237-241, 249-251
  and one-factorizations, 211-212
  skew-transversal, 249-251
  symmetric, 211-212, 238, 239-241, 251
  and triple systems, 237-241, 249-251
Line, 70
  concurrent, 71
Linked design, 4

M

MacNeish's theorem, 185
Matching, 209-210
Matrix
  conference, 145-150, 161
  congruent, 105
  Hadamard, 134-187
  identity, 22
  incidence, 7, 22
  permutation, 11
  (0,1), 22
Maximal arc, 88, 233-234
  exterior point, 234
  external line, 234
  secant, 234
  tangent, 234
Mean, 24, 295-296
Monic polynomial, 39
Monomial, 163
Multiplicative group, of finite field, 42-44

Subject Index

Mutually orthogonal Latin
  squares, 183, 192-193,
  204-208

N

non-Desarguesian plane, 92-94
Normalized conference matrix,
  145
Normalized skew-Hadamard
  matrix, 151
Nucleus, 89

O

One-factor, 210
One-factorization, 4, 110, 209-
  235, 244-245, 264
  automorphism, 216, 230
  automorphism group, 230
  of complete bipartite graph, 211
  of complete graph, 214-233
  division, 216-217, 223, 224-
    228, 232
  isomorphism, 215-216, 223-233
  of K_6, 214-215, 233-234
  of K_8, 220-223
  and Latin squares, 211-212
  orthogonal, 264
  perfect, 220-223
  and projective planes, 233-235
  and Room squares, 264
  total number, 215, 223
  and triple systems, 244-245
Orthogonality, 134, 135, 137,
  182-208, 264
Oval, 88-90
  nucleus, 89
  types I and II, 88

P

Pairwise balanced design, 14-22,
  187, 255-261, 284-285
  block, 14
  closure, 284-285
  frequency, 17
  index, 14
  and Latin squares, 187
  replication number, 17
Parameter, 4
Parallel class, 109, 254-258
Patterned starter, 213, 220, 223
PB (see Pairwise balanced design)
PB2-design, 115
Perfect one-factorization,
  220-223
Permutation matrix, 11
PG (see Finite projective
  geometry, Finite projective
  plane)
PG(d, n), 121
Point, 4, 70
  collinear, 71
Polynomial, 38-40
  monic, 39
Prime, in a finite affine plane, 76
Primitive element, 44
Primitive root, 44
Principal fiber, 285, 293-300
Profiles, 174-178
Projective plane, 233-234
  and one-factorizations, 233-234
  of parameter, 4, 233-234
Proper representation, 49

Q

Quadrangle, 81
  diagonal of, 81

[Quadrangle]
  diagonal point of, 81
Quadratic
  character, 44-46
  element, 44, 64, 124
Quasi-residual, 96-98
Quasi-symmetric balanced incomplete block design, 143
Quaternion, 152

R

Regular design, 5
Regular Hadamard matrix, 156
Replication number, 6, 17
Residual design, 95-98, 105
Resolution, 109
Resolvability, 109-117, 254-262
Resolvable design, 109-117, 254-262
(r, l)-design, 21
Room square, 263-282
  existence, 273-276
  isomorphism, 265
  and one-factorizations, 264
  skew, 269
  special, 277-278, 281
  standardized, 265
  subsquare, 269-273, 276-281

S

SBIBD (see Symmetric balanced incomplete block design)
SDR (see System of distinct representatives)
Secant, 88, 234
Self-orthogonal Latin square, 187, 192
SGDD (see Simple group-divisible design)
Simple, 249-251
  triple system, 249-251

Simple group divisible design, 192, 198, 258-261
Singer difference set, 60, 83-88
Skew Hadamard matrix, 145, 151-152
  normalized, 151-152
Smith class, 170-171
Smith normal form, 166
Spouse-avoiding mixed-double tournaments, 198-204
  sharply resolvable, 198
  symmetric, 203
SR (see Spouse-avoiding mixed-double tournaments)
Standardized Hadamard matrix, 136
Standardized Latin square, 183
Starter, 212-214, 220, 223, 265-269, 278
  adder for, 266-267, 269
  equivalence, 214
  patterned, 213, 220, 223
  and Room squares, 265-269, 279
  strong, 267-269, 278
Starter sequence, 204
Steiner triple system, 236, 238-239, 241-249, 252-262
  cyclic, 252-254
  resolvable, 254-262
  subsystem, 241-249
Subdesign, 7
Subgraph, 210
  spanning, 210
Subregular Hadamard matrix, 159
Subsquare
  of Latin square, 181-182
  of Room square, 269-273, 276-281
Subtournament, 202
Sums of squares, 46-53
  two square theorem, 50
  four square theorem, 53, 99
Supplementary difference sets, 62-69

Subject Index

Symmetric, 6, 29
Symmetric balanced incomplete
 block design, 26-29, 95-108,
 113-116, 128, 156-157, 159
Symmetric conference matrix,
 145
Symmetric t-design, 131
System of distinct representatives,
 10-13, 181
System of pairs, 4

T

$T(\lambda, v)$ (see Triple system)
Tangent, 88, 127, 234
t-Design, 32-34, 118-133,
 285-293
 derivative, 33
 generalized, 286-293
 trivial, 121
3-Design, 121-126, 131-133
Transpose equivalent, 163
Transversal, 183, 192
Transversal design, 192-193
Transversal system (see Transversal design)
Treatment, 5
Triangle, 127
Triple system, 236-262
 cyclic, 252-254

[Triple system]
 Kirkman, 254-262
 simple, 249-251
 Steiner, 236, 238-239, 241-249,
  254-262
 subsystem, 241-249
Triplet, 127
Trivial design, 6, 121
Trivial t-design, 121
Two square theorem, 50

V

Variance, 24, 295-296
Variance method, 24
Variety, 5
Vertex, 4, 210
 degree, 210

W

Weight of Hadamard matrix, 165
Witt cancellation law, 106

Z

Zero element, 36
$(0,1)$-matrix, 22